Vegetable Genomics and Breeding Research

Vegetable Genomics and Breeding Research

Guest Editors

Xiao Yang
Feng Yang
Bin Liu

Basel • Beijing • Wuhan • Barcelona • Belgrade • Novi Sad • Cluj • Manchester

Guest Editors
Xiao Yang
Institute of Urban Agriculture
Chinese Academy of
Agricultural Sciences
(IUA-CAAS)
Chengdu
China

Feng Yang
Rice and Sorghum
Research Institute
Sichuan Academy of
Agricultural Sciences
Deyang
China

Bin Liu
Hami-Melon Research Center
Xinjiang Academy of
Agricultural Sciences
Urumqi
China

Editorial Office
MDPI AG
Grosspeteranlage 5
4052 Basel, Switzerland

This is a reprint of the Special Issue, published open access by the journal *Horticulturae* (ISSN 2311-7524), freely accessible at: https://www.mdpi.com/journal/horticulturae/special_issues/6302NZ97CA.

For citation purposes, cite each article independently as indicated on the article page online and as indicated below:

Lastname, A.A.; Lastname, B.B. Article Title. *Journal Name* **Year**, *Volume Number*, Page Range.

ISBN 978-3-7258-7136-0 (Hbk)
ISBN 978-3-7258-7137-7 (PDF)
https://doi.org/10.3390/books978-3-7258-7137-7

Contents

horticulturae

Editorial

Recent Advances in Vegetable Genomics and Breeding Research

Xiao Yang [1], Feng Yang [2,*] and Bin Liu [3,*]

1 Institute of Urban Agriculture, Chinese Academy of Agricultural Sciences, Chengdu National Agricultural Science and Technology Center, Chengdu 610213, China; yangxiao@caas.cn
2 Rice and Sorghum Research Institute, Sichuan Academy of Agricultural Sciences, Deyang Branch, Deyang 618099, China
3 Biological Breeding Laboratory, Xinjiang Uygur Autonomous Region Academy of Agricultural Sciences, Urumchi 830091, China
* Correspondence: yangfeng@scrsri.cn (F.Y.); liu.bin@xaas.ac.cn (B.L.)

Vegetables are of critical importance to the human diet, providing the essential vitamins, minerals, and dietary fiber necessary for sustaining health [1]. In light of the rising global population and increasing demand for nutritious vegetables, there is an urgent need to enhance vegetable production, quality, and resilience to environmental challenges. Genomics and breeding research have emerged as powerful tools in achieving these objectives by unraveling the genetic bases of important traits and developing novel technologies and applications to breed new vegetable varieties with desirable characteristics.

This Special Issue, entitled "Vegetable Genomics and Breeding Research", comprises nine insightful and original research and review articles that aspire to shed light on recent advances in understanding the genetic mechanisms behind quality and stress resistance traits in vegetables and the application of cutting-edge breeding methodologies for vegetable crops.

A key topic in this collection is the advance of modern vegetable breeding technologies, with a focus on ploidy manipulation, haploid technology, and male sterility, as well as their applications in vegetable breeding. Fomicheva et al. (2025) thoroughly reviewed genome doubling in vegetable crops and its potential use in obtaining doubled haploids, which have the potential to be used to produce polyploids and helping to overcome interspecific hybrid sterility and improve agricultural traits [2]. Deng et al. (2025) performed whole-genome identification of DMP-family genes in Solanaceous vegetables, identifying *CaDMP8* (pepper) and *SmDMP8* (eggplant) as potential targets for developing haploid induction lines, which is critical to accelerating the genetic improvement of Solanaceous crops [3]. Addressing the practical challenges of polyploidization in vegetable breeding, El-Mahrouk et al. (2025) investigated and validated effective morphological markers for the early identification of diploid and tetraploid black cumin (*Nigella sativa*) seedlings, which offers a cost-effective and accessible tool for polyploid breeding [4]. The use of male sterility has been recognized as an effective strategy for controlling cross-pollination and accelerating the development of hybrids. Chikh-Rouhou et al. (2025) reviewed the current status of molecular markers linked with male sterility and their regulation of genetic and molecular mechanisms in onions, and highlighted the significant role male sterility plays as a transformative tool facilitating efficient hybrid seed production [5].

Due to increasing consumer demand for high-quality, nutrient-dense fresh vegetables, improving the quality of vegetables is of great importance. Yang et al. (2025) identified *BrC4H3*, *BrF3H1*, and *BrCHS1* as the key genes controlling the biosynthesis of purple leaf

 https://doi.org/10.3390/horticulturae11121520

traits in pak choi, which provided the genetic basis for breeding anthocyanin-enhanced pak choi cultivars [6]. Moreover, increasing the resilience of vegetables to biotic and abiotic stress is another major focus due to growing demand from growers. For instance, Zhao et al. (2025) performed pan-genome analysis to identify the *CsTRM* gene family and analyze its expression patterns in response to abiotic and biotic stresses in cucumber. Their results indicate that *CsTRM14* responds to salt stress, powdery mildew, gray mold, and downy mildew, while *CsTRM21* plays a significant role in regulating abiotic and biotic stress resistance; findings that provide a foundation for the further exploration of the potential role *TRMs* may play in stress resistance in cucumbers [7]. Oble et al. (2025) revealed that *SmPTI6* plays a significant role in the resistance of eggplant to *Phytophthora capsica* attacks, which insights will assist in the development of resistant eggplant cultivars [8]. Zhang et al. (2025) identified five hub regulatory genes that coordinate the defense response of watermelon against *Fusarium oxysporum* f. sp. *niveum*, which provided novel disease-resistant genes for breeding cultivars of watermelon resistant to Fusarium wilt [9]. In addition, Yuan et al. (2025) found that the transcription factor BrNAC19 positively regulates the heat stress response in Chinese cabbage by activating the antioxidant genes *BrCSD1* and *BrCAT2*, thereby mitigating the accumulation of reactive oxygen species (ROS) under heat stress, which exhibited potential as target genes for increasing heat stress tolerance [10].

In conclusion, this Special Issue provides valuable insights into the molecular bases of important quality and resistance traits in vegetables, which underpin the implementation of new breeding technologies to facilitate the development of new cultivars. The development of multi-omics approaches, artificial intelligence, speed breeding, and molecular breeding technologies, along with their applications in vegetable breeding, will enhance the efficiency, precision, and adaptability of vegetable breeding in the future, which promises address the needs of consumers, growers, and the global climate challenge.

Author Contributions: Writing—original draft preparation, X.Y.; writing—review and editing, X.Y., F.Y. and B.L. All authors have read and agreed to the published version of the manuscript.

Funding: This research received no external funding.

Acknowledgments: The authors gratefully acknowledge their colleague, Lin Ouyang, for her assistance in preparing the manuscript. They would also like to express their profound gratitude to all contributors and reviewers for their insightful contributions and the editor team of this Special Issue for their invaluable support.

Conflicts of Interest: The authors declare no conflicts of interest.

References

1. Zhang, L.; Huang, T.; Zhang, Q.; Wei, S.; Contreras, V.H.E.; Peng, J.; Song, B.; Li, Y.; Yang, Q.; Yang, X. Plant Factory Technology as A Powerful Tool for Improving Vegetable Quality: Lettuce as An Application Example. *Veg. Res.* **2024**, *4*, e017. [CrossRef]
2. Fomicheva, M.; Kulakov, Y.; Alyokhina, K.; Domblides, E. Spontaneous and Chemically Induced Genome Doubling and Polyploidization in Vegetable Crops. *Horticulturae* **2024**, *10*, 551. [CrossRef]
3. Deng, X.; Zhong, W.; Liu, B.; Shen, X.; Ren, Z.; Lu, Y.; Wang, X.; Ouyang, B. Identification of DMP Family Members in Solanaceous Vegetables Potentially Involved in Haploid Induction. *Horticulturae* **2025**, *11*, 1329. [CrossRef]
4. El-Mahrouk, M.E.; Maamoun, M.K.; Saifan, S.; Bayoumi, Y.A.; El-Ramady, H.; Abdalla, N. Exploring Polyploidization in *Nigella sativa* L.: An Applicable Strategy Towards Crop Improvement. *Horticulturae* **2025**, *11*, 1122. [CrossRef]
5. Chikh-Rouhou, H.; Singh, S.; Priyadarsini, S.; Mallor, C. Onion Male Sterility: Genetics, Genomics and Breeding. *Horticulturae* **2025**, *11*, 539. [CrossRef]
6. Yang, Q.; Huang, T.; Zhang, L.; Yang, X.; Zhang, W.; Chen, L.; Jing, Z.; Li, Y.; Yang, Q.; Xu, H.; et al. Integrated Phenotypic Physiology and Transcriptome Analysis Revealed the Molecular Genetic Basis of Anthocyanin Accumulation in Purple Pak-Choi. *Horticulturae* **2024**, *10*, 1018. [CrossRef]

7. Zhao, L.; Wang, K.; Wang, Z.; Chu, S.; Chen, C.; Wang, L.; Ren, Z. Pan-Genome Analysis of *TRM* Gene Family and Their Expression Pattern under Abiotic and Biotic Stresses in Cucumber. *Horticulturae* **2024**, *10*, 908. [CrossRef]
8. Obel, H.O.; Zhou, X.; Liu, S.; Yang, Y.; Liu, J.; Zhuang, Y. Transcriptomic Analysis of Resistant and Susceptible Eggplant Genotypes (*Solanum melongena* L.) Provides Insights into *Phytophthora capsici* Infection Defense Mechanisms. *Horticulturae* **2025**, *11*, 1026. [CrossRef]
9. Zhang, C.; Fang, X.; Zhang, J.; Wang, X.; Liu, Z.; Liu, S.; Song, Z.; Gao, P.; Luan, F. Weighted Gene Co-Expression Network Analysis Identifies Hub Genes Governing Resistance to *Fusarium oxysporum* f. sp. *niveum* in Watermelon (*Citrullus lanatus*). *Horticulturae* **2025**, *11*, 625. [CrossRef]
10. Yuan, S.; Yong, X.; Lu, Y.; Lei, Y.; Li, W.; Shi, Q.; Yao, X. The Transcription Factor BrNAC19 Acts as a Positive Regulator of the Heat Stress Response in Chinese Cabbage. *Horticulturae* **2024**, *10*, 1236. [CrossRef]

horticulturae

MDPI

Review

Spontaneous and Chemically Induced Genome Doubling and Polyploidization in Vegetable Crops

Maria Fomicheva [1,*,†], Yuri Kulakov [1,2,†], Ksenia Alyokhina [1] and Elena Domblides [1,*]

1 Federal State Budgetary Scientific Institution Federal Scientific Vegetable Center (FSBSI FSVC), VNIISSOK, 143072 Moscow, Russia; ykulakov12@yandex.ru (Y.K.); xenia.aliokhina@yandex.ru (K.A.)

2 Department of Biotechnology, Russian State Agrarian University—Moscow Timiryazev Agricultural Academy, Timiryazevskaya Street, 49, 127434 Moscow, Russia

* Correspondence: maria.fomicheva.1@yandex.ru (M.F.); edomblides@mail.ru (E.D.)

† These authors contributed equally to this work.

Abstract: Plant ploidy manipulation is often required for breeding purposes. However, there is no comprehensive review covering genome doubling in vegetable crops despite the abundance of data for a large number of vegetable species. Similar to other species, genome doubling is required in vegetable crops to obtain doubled haploids (DHs). It is also utilized for the production of polyploids to overcome interspecific hybrid sterility and improve agricultural traits. Spontaneous haploid genome duplication (SHGD) occurs in many Apiaceae, Brassicaceae, Cucurbitaceae, and Solanaceae crops, allowing for the laborious treatment with antimitotic agents to be bypassed. SHGD mechanisms are not fully understood, but existing data suggest that SHGD can occur via nuclear fusion, endoreduplication, or other mechanisms during microspore or ovule early embryogenic development. Other studies show that SHGD can occur at later developmental stages during extended plant growth in vitro or ex vitro, possibly due to the presence of phytohormones in the medium and/or diploid cell competitive advantage. For unresponsive accessions and species with rare SHGD, such as onion (*Allium cepa* L.) and beet cultivars (*Beta vulgaris* subsp. *vulgaris* L.), antimitotic agent treatment has to be applied. Antimitotic agent application efficiency depends on the treatment conditions, especially the agent concentration and exposure time. Also, plant developmental stage is critical for agent accessibility and plant survival. The existing methods can be used to further improve genome doubling methodology for major vegetable crops and other species.

Keywords: doubled haploid; DH technology; polyploidy; antimitotic compound; colchicine

1. Introduction

Plant breeding aims to improve cultivar yield and valuable compound abundance, for instance, sugar, antioxidant, and essential oil content. Another important goal of breeding is developing cultivars resistant to diseases, pests, and abiotic stressors, such as extreme temperature, salinity, and drought. Genome doubling is one of the tools required to speed up breeding process and increase genetic diversity. It is required for doubled haploid (DH) production. It is also utilized for doubling genome in interspecific hybrids and obtaining polyploids with improved agricultural traits. The methods of chromosome doubling are covered in many reviews for cereal and industrial, medicinal, and ornamental crops, but vegetable crops have not been discussed extensively so far [1–8].

Conventional breeding requires self-pollination for multiple generations, which makes this method extremely time-consuming. Many species have severe inbreeding depression [9] or self-incompatibility [10], which additionally complicates the generation of pure lines.

DH technologies allow for the production of fully homozygous pure lines in one generation, which immensely accelerates breeding. DH lines are produced when haploid cells, microspores, or ovules are stimulated to switch toward a sporophytic developmental route. The resulting haploid regenerants need to double their genome to become diploid plants with two identical copies of genetic material [11,12]. SHGD induction or doubling with antimitotic agents are required for the successful inclusion of regenerant plants in a breeding program because haploid plants are sterile or have low seed productivity [5]. Similarly, if mixoploids or polyploids are generated, they can be of limited value for breeding. Therefore, SHGD induction methods or antimitotic treatment protocols are crucial for DH technology's successful implementation. Detailed reviews provide protocol development guidelines for haploid genome doubling in many species [4–6]. However, there is no comprehensive review covering genome doubling in vegetable crops.

Vegetables play a crucial role in human nutrition, providing dietary fibers, vitamins, minerals, and other essential nutrients [13–15]. DH technologies are being developed and improved to facilitate vegetable crop breeding. Vegetable crops are less responsive to DH technologies compared to cereals and rapeseed (*Brassica napus* L.). Therefore, a lot of data on doubled haploids, including the putative mechanisms of SHGD, were obtained in cereals and rapeseed [6]. Recently, a large degree of progress was made in the development of DH technologies for vegetables [15]. However, problems with the induction of embryogenic growth in many vegetable crops, especially Solanaceae, and the limited SHGD, especially in Amaranthaceae and Amaryllidaceae, limited the production of sufficient numbers of DH plants for breeding programs [9,16,17].

Another problem faced by breeders is introducing new valuable traits. Polyploidization produces varieties with new characteristics that could not be achieved by other means. This method is exploited in ornamental crops [1,7,8], medicinal plants [1,2], and industrial crops [3] because polyploids often have increased plant size and vigor, improved decorative traits, or increased secondary metabolite synthesis. Polyploidization is used for vegetable breeding to improve agricultural traits as well. One of the commercial successes of this method is seedless watermelon hybrids produced from diploid and tetraploid parent lines [18]. Genome doubling is also applied to overcome interspecific hybrid sterility [19]. Broadening the genetic diversity of breeding material can be achieved via hybridization with related species. This method is commonly used to introduce disease, herbicide, and abiotic stress resistance, as well as cytoplasmic sterility genes and other traits from other species. However, the obtained hybrids are often sterile, but genome duplication with antimitotic compounds can restore fertility.

SHGD and colchicine treatments exert stress on the genome of treated plants. The skim sequencing of wheat (*Triticum aestivum* L.) showed that various chromosomal aberrations, including deletions, duplications, and aneuploidy, can happen due to genomic shock during DH production [20]. These alterations can affect plant fitness and agronomic performance. In addition, DNA methylation patterns change as a result of spontaneous and artificial doubling. Most differential DNA methylation occurred at random sites [21]. Colchicine treatment changed the expression of genes involved in hormone signal transduction, metabolism, cytoskeleton control, and others [22]. More studies are needed to identify how long these changes last and what their consequences are. Possibly, the development of less-stressful protocols of spontaneous and artificial genome doubling will be beneficial.

Genome doubling is an indispensable tool for a multitude of applications in plant breeding. In this review, we discuss SHGD incidence and mechanisms, as well as antimitotic treatment protocols for different vegetable crops. Emphasis was placed on the most recent papers and literature sources that provided well-described and statistically tested data with controls and a large number of tested plants. Very limited data were available for some species; therefore, any available information was analyzed for them. The critical treatment information was unified (concentrations in g/L, doubling rate in %) and summarized in a table (Table 1).

2. Spontaneous Genome Doubling

The term "spontaneous doubling" is used to differentiate from doubling induced by chemical treatment. SHGD is extremely advantageous for DH production as it allows for the laborious and costly genome doubling with antimitotic compounds to be omitted.

2.1. Spontaneous Haploid Genome Doubling in Vegetable Crops

SHGD occurs at different frequencies in different species. Moreover, it can vary greatly within a species between different cultivars. For instance, in cereal crops, the percentage of doubling in barley (*Hordeum vulgare* L.) is up to 87% [23], in rice (*Oryza sativa* L.) it is up to 72% [24], in wheat (*T. aestivum* L.) it is up to 50% [25], and in maize (*Zea mays* L.) genotypes it is up to 0–21.4% [26,27].

SHGD can be documented in vegetable crop plants generated by DH technologies (Figure 1). For the Amaranthaceae vegetable crop red beet (*Beta vulgaris* subsp. *vulgaris* L.), the SHGD rate varied in different studies. Some authors observed no incidence of SHGD [28], while others obtained up to 70% diploid plants [29] or tetraploid regenerants [30]. The close relative of red beet, sugar beet (*B. vulgaris* subsp. *vulgaris* (var. *saccharifera*)), had gynogenic regenerants with a haploid chromosome set, but the root tips had endopolyploid cells [31]. In another study, a 10% SHGD rate was observed for sugar beet [32].

For the Amaryllidaceae family, SHGD was reported in onion (*Allium cepa* L.) during gynogenesis in vitro, but the doubling efficiency was relatively low. When a large number of regenerants (about 100) were obtained, the doubling efficiency could be more accurately assessed, and it was determined to be about 10% [33–35].Various SHGD rates were reported in the Apiaceae family. In some studies, no doubling occurred, as reported by Hu et al., who showed that among 18 carrot (*Daucus carota* L.) plants obtained from the embryoids and calli, 16 plants were haploid (2n = 9), and the other 2 plants were aneuploids (2n = 10 and 11) [36]. In another study, regenerants obtained in a culture of isolated microspores had an SHGD rate ranging from 17.5% to 63.6% depending on carrot accession [37]. A total of 90% of carrot regenerants obtained from another culture were diploid [38]. The flow cytometric analysis of carrot plants obtained in an in vitro culture of unfertilized ovules in carrot showed that 97.7% of the regenerants were diploid [39].

About 50% of the dill (*Anethum graveolens* L.), caraway (*Carum carvi* L.), carrot (*D. carota* L.), fennel (*Foeniculum vulgare* Mill.), lovage (*Levisticum officinale* Koch.), and parsnip (*Pastinaca sativa* L.) regenerants obtained from microspore culture spontaneously doubled their genome. However, regenerant images and ploidy evaluation [40].

SHGD is widely observed in the Brassicaceae family. For instance, besides haploid plants, mixoploid, diploid, and tetraploid cabbage (*Brassica oleracea* var. *capitata* L.) regenerants produced from microspores can be observed as assessed by flow cytometry, guard cell chloroplast, and chromosome counting (Figure 1). SHGD occurred in 0–76.9% of cabbage (*Brassica oleracea* var. *capitata* L.) regenerants, in 52.2–100% of broccoli (*B. oleracea* var. *italica* L.) [41], in 50–70% plants of different *B. rapa* L. accessions [42–45], and in more than 60% of rapeseed (*B. napus* L.) regenerants [46].

In Cucurbitaceae species, for instance, in a summer squash (*Cucurbita pepo* L.) ovary/ovule culture, most regenerants doubled their genome. Although, occasional haploid, mixoploid, or polyploid regenerants were observed as well [47]. In styrian oil pumpkin (*Cucurbita pepo* ssp. *pepo* var. *styriaca*), most plants were also diploid, with rare occurrences of n, 3n, and 4n plants [48].

Figure 1. Obtaining cabbage (*B. oleracea* var. *capitata* L.) doubled haploids in microspore culture in vitro with subsequent regenerant ploidy analysis. (**A**) Microspores were isolated from the collected flower buds under sterile conditions. (**B**) Embryoids growing from microspores in liquid NLN medium. (**C**) Embryoids were transferred to solid MS medium. (**D**) Embryoids regenerated into plants. (**E**) Rooted plants were adapted to ex vitro conditions. (**F**) The adapted plantlets were grown under field conditions. (**G–G″**) The ploidy was analyzed with a flow cytometer (Beckman Coulter, USA). Nuclei were isolated from plant leaves in Galbraith lysis buffer and stained with propidium iodide. The analyzed sample (green peaks) was compared with a control diploid cabbage (red peaks). Mixoploid (**G**), spontaneously doubled diploid (**G′**), and tetraploid (**G″**) plants were observed. (**H–H″**) The ploidy estimation by counting the chloroplast number in stomatal guard cells. Stomatal guard cells in phase contrast (top images) and chloroplast red autofluorescence (bottom images) were imaged. Haploid (**H**), spontaneously doubled diploid (**H′**), and tetraploid (**H″**) plants can be distinguished by the number of chloroplasts that is higher in higher ploidy plants. (**I**) The direct counting of chromosomes in DAPI-stained spontaneously doubled diploid (2n = 2x = 18) and tetraploid (2n = 4x = 36) samples.

Solanaceae crops also demonstrate genome doubling in DH regenerants, although the doubling rate often differs greatly not only between different species but also within the same species between different genotypes or even between different plants within the same genotype [49]. The frequency of SHGD in bell pepper (*Capsicum annuum* L.) plants was 14%–51% [50], 14.3–65.8% [51], 30–55% [52,53], 32.6% [54], 33% [55], and 65% [56]. In hot pepper (*C. annuum* L.), the SHGD rate was 16.3% [57]. In Indonesian hot pepper (*C. annuum* L.), spontaneous doubling was 14–33% in four accessions and 47% and 51% in the other two pepper accessions [50]. The analysis of cayenne pepper (*C. frutescens* L.)

regenerants showed that about 40% of plants were diploids and 8% were mixoploids [58]. Interspecific hybrids, *C. frutescens* × *C. annuum* and *C. frutescens* × *C. chinense* had a lower haploid embryoid regeneration rate compared to *C. annuum*, but the SHGD was 50% and 80%, respectively [49]. In the genus *Physalis*, a high level of spontaneous diploidization (up to 72%) was observed in tomatillo (*Physalis ixocarpa* Brot.) (2n = 2x = 24) [10,59]. At the same time, for the tetraploid species cape gooseberry (*Physalis peruviana* L.) (2n = 4x = 48), a fairly low doubling level was documented (28%), and antimitotic treatment was required to double the genome in haploid regenerants [60].

Tomato (*Solanum lycopersicum* L.) is one of the most recalcitrant crops for DH technologies. Unlike most other members of the Solanaceae family, tomato is unresponsive to all the existing DH methods available in the literature. Therefore, it is difficult to make conclusions about SHGD for tomato. The largest number of plants was obtained in the work of Zagorska et al., 2003, who analyzed ploidy in 700 androgenic regenerants. They concluded that 21,5% were haploid, 11.3% were diploid, and the rest were mixoploid [61].

Diploid plants obtained from the anther or ovule cultures can originate from surrounding somatic tissues. In addition, some regenerant plants can come from unreduced gametes that formed via meiotic restitution [62]. Therefore, regenerant origin has to be validated by molecular markers to select only true DHs. The reported SHGD rate may be affected by the erroneous inclusion of plants obtained via the abovementioned mechanisms.

A large number of vegetable species described above undergo spontaneous genome doubling. However, some accessions have an insufficient SHGD rate, and some species, including onion and beet, are recalcitrant to spontaneous doubling induction. Despite the challenges of the DH technologies in vegetable crops, doubled DH lines are included in breeding programs. Examples of the successful use of spontaneously doubled DH lines include the F1 hybrid of cabbage 'Nataly', kohlrabi 'Dobryniya', sweet pepper 'Mila', 'Nataly' and 'Gusar', pumkin 'Vega', and a carrot variety 'Sonata' [63].

2.2. SHGD Mechanisms

A list of factors influence spontaneous doubling induction, including genetic factors and microspore or anther temperature stress treatments. In cereal crops, mannitol or 2-hydroxynicotinic acid can also be used as a stress treatment [62]. In addition, spontaneous doubling can occur after a long-term regenerant subculturing in vitro or ex vitro.

Four possible mechanisms of genome doubling were proposed: nuclear fusion, endoreduplication, C-mitosis, and endomitosis [6,27,62]. Endoreduplication is a process when one or several rounds of DNA replication happen without mitosis. It is widely observed in angiosperms at different stages of development [64]. One of the critiques to this plausible mechanism is that endoreduplication is generally seen as a terminal stage of cell differentiation served to obtain higher metabolic competences and/or cell size, which differs from the processes observed in developing haploids [27].

The nuclear fusion of two haploid nuclei is another possible mechanism. Nuclear fusion was experimentally documented in a barley microspore culture (*H. vulgare* L.) by live cell imaging [65]. More studies in different taxa are needed to further confirm this mechanism's applicability for SHGD.

Endomitosis happens when mitotic stages occur inside the nuclear membrane without spindle formation and daughter cell separation. However, endomitosis is rarely observed in angiosperms. C-mitosis is an artificially aborted cell division via the disruption of mitotic spindle by colchicine or other antimitotic agents [27,62], which is discussed in Section 2 of this review.

Besides haploid and doubled haploid plants, triploid and polyploid regenerants can often be observed in many crops, including *Brassica* species, *C. annuum* L., *D. carota* L. [4], and *C. pepo* L. [47,48]. Possibly, triploid plants can form due to the fusion of diploid and haploid nuclei. Tetraploid and higher ploidy plants are likely to result from more than one round of doubling [6]. As it was mentioned before, different cultivars can have a big difference in the incidence of SHGD. The genetic mechanisms that make some cultivars more prone to SHGD are not clear, but the research conducted on maize showed one major and a few minor QTLs that are associated with spontaneous genome doubling. The possible candidate gene associated with the reported QTLs is the absence of the first division (*afd1*) gene that affects the first meiotic division, resulting in a single equational division. Another candidate is the formin-like protein 5 that affects actin cytoskeleton [66]. Studies of other species, including vegetable cultivars, are required for the elucidation of the genetic mechanisms controlling SHGD. The development of SHGD-associated markers would simplify the introgression of this valuable trait into elite germplasm and increase DH production efficiency.

2.3. SHGD Timing during Development

Spontaneous genome doubling can occur at different stages of microspore development [6,62]. The applied stress pretreatment leads to cytoskeleton perturbation. It disrupts mitotic spindle or cell wall formation that results in SHGD. It was demonstrated that to obtain the best results, the stress treatment has to be applied specifically at late uninucleate to early binucleate stages. Microspores isolated at early uninucleate stages predominantly resulted in haploid regenerants, while the binucleate microspores produced more doubled haploid and polyploid regenerants [6]. Stress application at later stages can increase the incidence of triploid and polyploid regenerants [62].

Spontaneous chromosome doubling can occur at later developmental stages as well. Yuan et al. reported that the long-term subculturing of cabbage or broccoli haploids on MS-2 medium with 0.1 mg/L NAA and 0.2 mg/L 6-BAP led to a gradual increase in the number of plants with a doubled chromosome set. After one or more years in tissue culture, most of the cabbage or broccoli haploids turned into DHs or mixoploid plants [41]. Similar results were observed in haploid pepper (*C. annuum* L.) plants grown ex vitro for 6 years. The plants were rejuvenated by cutting off the shoots and allowing young shoots to regrow. The authors reported that out of 12 plants, 1 plant had all diploid shoots, 7 plants had both haploid and diploid shoots, 2 plants remained haploid, and 2 plants died [67]. For plants that change their ploidy after long-term growth in vitro or ex vitro, it can be speculated that spontaneous endoreduplication or doubling via other mechanisms can happen in some cells. Subsequently, doubled haploid cells could outcompete haploid cells. However, this hypothesis requires experimental testing. Possibly, culture media hormones can affect the doubling rate. For instance, auxins are involved in the transition from the mitotic cycle to endoreduplication [68]. For instance, it was shown that NAA caused genome duplication in sugar beet [69]. 2,4-D is also known to increase plant ploidy, as shown, for example, in an orchid tissue culture [70]. The cucumber embryogenic callus established from immature embryos on media supplemented with 6-BAP, NAA, and 2,4-D regenerated not only into diploid (57%) but also tetraploid (18%), octoploid (4%), and mixoploid (2n/4n—4% and 4n/8n—17%) regenerants, as tested by flow cytometry [71].

Currently, the SHGD mechanisms are not fully understood. More research is needed to definitively demonstrate how SHGD occurs and what molecular mechanisms underlie these processes. Understanding the mechanisms of spontaneous doubling can vastly improve the genome doubling rate and the efficiency of DH technology.

3. Chemically Induced Genome Doubling Protocols

When SHGD does not occur or happens at a low frequency, genome doubling can be induced with antimitotic compounds. Currently, a number of antimitotic agents are used to induce genome duplication, including a natural alkaloid colchicine (C22H25NO6) and herbicides, such as trifluralin (C13H16F3N3O4), oryzalin (C12H18N4O6S), pronamide (C12H11Cl2NO), and amiprofosmethyl (APM) (C11H17N2O4PS). Alcohol n-butanol (C4H9OH) is also utilized as an antimitotic agent. The listed antimitotic compounds have a similar mechanism of action. They perturb the microtubule cytoskeleton that prevents mitotic spindle formation and chromosome separation. As a result, the cell receives a doubled chromosome set [4,72,73].

The antimitotic compound choice, concentration and treatment time, supplementary compounds, and the explants used for treatment are critical for successful genome doubling. Different approaches for chemically induced genome doubling are covered below.

3.1. Antimitotic Agent Choice

Colchicine is one of the most widely used antimitotic agents (Table 1). Colchicine is typically applied at concentrations of 0.05 to 5 g/L, but in some cases, it may be outside this range. For example, Gurel S. et al. (2021) treated sugar beet (*B. vulgaris* subsp. *vulgaris* (var. *saccharifera*)) with 20 g/L colchicine, while Vasilchenko et al. (2018) used it at a 4000 times smaller concentration (0.005 g/L) on the same crop [16,74]. Hence, the antimitotic agent should be tested in a wide range of concentrations to determine the most effective and cost-efficient treatment protocol. Colchicine must be handled with caution with the use of protective clothing and gloves due to its toxicity [75]. In addition, colchicine is a light-sensitive compound; therefore, it has to be used under limited or no light conditions [76]. For instance, plant apices with applied cotton balls soaked in colchicine should be covered with foil to prevent colchicine degradation.

Trifluralin is another common substance used to increase the ploidy level. It is a pre-emergence herbicide that prevents seed germination. Trifluralin has low acute toxicity, but it has been classified as a group C, possible human carcinogen [77]. Similar to colchicine, it has to be protected from light due to its light sensitivity [72]. Trifluralin is typically used at 3.35×10^{-4} [78] to a 0.1 g/L concentration. However, concentrations close to 0.1 g/L or above have led to plantlet death [79,80]. Trifluralin was successfully used in a list of species, including onion (*A. cepa* L.) [81], rapeseed (*B. napus* L.) [78,82], sugar beet (*B. vulgaris* subsp. *vulgaris* (var. *saccharifera*)) [79,83], rice (*O. sativa* L.) [5], maize (*Z. mays* L.) [84], and others (Table 1).

Oryzalin is used to increase ploidy in many horticultural and ornamental species. For genome doubling purposes, oryzalin is used at concentrations ranging from 4×10^{-4} [79] to 0.1 g/L. Similar to trifluralin, high oryzalin concentrations (0.1 g/L or above) are also detrimental for plant survival [79,85]. Oryzalin was tested in rapeseed (*B. napus* L.) [82], sugar beet (*B. vulgaris* subsp. *vulgaris* (var. *saccharifera*)) [79], onion (*A. cepa* L.) [34,81,86], maize (*Z. mays* L.) [84], and other species (Table 1).

Amiprofos-methyl (APM) is an antimitotic agent that destroys mitotic spindle or induces multipolar spindle division [4]. Successful genome doubling using APM was reported for a list of species, including rapeseed (*B. napus* L.) [82], wheat (*T. aestivum* L.) [87], and maize (*Z. mays* L.) [84,88,89]. For vegetable crops, it shows promising results in onion (*A. cepa* L.) [81,86,90–93].

The use of herbicide pronamide is limited to a small number of studies. In maize, 0.0026 g/L pronamide resulted in about a 60% doubling rate [84]. In sugar beet and fodder beet, 0.004 g/L pronamide induced 2% doubling [79]. Rapeseed treatment with pronamide increased the number of plants with doubled genome (52%) compared to untreated control (36.4%), but it was less effective than colchicine treatment (69.6%) [94] (Table 1).

N-butanol is a primary 4-carbon alcohol that induces cortical microtubule depolymerization [95,96]. A drastic increase in the microspore embryogenesis rate after n-butanol treatment was observed in wheat (*T. aestivum* L.) [97] and certain barley cultivars (*Hordeum vulgare* L.) [98]. N-butanol has some positive effect on embryogenesis in maize (*Z. mays* L.) [95,99,100] and pepper (*C. annuum* L.) [101]. A further improvement in the n-butanol treatment regimen and testing of this compound in other species is still needed.

The efficacy of an antimitotic agent treatment critically depends on a number of factors that determine the toxic effect of the antimitotic compound, as well as the duplication and the aberrant ploidy rate. The agent concentration, supplementary components, the application method, and other contributing factors are discussed below.

Table 1. Vegetable crop and sugar beet (*Beta vulgaris* L.) genome doubling approaches.

The Crop	Ploidy before Treatment	The Best Doubling Efficiency	Application Method	Antimitotic Agent	Treatment Time	Growth Conditions after Treatment	Practical Results	References
Amaranthaceae								
Sugar beet (*Beta vulgaris*)	n	–	Seedlings with 3–7 leaves	5 g/L colchicine	5 min 2 times	MS, 10 g/L sucrose, 6.5 g/L Gelrite, 0.002 g/L kinetin	–	[16]
	n	91.3%	Microclones	0.0005 g/L colchicine	48 h	MS, GA, 6-BAP, kinetin—0.0002 g/L each	Selection of new lines with sterile cytoplasm (confirmed by PCR and RFLP analysis). Four lines were selected for breeding purposes (the plant qualities are not described)	[74]
	n	29.1%	Shoots higher than 1 cm with roots removed	0.05, 0.1, 0.15 or 0.5 g/L colchicine	12, 24, 36 or 48 h	MS, 0.001 g/L 6-BAP, solid or liquid media	–	[83]
		20.7%		0.017, 0.0034 or 0.005 g/L trifluralin				
	n	4.7% (0.03 g/L)	Ovules. Antimitotic agent with 1.5% DMSO	0–0.09 g/L APM	5 h	Liquid culture medium	–	[79]
		2.8% (0.00035 g/L)		0–0.1 g/L oryzalin				
		2.0% (0.003 g/L)		0–0.1 g/L pronamid				
		2.0% (0.003 g/L)		0–0.1 g/L trifluralin				
	n	60% (4 g/L, 2.5 h)	Ovules after 7 days in culture	4–60 g/L colchicine	0.08—5 h	Induction medium	–	[102]
	n	64% (0.046 g/L APM, 5 h)	Ovules after 10 days culture	0.006, 0.046, 0.092 g/L APM + 15 g/L DMSO in liquid induction medium	2, 5, 18 h	Culture medium	–	[103]
	n	19%	Roots of regenerant plants	3 g/L colchicine	24 h	Plants were planted in soil	Three lines that exceeded diploid control for yield and sugar concentration were produced (for instance, BTS 40 DH line—377.7 g and 19.4%, control diploid—204.2 g and 18.4% root weight and sugar content, respectively)	[104]
	n	8.4% 2n, 3.4% 4n, 1.1% 8n, 33% mixoploids	Meristem of plants at the 6–8-leaf stage	1 g/L colchicine with 20 g/L DMSO	72 h	–	–	[105]

Table 1. *Cont.*

The Crop	Ploidy before Treatment	The Best Doubling Efficiency	Application Method	Antimitotic Agent	Treatment Time	Growth Conditions after Treatment	Practical Results	References
				Amaryllidaceae				
Garlic (*A. sativum*)	2n	–	Basal plates in liquid B5 with 20 g/L DMSO and colchicine	2.5–7.5 g/L colchicine	36–72 h	B5, 0.0008 g/L 2iP, 0.0001 g/L NAA	4n plants leaves had 3 times larger area. Allicin content increased by 30%	[106]
Onion (*Allium cepa*)	n	–	Basal explant in MSO (MS based) medium with colchicine	0.4 g/L colchicine	48 h	1/2 BDS, 30 g/L glucose and 7.0 g/L agar, pH 6.0	–	[9]
	n	36.7%	Embryos in medium with APM	0.015 g/L APM	48 h	1/2 BDS, 30 g/L glucose, 7.0 g/L agar, pH 6.0	–	[93]
	n	36%	Intact or cut longitudinally into halves basal explants in BDS medium with antimitotic agents	0.003; 0.006; 0.009 g/L APM	72 h	1/2 BDS or M4, 30 g/L glucose and 0 or 7.0 g/L agar, pH 6.0	Fertility/fecundity recovered in some genotypes, but not others. Lines with uniform bulb shapes were produced	[90]
		40–46% 2n, 18–30% mixoploids (0.2–0.4 g/L colchicine in liquid medium, 48 h		0.01–0.4 g/L colchicine	24, 48, 72 h			
	n	32.8% (0.4 g/L)	Whole basal explants	0.3–0.4 g/L colchicine	–	MSO, 30 g/L sucrose	Fertility/fecundity recovery	[86]
		15.8% (0.03 g/L)		0.017–0.051 g/L oryzalin				
		25% (0.03 g/L, 0.045 g)		0.015–0.045 g/L APM				
	n	100% (0.25 g/L, 48 h)	in vitro plants	0.25 or 0.5 g/L colchicine	24 and 48 h	1/2 MS, 40 g/L sucrose, 7.5 g/L agar, pH 5.8	–	[92]
		57.7% (0.015 g/L 48 h)		0.015–0.06 g/L APM				
	n	46% (0.01 g/L, 72 h)	Root tips, shoot apex	0.01; 0.1 g/L colchicine	24 and 72 h	M3 medium	–	[91]
	n	up to 65.7% (1 g/L)	in vitro plants	0.25–5 g/L colchicine	24 h	MS	–	[34]
		up to 57.1% (0.0175 g/L)		0.0035–0.07 g/L oryzalin				
	n	44%	Two-month-old haploid plants on BDS basic medium with colchicine	0.015 g/L colchicine	72 h	The plants were grown in peat blocks	–	[107]

Table 1. *Cont.*

The Crop	Ploidy before Treatment	The Best Doubling Efficiency	Application Method	Antimitotic Agent	Treatment Time	Growth Conditions after Treatment	Practical Results	References
Onion (*Allium cepa*)	n	47.1% 2n, 29,4% 29,4% another ploidy (0.017 g/L 72 h)	Gynogenic embryos were plated on the media with antimitotic agents in the dark	0.0017 g/L; 0.017 g/L trifluralin in acetone	24 and 72 h	Medium R1: 1/2 BDS, 30 g/L glucose and 7.0 g/L agar, pH 6.0; medium R2: BDS, 40 g/L sucrose, 7.0 g/L agar, pH 6.0	–	[81]
		47.1% 2n, 8.8% other ploidy (24 h)		0.012 g/L oryzalin in acetone				
		34.8% 2n, 17.4% other ploidy (72 h)		0.015 g/L APM in methanol				
		35.3% 2n, 5.9% other ploidy (0.05 g/L 72 h)		0.001 g/L; 0.05 g/L colchicine with 20 g/L DMSO				
	n	Up to 38% 2n (0.015 g/L, liquid media, 24 h)	Embryos in liquid or solid media with APM	0.008 g/L; 0.015 g/L APM	24 and 72 h	1/2 BDS, 15 g/L glucose	–	[108]
Onion (*A. fistulosum* × *A. cepa*)	2n	51.4% 4n (10 g/L)	Callus in liquid EDS on a shaker at 60 rpm	5–20 g/L colchicine	36–72 h	solid BDS medium with 8 g/L agar	Five 4n, likely amphidiploid plants were obtained and adapted to field conditions for future breeding	[109]
Onion (*A. cepa* × *A. vavilovii*, *A. cepa* × *A. nutans*, *A. cepa* × *A. schoenoprasum*)	3n *	–	Meristems in vitro	0.01 g/L colchicine	–	–	An increased vegetative mass (no quantitative data) and resistance to downy mildew was observed in interspecific hybrids	[110]
Persian shallot (*A. hirtifolium*)	2n	died after treatment	Basal plates in liquid MS with 10 g/L DMSO and antimitotic agent on a shaker at 100 rpm	3–7 g/L colchicine	24–48 h	MS, 30 g/L sucrose, 8 g/L agar, 0.001 g/L 6-BAP, 0.0005 g/L NAA	Increased total phenolic compound and allicin content by 27 and 15%, respectively[111]	[111]
		died after treatment		0.01–0.04 g/L oryzalin	4–8 h			
		35% 4n (5 g/L, 36 h); 31.82% mixoploids (3 g/L, 36 h)	Basal plates on solid MS with 8 g/L agar with antimitotic agents	3–7 g/L colchicine	24–48 h			
		45.45% 4n (0.01 g/L, 8 h); 16.9% mixoploids (0.04 g/L)		0.01–0.04 g/L oryzalin	4–8 h			

Table 1. *Cont.*

The Crop	Ploidy before Treatment	The Best Doubling Efficiency	Application Method	Antimitotic Agent	Treatment Time	Growth Conditions after Treatment	Practical Results	References
				Apiaceae				
Ajowan (*Trachyspermum ammi*)	2n	11.53% 4n (0.5 g/L, 24 h)	Seeds	0.24–0.5 g/L colchicine	6–48 h	liquid MS	Thymol in essential oil increased by 39% in 4n plants. The increase was observed in plant organ sizes with the largest difference in plant height (over 2 times)	[112]
American wild carrot (*Daucus pusillus*)	n	–	Cut umbrellas before flowering in the green bud phase	1 g/L colchicine with 20 g/L DMSO	20 h	MSm, 0.0002 g/L 2,4-D, 6–8 g/L agar	–	[110]
Caraway (*Carum carvi*)	n	–	Root system of in vitro plants	0.4 g/L colchicine with 10 drops/L DMSO	24 h	Containers with soil, plants were covered to maintain high humidity	–	[113]
Carrot (*Daucus carota* L.)	n	–	Microspores	0.5 g/L colchicine	24 h	B5, 0.0001 g/L 2,4-D, 0.0001 g/L NAA, 100 g/L sucrose	–	[114]
Carrot (*D. carota* L.)	n	–	Microspores	0.5 g/L colchicine	48 h	NLN, 0.0001 g/L 2,4-D, 0.0001 g/L NAA, 130 g/L sucrose	–	[115]
Fennel (*Foeniculum vulgare*), Dill (*Anethum graveolens*)	n	–	Root system and crowns	3.4 g/L colchicine	1.5 h	Containers with soil, plants were covered to maintain high humidity	–	[116]
Parsley (*Petroselinum crispum* L.)	2n	about 30% (0.5, 24 h)	Seeds ex vitro on a rotary shaker at 120 rpm	0.25, 0.5, 1, 2 g/L colchicine	8–48 h	Seeds were planted in pots with soil mixture	Plant height is increased by 42% and leaf length is increased by 64% in 4n plants	[117]
		100% (1 g/L, 24 h)	Plant nodes in vitro on a rotary shaker at 120 rpm			MS with 0.001 g/L 2,4-D		

Table 1. *Cont.*

The Crop	Ploidy before Treatment	The Best Doubling Efficiency	Application Method	Antimitotic Agent	Treatment Time	Growth Conditions after Treatment	Practical Results	References
				Brassicaceae				
Broccoli (*Brassica oleracea* var. *italica*)	n	50% (2 g/L, 6 h), 66.7% (1 g/L, 12 h)	In vitro seedling roots were trimmed to 1–2 cm length and immersed in colchicine	0.5, 1, 2, 4 g/L colchicine with 20 g/L DMSO	6–12 h	Containers with soil, plants were covered to maintain high humidity	Genome doubling recovered fertility (fertile DH and partially fertile mixoploids), high colchicine toxicity for broccoli was shown (from 50 (1 g/L, 6 h) to 100% (4 g/L, 9–12 h)	[41]
White cabbage (*B. oleracea* var. *capitata* L.)		50% (2 g/L, 9 h)			3–12 h			
B. oleracea × leaf mustard *B. juncea*	2n ***	Best results for 1.5, 2 g/L	Rooted ex vitro cuttings axillary meristems were soaked in colchicine solution and covered with foil	0.5, 1, 1.5, 2, 2.5 g/L colchicine	Two treatments/day for 3 days	–	Genome doubling recovered fertility (7–84% pollen fertility), an increased main stem thickness and leaf size, more compact inflorescences, different leaf texture and margins compared to ABC hybrids and parents (no quantitative data)	[118]
Chinese cabbage (*B. rapa*) × white cabbage (*B. oleracea*), rapeseed (*B. napus*)/leaf mustard (*B. juncea*) × Chinese cabbage (*B. rapa*)	2n **	–	Plantlets in vitro	1 g/L colchicine	4 h	Plants were planted in coco-peat	The white cabbage with orange/yellow inner leaves (no quantitative data) and Chinese cabbage with an increased anthocyanin content (increased from 0 to 4.7 mg/g)	[119]
				Cucurbitaceae				
Cucumber (*Cucumis sativus*)	n	–	Cuttings with 2 axillary buds in E20H8 medium with colchicine	0.2 g/L colchicine	48 h	–	Genome doubling recovered fertility. A 40–80% mortality rate from colchicine treatment.	[120]
Cucumber (*C. sativus*)	n	24%	Haploid plants were on CBM basic medium with colchicine	0.2 g/L colchicine	96 h	The roots were rinsed in water. Then, the plants were grown in peat blocks	–	[107]
Melon (*Cucumis melo*)	n	–	The main apical stem	colchicine	2 h	–	–	[121]

Table 1. *Cont.*

The Crop	Ploidy before Treatment	The Best Doubling Efficiency	Application Method	Antimitotic Agent	Treatment Time	Growth Conditions after Treatment	Practical Results	References
Pumpkin (*Cucurbita pepo*)	n	–	Plantlets on a shaker at 120 rpm	5 g/L colchicine	12 h	–	–	[122]
			Apical shoot	10 g/L colchicine	3 1 h treatments per day			
Watermelon (*Citrullus lanatus*)	2n	Over 60% (colchicine, ethalfluralin, oryzalin at the highest concentration, 9 days)	Shoot buds of in vitro plants in MS with antimitotics on a shaker	0.2, 0.4, 0.6, 0,8 g/L colchicine. Ethalfluralin, oryzalin, cobex, amex (25, 50, 75, 100 µM/L)	3, 6, 9 days	MS, 30 g/L sucrose, 7 g/L agar with 0.00225 g/L 6-BAP	–	[123]
	2n	–	Seedlings were immersed in colchicine aqueous solution	2 g/L colchicine	6 days	–	4n rootstocks are more tolerant to salt stress (minor withering at 300 mM NaCl, no quantitative data) due to lower Na+ /K+ ratio, higher photosynthetic capacity, antioxidant enzyme activity, and osmoregulatory gene expression	[124]
				Solanaceae				
African nightshade (*Solanum nigrum ssp. villosum*)	2n ****	about 10% (0.1 g/L)	Ex vitro seedlings at cotyledonary stage were sprayed with colchicine solution and covered with polyethylene sheets	0.1, 0.5, 2.5 g/L colchicine in 1 mL DMSO and 0.1 mL Tween-20	7 d, sprayed once a day	–	–	[125]
Cape gooseberry (*Physalis peruviana*)	n ****	over 60% (2 g/L, 2 h)	Excised axillary buds were immersed in colchicine solution in the dark	2, 4, 6 g/L colchicine with 20 g/L DMSO	2 h, 4 h, 6 h	MS, 0.1 mg/ L IBA	Recovery of fertility and seed production	[60]
Chili (*C. annuum*); Ancho chili × habanero chili (*C. annuum* × *C. chinense*)	n	–	Seedlings in vitro/ex vitro. Each seedling was immersed to the base of the stem in a bottle with colchicine solution	5 g/L colchicine	8 h	Pots with sterile substrate	Maintainers of cytoplasmic male sterility resistant to *Phytophthora capsici* Leo. and to the geminiviruses PepGMV and PHYVV were obtained	[126]

Table 1. *Cont.*

The Crop	Ploidy before Treatment	The Best Doubling Efficiency	Application Method	Antimitotic Agent	Treatment Time	Growth Conditions after Treatment	Practical Results	References
Chili (*C. annuum*)	2n	–	Seedlings at the 4-leaf stage	3 g/L colchicine	9 or 12 h for 2 days or 8 h for 3 days	–	4n plants exhibit gigas characters when compared to the 8n (leaf area 1.8 times larger; fruit diameter 2.7 times larger; flower diameter 1.6 times larger), except for the anthers, which are large, thick, sometimes deformed and coalesce with the corolla in the 8n plants. 8n plants were less vigorous.	[127]
Eggplant (*Solanum melongena*)	n	–	Plants in vitro	5–10 g/L colchicine	2 h (5 g/L); 1 h (10 g/L)	–	–	[17]
		25% more compared to SHGD	Plant axillary buds ex vitro	5 g/L colchicine in lanolin paste	48 h			
	n	50–70%	Remove the apical and axillary buds from plantlets. Apply colchicine to secondary axillary buds in the dark	5 g/L colchicine in lanolin paste	48 h	Greenhouse conditions; remove shoots produced by untreated buds	–	[128]
	n	35%	2–3 leaf plantlets on 1/2 MS with colchicine	0.6 g/L colchicine	72 h	The plants were grown in peat blocks	–	[107]
	n	100%	Colchicine solution was applied to axillary buds with a piece of cotton in the dark	5 g/L	2 h	–	–	[129]
				10 g/L colchicine	1 h			
	n	–	Plantlets in vitro	5 g/L	2 h	Hormone-free medium R	–	[130]
	n	–	Plantlets roots in vitro/ex vitro	0.1 g/L colchicine	4 h	The plants were acclimated and transferred to the greenhouse	–	[131]
Paprika (*C. annuum*)	n	95% on average	In vitro plantlets at 2–3-leaf stage in plain V3 medium with colchicine	10 g/L colchicine	1 h	Plastic pots containing 1:1 mixture of non-sterilized peat and sandy soil mix	–	[107,132]
Pepper (*C. annuum*)	n	50–95%	In vitro plantlets in V3 medium with colchicine	0.05–0.4 g/L colchicine	96–144 h	–	–	[133]

Table 1. *Cont.*

The Crop	Ploidy before Treatment	The Best Doubling Efficiency	Application Method	Antimitotic Agent	Treatment Time	Growth Conditions after Treatment	Practical Results	References
Pepper (*C. annuum*)	n	up to 100%	Axillary buds of ex vitro plants were covered with cotton soaked in colchicine and wrapped with foil	5 g/L colchicine	12 h	Colchicine applications were continued until chromosome doubling was achieved	–	[134]
Pepper (*C. annuum*)	n	–	Axillary buds of ex vitro plants were covered with a piece of cotton soaked in colchicine in the dark	0.5 g/L colchicine	2 h	–	–	[135]
Pepper (*C. annuum*)	n	57.6% and 47.3% 33% SHGD	Anthers	3 g/L colchicine	30 days	MS with supplements	Addition of colchicine to culture medium resulted with positive effects on viability of embryo, regeneration, and growth into full developed plantlets; haploid plants are smaller, plant viability, in vitro regeneration, development, and growth is lower compared to diploid donor plants	[55]
Pepper (*C. annuum*)	n	25–27% 2n, 29–55% mixoploids	The apical shoot fragment with 2–3 leaves in vitro	0.2, 0.4 g/L g/L colchicine	6–9 days	MSm, 30 g/L sucrose, 8 g/L agar, pH 5.8	Four weeks after the second colchicine treatment the growth disorder was observed, which proved to be directly proportional to the time of explant incubation on the colchicine-containing MS medium; after six days of colchicine treatment, 38% of the plants regenerated, and after nine days, their number decreased to 33%	[136]
Spice and bell pepper (*C. annuum*)	n	75%	4–6 leaf stage in vitro plantlets	3 g/L colchicine with 1.5 g/L DMSO	3 h	The roots were rinsed in water. Then, the plants were grown in peat blocks	–	[137]

Table 1. *Cont.*

The Crop	Ploidy before Treatment	The Best Doubling Efficiency	Application Method	Antimitotic Agent	Treatment Time	Growth Conditions after Treatment	Practical Results	References
F_2 hybrids *C. annuum* L. (cv. Zdorovie) × *C. chinense* Jacq; BC_2 of *C. annuum* L. (cv. Zdorovie) × *C. chinense* Jacq × cv. Zdorovie × cv. Zdorovie	n/2n	-	Apical meristem ex vitro plants	5 g/L colchicine with Tween-20 (1 drop/100 mL)	-	Plants were covered with polyethylene film to reduce evaporation	Fast generation of homozygous lines combining *C. annuum* L. and *C. chinense* Jacq traits. 17.5% were fertile and set seeds; fruit mass was 3 times larger than in *C. chinense*	[138]
Bell pepper, capia, charleston, and green types (*C. annuum*)	n	-	Ex vitro plants	5 g/L colchicine	2 h	Self-pollination was performed in at least 3–4 flowers. Three or four fruits were harvested from each self-pollinated plant, and their seeds were removed. Then, the seeds were labeled and packed.	Three homozygote pure lines with agronomically valuable traits, including Me1, Me3, N and Me7 nematode resistance genes, were obtained. Also, one homozygote pure line containing Me1 was found	[139]
Tomatillo (*Physalis ixocarpa*)	n	50% 2n, 50% 4n	Excised apical axillary buds from in vitro regenerants were inoculated on MS medium with colchicine	0.5, 1 g/L colchicine	2–6 days	MSD medium	–	[59]

Notes: * *A. cepa* 2n = 2x = 16; *A. vavilovii, A. nutans, A. schoenoprasum* are natural tetraploids 2n = 4x = 32. ** *B. rapa* 2n = 2x = 20, *B. oleracea* 2n = 2x = 18, *B. napus* 2n = 2x = 38, *B. juncea* 2n = 2x = 36. *** *B. oleracea* 2n = 2x = 18, *B. juncea* 2n = 2x = 36. **** the species is a natural tetraploid 2n = 4x: *S. nigrum* ssp. *villosum* 2n = 4x = 48; *Physalis peruviana* L. 2n = 4x = 48.

3.2. Antimitotic Agent Application Method

Various modifications of antimitotic agent treatment were used in the literature. Antimitotic agents can be added at different stages of plant development that can affect plant survival and the efficiency of genome doubling (Table 1). Antimitotic agents can be added to isolated microspores [78,82,94,101,140,141] and ovules [79,83,102,103] with subsequent transfer to antimitotic free medium. If doubling occurs at one of the first of the microspore mitotic divisions, mixoploids may not form [6,141]. However, antimitotic treatment can negatively affect microspore survival and normal development [4]. Regenerated embryoids [81,142,143] or entire plantlets [17,92,122,144], apices [91,122], or roots [41,91,113,116] of regenerated plants can be used for treatment too.

Alternatively, plants can be treated via the injection of antimitotic agents into the plantlet stems. However, this protocol is laborious and can lead to the formation of mixoploids, low survival, and seed production [41,140,145–147].

For polyploidy induction from diploid plants, the antimitotic treatment of seeds, especially pre-germinated seeds with emerging roots, can be the most effective way to double the genome [7,112,117].

3.3. Antimitotic Agent Supplementary Compounds

Antimitotic treatment efficiency can be improved with the addition of supplementary compounds. For instance, to increase the antimitotic compound solubility, stock solutions are prepared in different solvents. Trifluralin is dissolved in acetone, APM is dissolved in methanol, and oryzalin is dissolved in 1 M NaOH, 70% ethanol, or acetone. Colchicine can be dissolved in 96% ethanol, water, culture medium, or 2% dimethyl sulfoxide (DMSO) [1,148]. DMSO is used to increase cell permeability for antimitotic compounds. However, it also exerts a toxic effect on the cells. The application of colchicine with the addition of 2 or 4% DMSO decreased the survival rate compared to colchicine dissolved in water, but the doubling rate increased [144,149]. The addition of surfactants such as Tween 20, Teepol, or other detergents also increases cell penetration by antimitotic compounds [1,144,150].

Cotton balls, lanolin paste, glycerol, or agar can be used to localize the antimitotic solution [7,151]. Caffeine can be a promising supplement to reduce the number of albino regenerants, as shown in wheat. A 0.5 mM caffeine treatment for 24 h significantly increased the fraction of normal regenerants in two of the six spring wheat crosses [152]. Caffeine could be tested in other species, including vegetable crops, to study if it can improve survival or reduce the number of albino plants.

3.4. Antimitotic Agent Exposure Time

Antimitotic exposure time is one of the key factors determining the success of genome doubling. It depends on the cell cycle length and the accessibility of the compound. If the treatment is too short, only a small fraction of cells that enter cell division during the treatment will double their genome. As a result, no ploidy changes occur or mixoploid plants form. For better results, actively growing plants [6] and the sufficient treatment time to have most cells enter mitosis must be used. On the other hand, an excessive antimitotic treatment is toxic and leads to cell death [123]. Since the cell cycle time and sensitivity to the antimitotic compound depend on genotype and a number or external factors, the treatment time is determined empirically in different studies. The most commonly used exposure time range is 3–72 h (Table 1). After the treatment, the plant material is transferred to antimitotic-free medium, or the antimitotic is washed off from the roots or apices.

4. Chemically Induced Haploid Genome Doubling in Vegetable Crops

The successful genome doubling protocol depends on multiple factors and has to be adjusted for the species and genotype of interest (Figure 2). However, published data can be a good starting point for the experimental design (Table 1). Genome doubling protocols for vegetable crops are discussed below.

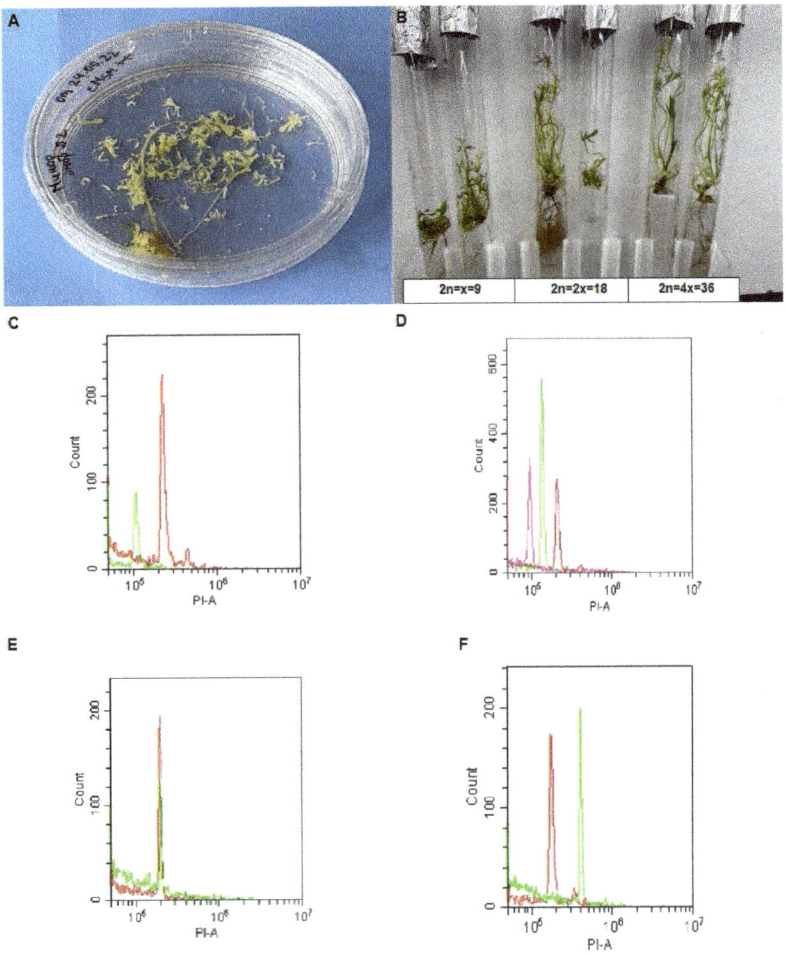

Figure 2. Carrot (*D. carota* L.) haploid regenerant artificial genome doubling with colchicine. (**A**) Carrot embryoids growing from microspores in liquid MSm medium. (**B**) Plants regenerated from embryoids were treated with 0.5 g/L colchicine for 48 h. Untreated haploid control tubes are shown on the left, the colchicine-treated plants that underwent one or two rounds of genome duplication are shown in the center and on the right, respectively. (**C–F**) The cytometric analysis of nuclei isolated from carrot leaves and stained with propidium iodide. The analyzed sample (green peak) was compared with a control diploid plant (red peak) or a control haploid plant (purple peak). (**C**,**D**) Haploid plants (**C**) were used for colchicine treatment; aneuploid plants (**D**) were discarded. (**E**,**F**) The colchicine-treated plants were analyzed by flow cytometry 6 weeks after treatment. Out of 8 treated plants, 2 (25%) became diploid (**E**) and 6 (75%) became tetraploid (**F**).

There are no data on the efficiency of red beet (*B. vulgaris* subsp. *vulgaris*) (the Amaranthaceae family) colchicine treatment. However, there are a number of papers on sugar beet (*B. vulgaris* subsp. *vulgaris* (var. *saccharifera*)) genome doubling. APM treatment yielded 64% genome doubling [103]. In other studies, treatment with different antimitotic compounds produced 2–4.7% doubling [79], 8.4% doubling [105], 19% [104], or doubling efficiency was not reported [16,74,83,102].

Low rates of spontaneous doubling in onion (*A. cepa* L.) (the Amaryllidaceae family) make an antimitotic treatment a necessary step. The application of an antimitotic agent was attempted at different stages of development, including embryos [81,93], plantlets in vitro [34,86,90,92], shoot apices, and root tips [91]. The least laborious approach was the antimitotic treatment of embryos [153]. Trifluralin, oryzalin, and amiprofos-methyl (APM) were used successfully in many studies [34,81,86,90,92,93] (Table 1). Genome doubling efficiency was ranging from over 30% [81,86,90,93], 46% [91], 57.1%, and 65.7% [34] to 100% [92].

To date, there are few studies on the efficiency of induced chromosome doubling in the Apiaceae family. The most commonly used method for genome doubling in Apiaceae species is colchicine treatment (Table 1). For carrot (*D. carota* L.), a 0.5 g/L colchicine treatment was used for the in vitro culture of isolated microspores for 24 h [114] or 48 h [115]. In caraway (*Carum carvi*) [113], fennel (*Foeniculum vulgare*) and dill (*Anethum graveolens*) roots were treated with colchicine [116]. No data on the doubling efficiency were presented by the authors. In our preliminary studies, we treated eight carrot haploid plants produced from microspores with 0.5 g/L colchicine for 48 h (Figure 2A,B). Two plants (25%) became diploid, and six regenerants became tetraploid, as assessed by flow cytometry (Figure 2C–F).

Brassicaceae crops often have a high spontaneous doubling rate and do not need antimitotic treatment. If needed, antimitotic treatment can be applied. For instance, an over 50% doubling rate was observed when the haploid seedling roots of cabbage (*B. oleracea* var. *capitata*) and broccoli (*B. oleracea* var. *italica*) were treated with colchicine. However, colchicine negatively affected plant survival [41].

In *Cucurbita* spp., Kurtar (2018) doubled winter squash (*Cucurbita maxima* Duch.) and pumpkin (*Cucurbita moschata* Duch.) haploid regenerants obtained in the anther culture using repeated 1% colchicine one-hour treatments of ex vitro shoot tips. The best reported efficiency was 93.3%. However, the actual data showing plant regeneration from microspores and ploidy measurement data were not shown [154]. Colchicine doubling protocols were tested in melon [121], pumpkin [122], and cucumber [120], but the doubling efficiency was not reported.

In Solanaceae crops, antimitotic treatment protocols were tested in a number of species. In pepper (*C. annuum* L.), the colchicine treatment of anther culture increased the doubling rate by 14.1–17.3% compared to SHGD. In addition, in the presence of colchicine, the embryo formation increased by 36.8% or 89.5%, depending on the media, compared to the control [55]. In another study, pepper (*C. annuum* L.) haploid plant treatment with colchicine led to a 25–27% incidence of diploids and 29–55% rate of mixoploids [136]. The obtained DH lines are successfully used for breeding. For instance, DH pepper (*C. annuum* L.) lines resistant to bacterial spot (*Xanthomonas campestris* pv. *vesicatoria*) [155], *Verticillium dahliae* Kleb. [156], or carrying nematode resistance genes (Me1, Me3, N, and Me7) [139] were obtained. Elite homozygous sweet pepper [156,157], long sweet pepper [158], and four minipaprika [159] lines recommended for variety testing or new approved varieties with improved fruit qualities and productivity were created using DH technology.

In eggplant (*Solanum melongena* L.), an in vitro antimitotic treatment was more effective, yielding from 35% [107] to 100% plants with doubled genome [128,129]. In vitro treatment also saved time, allowing for earlier doubled haploid plant development and blooming [128]. The colchicine treatment of ex vitro plants in eggplant was also used in some studies. Axillary bud ex vitro treatment increased the doubling rate by 25% compared to SHGD in the untreated control [17]. In another study, eggplant ex vitro plant axillary buds were treated with colchicine in lanolin paste to minimize the evaporation of the antimitotic compound. A total of 50–70% of treated plants became diploid [128]. However, a similar treatment of tomatillo (*P. ixocarpa* Brot.) was considered unsuccessful. The best protocol was to excise the apical or axillary buds from the regenerants and place them on MS medium with 0.05 or 0.1% colchicine. The treatment yielded 50% diploids and 50% tetraploids [59].

5. Artificial Polyploidization for Breeding Purposes

Polyploidization is a common event in the evolutionary history of angiosperms. Polyploidy was documented in over 80% of plant species. It is attributed to 2–4% of angiosperm speciation [160]. The mechanisms of how polyploidy affects phenotype are not fully understood, but it is likely to be a multitude of factors, including a larger cell size, increased heterozygosity level, gene dosage, and new epigenetic and genetic interactions [7].

Polyploidy can give an evolutionary advantage to a species. It can also be useful for plant breeders since polyploidy can improve plant agronomic traits. Many domesticated crops, including durum wheat (*Triticum turgidum subsp. durum*), oat (*Avena sativa* L.), millet (*Panicum miliaceum*), potatoes (Solanum tuberosum L.), cotton (*Gossypium hirsutum*), and sweet potato (*Ipomoea* batatas), are polyploid organisms [160]. Polyploidy can occur naturally due to cell division failure or the fusion of unreduced gametes or via artificial polyploidization by antimitotic agents. Polyploidy can increase the organ size, useful substance content, tolerance to stressors, diseases, insects, and other traits [7].

5.1. Chemically Induced Polyploidization in Vegetable Crops

Polyploidization is often utilized in vegetable crops. Polyploidy induction in *Cucurbitaceae* crops plays a major role in obtaining varieties with superior qualities. Triploid hybrids are produced by crossing diploid and tetraploid watermelons. This method was first developed by Kihara in 1951. Triploid watermelons produce seedless fruits with an increased yield and other favorable traits [161]. The colchicine, oryzalin, or ethalfluralin treatment of watermelon in vitro shoots yielded over 60% tetraploids [123].

In other species, polyploidization is also attempted with the goal of improving the agricultural value. Colchicine treatment of Katokkon pepper (*C. annuum* L.) seeds produced 50% of mixoploid plants [162]. Genome doubling in Apiaceae species was performed with the goal of increasing the size and improving the essential oil content. Parsley (*Petroselinum crispum* L.) [117] and ajowan (*Trachyspermum ammi*) [112] treatment with colchicine produced 100% and 11.53% tetraploids, respectively. Tetraploid parsley and ajowan had larger plant size (1.4 and 2 times larger, respectively). The stomata size was increased, while the stomata density decreased in both studies [112,117]. Tetraploid ajowan had a 39% larger thymol content in essential oil [112].

Garlic and shallot are propagated vegetatively; therefore, breeders have a high need to increase the genetic diversity of these crops. One of the approaches that is being taken into account is the production of tetraploids. Persian shallot (*Allium hirtifolium*) tetraploids produced with colchicine or oryzalin treatment had increased the total phenolic compound and allicin content by 27 and 15%, respectively [111]. Garlic (*Allium sativum* L.) tetraploid plants obtained by colchicine treatment had a three times larger leaf area, and their allicin content increased by 30% [106].

5.2. Chemically Induced Polyploidization in Interspecific Hybrids

Chromosome doubling is also used to restore the fertility of interspecific hybrids. Interspecific hybridization is widely used for the introgression of valuable traits, including resistance to diseases, herbicides, salinity, extreme temperature, and others. Interspecific hybrids are obtained by embryo rescue or natural seed setting [19]. However, most interspecific hybrids are sterile, and overcoming their sterility can be a non-trivial task. However, this is a necessary step to reproduce hybrid forms and perform backcrossing. One of the approaches to overcome sterility is genome doubling [163]. For example, colchicine treatment was used to restore fertility in Japanese bunching onion (*Allium fistulosum*) × onion (*A. cepa* L.) interspecific hybrids [110,164]. The genome doubling in *Allium cepa* × *A. fistulosum* hybrids was necessary for those plants to produce seeds. Plants grown from those seeds demonstrated robust shoots and inflorescences that were larger than their parent plants. They tolerated adverse environmental conditions well. They also produced 1.5 times more green mass that contained more sugars and vitamin C than *A. fistulosum* samples [110].

Interspecific onion hybrids (*A. cepa* L. (2n = 2x = 16) × *A. vavilovii* (2n = 4x = 32), *A. cepa* L. (2n = 2x = 16) × *A. nutans* (2n = 4x = 32), and *A. cepa* L. (2n = 2x = 16) × *A. schoenoprasum* (2n = 4x = 32)) meristems were treated with 0.01 g/L colchicine, which recovered fertility in over 60% of hybrid onion plants. By doubling the chromosome set of triploid forms (2n = 3x = 24), hexaploids (2n = 6x = 48) with fertile pollen were obtained. Hexaploid plants *A. cepa* L.× *A. nutans* (6x) were subsequently used as a mother plant for crossing with *A. cepa* L. (2x) to obtain fertile tetraploid interspecific hybrids. The resulting hybrid forms were perennial, wintered well, had an increased vegetative mass, and were resistant to downy mildew [110]. A similar approach was utilized in Brassicaceae crops. , crosses were treated with colchicine by soaking the axillary meristems, and resulting plants with a doubled genome were able to set seeds in contrast to the original plants. The best outcome was observed for the 1.5–2 g/L colchicine treatment. As a result, allohexaploid (AABBCC) crosses that do not exist in nature were obtained. These plants can be useful for combining valuable traits from A, B, and C *Brassica* genomes [118]. In another study, Chinese cabbage (*B. rapa* subsp. *pekinensis*) × white cabbage (*B. oleracea* var. *capitata*) and rapeseed (*B. napus*)/leaf mustard (*B. juncea*) × Chinese cabbage (*Brassica rapa* subsp. *pekinensis*) crosses grown in vitro were submerged in 0.1 g/L colchicine for 4 h. The treated plants were fertile and were used for subsequent backcrossing. This study allowed for plants with new traits to be produced; the white cabbage with orange/yellow inner leaves trait (probably due to increased β-carotene levels) was transferred from Chinese cabbage. Also, Chinese cabbage with increased anthocyanin content (4.7 mg/g) was obtained by interspecific hybridization with rapeseed or leaf mustard [119].

Interspecific hybridization in *Solanaceae* also allows for the transfer of valuable traits. For instance, *C. annuum* L. × *C. chinense* crosses allowed for the transfer of the Tomato spotted wilt virus (TSWV) resistance gene from *C. chinense*. In total, 40 DH regenerants were obtained from both donor plants. F2 *C. annuum* L. (variety Zdorovye) × *C. chinense* and BC2 *C. annuum* L. (variety Zdorovye) × *C. chinense* × Zdorovye × Zdorovye were generated. The colchicine treatment recovered fertility. The fruit shape was intermediate between the two species. The weight of fruits in plants obtained from F2 *C. annuum* L. (Zdorovye) × C. chinense was three times larger than in the parental form of *C. chinense* and almost two times smaller than the fruits of the parental form of *C. annuum* L. (Zdorovye). Plants obtained from BC2 *C. annuum* L. (Zdorovye) × C. chinense × Zdorovye × Zdorovye had fruits with a weight that was equal or slightly inferior to the fruits of the Zdorovye variety [138].

Altogether, polyploidization can be a useful tool to tackle a list of problems faced by a breeder, including improving cultivar characteristics and overcoming interspecific hybrid sterility.

6. Ploidy Determination Methods

The analysis of ploidy is essential for genome doubling experiments and for the determination of spontaneously doubled haploid regenerants. Plants with different ploidy often differ in terms of size and morphology, as well as the development of reproductive organs. Haploid plants usually have smaller flowers, abnormal ovary development, irregular and uneven anther development, as well as smaller leaves. Seed setting is used as a marker to distinguish DHs from haploid plants since the latter ones are mostly sterile [4].

The most accurate ploidy determination methods are chromosome counting and the flow cytometry of isolated and stained nuclei since they allow for a direct genome size assessment.

The flow cytometry of cell nuclei is one of the best options for ploidy assessment since it allows for the fast and precise examination of samples. Tens and even hundreds of samples per day can be tested, which is advantageous for large-scale breeding programs. Also, this method allows for ploidy determination at any stage of plant development in vitro and ex vitro. Flow cytometry is the only method that provides information about thousands of cells in the sample, allowing for mixoploid plant documentation [165].

For flow cytometric analysis, plant tissue (mostly leaves) is minced with a razor blade. The releasing nuclei are filtered to remove debris. Then, they are stained with a DNA-specific fluorescent dye, such as propidium iodide. The cytometer detects the fluorescence brightness of the nuclei. Since the dye stoichiometrically binds DNA, it allows for the differences in the DNA content between the standard and the analyzed sample to be distinguished (Figure 1G–G″, Figure 2C–F). Although rapid processing samples have made flow cytometry the most efficient approach for determining the ploidy, its use in many laboratories is still limited due to the high cost of equipment and challenges in mastering quality sample preparation and instrument management [165,166].

The classical method of ploidy determination is chromosome counting in stained cytological preparations (Figure 1I). The root tips or developing flower buds are fixed, macerated with enzymes, and stained with DNA dyes for chromosome visualization. Direct chromosome counting is a very reliable method for ploidy determination, but this method is extremely time-consuming and technically challenging [166,167].

Also, the number of stomata guard cells in the field of view, the size of the stomata, and the number of chloroplasts in the guard cells of the stomata can be used to distinguish plants with different ploidy. The smaller the number of stomata, the larger the stomata, and the larger the number of chloroplasts in stomatal guard cells, the more they are associated with higher ploidy (Figure 1H–H″). This method is widely used since it is fast and inexpensive, but it can be applied only to ex vitro plants raised under the same conditions [166,167].

7. Conclusions

Increasing plant ploidy is a critical step for many biotechnological methods used for new breeding material production. In this review, we discussed genome doubling in vegetable crops for different applications, including DH technologies, obtaining polyploids and overcoming interspecific hybrid sterility.

Vegetable crops are a challenging subject for DH technologies. During DH production, microspores or ovules switch their development to sporophytic growth. At the second step, genome doubling is induced in haploid regenerants. Vegetable crops often pose difficulties at one or both stages of DH production. Brassicaceae crops can produce a largely varying number of haploid embryoids depending on a genotype, but SHGD is very common for them [41]. As a result, spontaneously doubled regenerants from responsive genotypes can be included in breeding programs. In Apiaceae, Brassicaceae, Cucurbitaceae, and Solanaceae crops, one or both stages of DH technologies are still challenging [9,17,113,116,122]. Improving both stages would allow for the implementation of DH technologies in vegetable crop breeding on a routine basis.

Multiple factors have to be adjusted to maximize the genome doubling rate. The published protocols can be used as a basis for experimental design, but different treatment regimens should be tested for the best results. The antimitotic agent and supplementary compound choice and concentration, exposure time, and the method of application should be tested on a small population of the plants of interest or on plants that are readily available. For instance, plants from a similar cultivar grown from seeds or micropropagated in vitro can be used. If no doubling occurs, the increased antimitotic agent concentration, exposure time, or a different application method can be tested. If mixoploidy or higher-than-needed ploidy is achieved, the plants can be unusable for future breeding purposes. To avoid this, the antimitotic agent concentration or exposure time should be adjusted. Excessive concentrations of antimitotic agents or DMSO can lead to plant death. The genotype, the quality of the reagents, and the researcher technique can also affect the genome doubling results. When the best regimen is found, it can be applied to the limited breeding material.

Unfortunately, many published DH protocols do not report the antimitotic agent application details and the efficiency of the treatment (Table 1). Interspecific hybrid genome doubling protocols are also often omitted or not covered in full. The assessment of the chemically induced genome doubling efficiency can be complicated by spontaneously doubled plants. The best design for experiments targeted to determine the best antimitotic doubling protocols should include an untreated control, but this is often lacking in published studies. In addition, the number of treated plants is often insufficient for the protocol efficiency statistical assessment or comparison with other treatment regimens. This can happen because researchers often have a practical goal of obtaining plants with a doubled genome from a limited amount of breeding material. Well-designed genome doubling experiments would provide valuable information for other researchers for the use and further development of genome doubling protocols.

Author Contributions: Conceptualization, M.F. and E.D.; investigation, M.F., Y.K., K.A. and E.D.; resources, M.F. and E.D.; writing—original draft preparation, M.F., Y.K. and E.D.; writing—review and editing, M.F.; supervision, M.F. and E.D.; funding acquisition, M.F. All authors have read and agreed to the published version of the manuscript.

Funding: The work was funded by a grant from the Russian Science Foundation, grant number 23-76-01034 (https://rscf.ru/en/project/23-76-01034/ (accessed on 16 March 2024)).

Data Availability Statement: Not applicable.

Conflicts of Interest: The authors declare no conflicts of interest.

References

1. Dhooghe, E.; Van Laere, K.; Eeckhaut, T.; Leus, L.; Van Huylenbroeck, J. Mitotic Chromosome Doubling of Plant Tissues in Vitro. *Plant Cell Tissue Organ Cult.* **2011**, *104*, 359–373. [CrossRef]
2. Madani, H.; Escrich, A.; Hosseini, B.; Sanchez-Muñoz, R.; Khojasteh, A.; Palazon, J. Effect of Polyploidy Induction on Natural Metabolite Production in Medicinal Plants. *Biomolecules* **2021**, *11*, 899. [CrossRef] [PubMed]
3. Trojak-Goluch, A.; Kawka-Lipińska, M.; Wielgusz, K.; Praczyk, M. Polyploidy in Industrial Crops: Applications and Perspectives in Plant Breeding. *Agronomy* **2021**, *11*, 2574. [CrossRef]
4. Ahmadi, B.; Ebrahimzadeh, H. In Vitro Androgenesis: Spontaneous vs. Artificial Genome Doubling and Characterization of Regenerants. *Plant Cell Rep.* **2020**, *39*, 299–316. [CrossRef]
5. Hooghvorst, I.; Ribas, P.; Nogués, S. Chromosome Doubling of Androgenic Haploid Plantlets of Rice (*Oryza sativa*) Using Antimitotic Compounds. *Plant Breed.* **2020**, *139*, 754–761. [CrossRef]
6. Kasha, K.J. Chromosome Doubling and Recovery of Doubled Haploid Plants. In *Haploids in Crop Improvement II*; Don Palmer, C.E., Keller, W.A., Kasha, K.J., Eds.; Biotechnology in Agriculture and Forestry; Springer: Berlin/Heidelberg, Germany, 2005; pp. 123–152, ISBN 978-3-540-26889-5.
7. Manzoor, A.; Ahmad, T.; Bashir, M.A.; Hafiz, I.A.; Silvestri, C. Studies on Colchicine Induced Chromosome Doubling for Enhancement of Quality Traits in Ornamental Plants. *Plants* **2019**, *8*, 194. [CrossRef] [PubMed]
8. Touchell, D.H.; Palmer, I.E.; Ranney, T.G. In Vitro Ploidy Manipulation for Crop Improvement. *Front. Plant Sci.* **2020**, *11*, 517580. [CrossRef] [PubMed]
9. Alan, A.R. Doubled Haploid Onion (*Allium cepa* L.) Production Via In Vitro Gynogenesis. In *Doubled Haploid Technology: Volume 1: General Topics, Alliaceae, Cereals*; Segui-Simarro, J.M., Ed.; Methods in Molecular Biology; Springer: New York, NY, USA, 2021; Volume 1, pp. 151–169. ISBN 978-1-07-161315-3.
10. Escobar-Guzmán, R.; Ochoa-Alejo, N. Anther Culture of the Gametophytic Self-Incompatible Species *Physalis ixocarpa* Brot. In *Doubled Haploid Technology: Volume 2: Hot Topics, Apiaceae, Brassicaceae, Solanaceae*; Segui-Simarro, J.M., Ed.; Springer: New York, NY, USA, 2021; pp. 319–326. ISBN 978-1-07-161335-1.
11. Asif, M. *Progress and Opportunities of Doubled Haploid Production*; SpringerBriefs in Plant Science; Springer International Publishing: Heidelberg, Germany, 2013; Volume 6, ISBN 978-3-319-00731-1.
12. Seguí-Simarro, J.M.; Jacquier, N.M.A.; Widiez, T. Overview of In Vitro and In Vivo Doubled Haploid Technologies. In *Doubled Haploid Technology: Volume 1: General Topics, Alliaceae, Cereals*; Segui-Simarro, J.M., Ed.; Springer: New York, NY, USA, 2021; pp. 3–22, ISBN 978-1-07-161315-3.
13. Pivovarov, V.F.; Pyshnaya, O.N.; Gurkina, L.K. Vegetables are products and raw material for functional nutrition. *Vopr. Pitan.* **2017**, *86*, 121–127. [CrossRef]
14. Behera, T.; Devi, J.; Tiwari, J.; Singh, B. Vegetable Breeding: Status and Strategies. *Veg. Sci.* **2023**, *50*, 131–145. [CrossRef]
15. Thakur, P.; Kumari, N.; Kumar, A.; Sharma, P.; Chadha, S. Recent Advances in Development and Utilization of Double Haploids (DHs) in Economically Important Vegetable Crops. *Plant Cell Tissue Organ Cult.* **2023**, *156*, 15. [CrossRef]
16. Gurel, S.; Pazuki, A.; Aflaki, F.; Gurel, E. Production of Doubled Haploid Sugar Beet (*Beta vulgaris* L.) Plants Through Gynogenesis. In *Doubled Haploid Technology: Volume 3: Emerging Tools, Cucurbits, Trees, Other Species*; Segui-Simarro, J.M., Ed.; Methods in Molecular Biology; Springer: New York, NY, USA, 2021; pp. 313–323. ISBN 978-1-07-161331-3.
17. Mir, R.; Calabuig-Serna, A.; Seguí-Simarro, J.M. Doubled Haploids in Eggplant. *Biology* **2021**, *10*, 685. [CrossRef] [PubMed]
18. Jaskani, M.J.; Kwon, S.W.; Kim, D.H. Comparative Study on Vegetative, Reproductive and Qualitative Traits of Seven Diploid and Tetraploid Watermelon Lines. *Euphytica* **2005**, *145*, 259–268. [CrossRef]

19. Rogo, U.; Fambrini, M.; Pugliesi, C. Embryo Rescue in Plant Breeding. *Plants* **2023**, *12*, 3106. [CrossRef] [PubMed]
20. Shrestha, S.; Koo, D.-H.; Evers, B.; Wu, S.; Walkowiak, S.; Hucl, P.; Pozniak, C.; Fritz, A.; Poland, J. Wheat Doubled Haploids Have a Marked Prevalence of Chromosomal Aberrations. *Plant Genome* **2023**, *16*, e20309. [CrossRef] [PubMed]
21. Piskorz, E.W.; Xu, L.; Ma, Y.; Jiang, H. Doubled-Haploid Induction Generates Extensive Differential DNA Methylation in Arabidopsis. *J. Exp. Bot.* **2023**, *74*, 835–847. [CrossRef] [PubMed]
22. Zhou, K.; Fleet, P.; Nevo, E.; Zhang, X.; Sun, G. Transcriptome Analysis Reveals Plant Response to Colchicine Treatment during on Chromosome Doubling. *Sci. Rep.* **2017**, *7*, 8503. [CrossRef] [PubMed]
23. Hoekstra, S.; van Zijderveld, M.H.; Heidekamp, F.; van der Mark, F. Microspore Culture of Hordeum Vulgare L.: The Influence of Density and Osmolality. *Plant Cell Rep.* **1993**, *12*, 661–665. [CrossRef]
24. Cho, M.S.; Zapata, F.J. Plant Regeneration from Isolated Microspore of Indica Rice. *Plant Cell Physiol.* **1990**, *31*, 881–885. [CrossRef]
25. Castillo, A.M.; Allue, S.; Costar, A.; Alvaro, F.; Valles, M.P. Doubled Haploid Production from Spanish and Central European Spelt by Anther Culture. *J. Agric. Sci. Technol.* **2019**, *21*, 1313–1324.
26. Barnabás, B.; Obert, B.; Kovács, G. Colchicine, an Efficient Genome-Doubling Agent for Maize (*Zea mays* L.) Microspores Cultured in Anthero. *Plant Cell Rep.* **1999**, *18*, 858–862. [CrossRef]
27. Seguí-Simarro, J.M.; Nuez, F. Pathways to Doubled Haploidy: Chromosome Doubling during Androgenesis. *Cytogenet. Genome Res.* **2008**, *120*, 358–369. [CrossRef]
28. Zayachkovskaya, T.; Domblides, E.; Zayachkovsky, V.; Kan, L.; Domblides, A.; Soldatenko, A. Production of Gynogenic Plants of Red Beet (*Beta vulgaris* L.) in Unpollinated Ovule Culture In Vitro. *Plants* **2021**, *10*, 2703. [CrossRef]
29. Kiszczak, W.; Kowalska, U.; Burian, M.; Podwyszyńska, M.; Górecka, K. Influence of Polyamines on Red Beet (*Beta vulgaris* L. Ssp. Vulgaris) Gynogenesis. *Agronomy* **2023**, *13*, 537. [CrossRef]
30. Górecka, K.; Krzyżanowska, D.; Kowalska, U.; Kiszczak, W.; Podwyszyńska, M. Development of Embryoids by Microspore and Anther Cultures of Red Beet (*Beta vulgaris* L. subsp. vulgaris). *J. Cent. Eur. Agric.* **2017**, *18*, 185–195. [CrossRef]
31. Bossoutrot, D.; Hosemans, D. Gynogenesis in *Beta vulgaris* L.: From in Vitro Culture of Unpollinated Ovules to the Production of Doubled Haploid Plants in Soil. *Plant Cell Rep.* **1985**, *4*, 300–303. [CrossRef]
32. Lux, H.; Herrmann, L.; Wetzel, C. Production of Haploid Sugar Beet (*Beta vulgaris* L.) by Culturing Unpollinated Ovules. *Plant Breed.* **1990**, *104*, 177–183. [CrossRef]
33. Alan, A.R.; Mutschler, M.A.; Brants, A.; Cobb, E.; Earle, E.D. Production of Gynogenic Plants from Hybrids of *Allium cepa* L. and *A. roylei* Stearn. *Plant Sci.* **2003**, *165*, 1201–1211. [CrossRef]
34. Geoffriau, E.; Kahane, R.; Bellamy, C.; Rancillac, M. Ploidy Stability and in Vitro Chromosome Doubling in Gynogenic Clones of Onion (*Allium cepa* L.). *Plant Sci.* **1997**, *122*, 201–208. [CrossRef]
35. Jakše, M.; Hirschegger, P.; Bohanec, B.; Havey, M.J. Evaluation of Gynogenic Responsiveness and Pollen Viability of Selfed Doubled Haploid Onion Lines and Chromosome Doubling via Somatic Regeneration. *J. Am. Soc. Hortic. Sci.* **2010**, *135*, 67–73. [CrossRef]
36. Hu, K.L.; Matsubara, S.; Murakami, K. Haploid Plant Production by Anther Culture in Carrot (*Daucus carota* L.). *J. Jpn. Soc. Hort. Sci.* **1993**, *62*, 561–565. [CrossRef]
37. Li, J.-R.; Zhuang, F.-Y.; Ou, C.-G.; Hu, H.; Zhao, Z.-W.; Mao, J.-H. Microspore Embryogenesis and Production of Haploid and Doubled Haploid Plants in Carrot (*Daucus carota* L.). *Plant Cell Tissue Organ Cult.* **2013**, *112*, 275–287. [CrossRef]
38. Kiszczak, W.; Krzyżanowska, D.; Strycharczuk, K.; Kowalska, U.; Wolko, B.; Górecka, K. Determination of Ploidy and Homozygosity of Carrot Plants Obtained from Anther Cultures. *Acta Physiol. Plant* **2011**, *33*, 401–407. [CrossRef]
39. Kiełkowska, A.; Adamus, A. In Vitro Culture of Unfertilized Ovules in Carrot (*Daucus carota* L.). *Plant Cell Tissue Organ Cult.* **2010**, *102*, 309–319. [CrossRef]
40. Ferrie, A.M.R.; Bethune, T.D.; Mykytyshyn, M. Microspore Embryogenesis in Apiaceae. *Plant Cell Tissue Organ Cult.* **2011**, *104*, 399–406. [CrossRef]
41. Yuan, S.; Su, Y.; Liu, Y. Chromosome Doubling of Microspore-Derived Plants from Cabbage (*Brassica oleracea* var. *capitata* L.) and Broccoli (*Brassica oleracea* var. *italica* L.). *Front. Plant Sci.* **2015**, *6*, 1118. [CrossRef]
42. Gu, H.H.; Zhou, W.J.; Hagberg, P. High Frequency Spontaneous Production of Doubled Haploid Plants in Microspore Cultures of *Brassica rapa* ssp. *chinensis*. *Euphytica* **2003**, *134*, 239–245. [CrossRef]
43. Niu, L.; Shi, F.; Feng, H.; Zhang, Y. Efficient Doubled Haploid Production in Microspore Culture of Zengcheng Flowering Chinese Cabbage (*Brassica campestris* L. ssp. *chinensis* [L.] Makino var. *utilis* Tsen et Lee). *Sci. Hortic.* **2019**, *245*, 57–64. [CrossRef]
44. Shumilina, D.V.; Shmykova, N.A.; Bondareva, L.L.; Suprunova, T.P. Effect of Genotype and Medium Culture Content on Microspore-Derived Embryo Formation in Chinese Cabbage (*Brassica rapa* ssp. *chinensis*) Cv. Lastochka. *Biol. Bull. Russ. Acad. Sci.* **2015**, *42*, 302–309. [CrossRef]
45. Zhang, Y.; Wang, A.; Liu, Y.; Wang, Y.; Feng, H. Improved Production of Doubled Haploids in *Brassica rapa* through Microspore Culture. *Plant Breed.* **2012**, *131*, 164–169. [CrossRef]
46. Prem, D.; Solís, M.-T.; Bárány, I.; Rodríguez-Sanz, H.; Risueño, M.C.; Testillano, P.S. A New Microspore Embryogenesis System under Low Temperature Which Mimics Zygotic Embryogenesis Initials, Expresses Auxin and Efficiently Regenerates Doubled-Haploid Plants in *Brassica napus*. *BMC Plant Biol.* **2012**, *12*, 127. [CrossRef]
47. Domblides, E.; Ermolaev, A.S.; Belov, S.N. Obtaining doubled haploids of Cucurbita pepo L. Vegetable crops of Russia. *Veg. Crops Russ.* **2021**, *4*, 11–26. [CrossRef]

48. Košmrlj, K.; Murovec, J.; Bohanec, B. Haploid Induction in Hull-Less Seed Pumpkin through Parthenogenesis Induced by X-Ray-Irradiated Pollen. *J. Am. Soc. Hortic. Sci.* **2013**, *138*, 310–316. [CrossRef]
49. Nowaczyk, P.; Olszewska, D.; Kisiała, A. Individual Reaction of Capsicum F2 Hybrid Genotypes in Anther Cultures. *Euphytica* **2009**, *168*, 225–233. [CrossRef]
50. Supena, E.D.J.; Suharsono, S.; Jacobsen, E.; Custers, J.B.M. Successful Development of a Shed-Microspore Culture Protocol for Doubled Haploid Production in Indonesian Hot Pepper (*Capsicum annuum* L.). *Plant Cell Rep.* **2006**, *25*, 1–10. [CrossRef]
51. Dumas De Vaulx, R.; Chambonnet, D.; Pochard, E. Culture in Vitro d'anthères de Piment (*Capsicum annuum* L.): Amélioration Des Taux d'obtention de Plantes Chez Différents Génotypes Par Des Traitements à + 35 °C. *Agronomie* **1981**, *1*, 859–864. [CrossRef]
52. Irikova, T.; Grozeva, S.; Rodeva, V. Anther Culture in Pepper (*Capsicum annuum* L.) in Vitro. *Acta Physiol. Plant* **2011**, *33*, 1559–1570. [CrossRef]
53. Keleş, D.; Pınar, H.; Ata, A.; Taşkın, H.; Yıldız, S.; Büyükalaca, S. Effect of Pepper Types on Obtaining Spontaneous Doubled Haploid Plants via Anther Culture. *HortScience* **2015**, *50*, 1671–1676. [CrossRef]
54. Gyulai, G.; Gémesné, J.A.; Sági, Z.S.; Venczel, G.; Pintér, P.; Kristóf, Z.; Törjék, O.; Heszky, L.; Bottka, S.; Kiss, J.; et al. Doubled Haploid Development and PCR-Analysis of F1 Hybrid Derived DH-R2 Paprika (*Capsicum annuum* L.) Lines. *J. Plant Physiol.* **2000**, *156*, 168–174. [CrossRef]
55. Comlekcioglu, N. Effect of Colchicine Addition to Culture Medium on Induction of Androgenesis in Pepper (*Capsicum annuum* L.). *Pak. J. Bot.* **2021**, *53*, 1001–1005. [CrossRef]
56. Dolcet-Sanjuan, R.; Claveria, E.; Huerta, A. Androgenesis in *Capsicum annuum* L.—Effects of Carbohydrate and Carbon Dioxide Enrichment. *J. Am. Soc. Hortic. Sci.* **1997**, *122*, 468–475. [CrossRef]
57. Kim, M.; Park, E.-J.; An, D.; Lee, Y. High-Quality Embryo Production and Plant Regeneration Using a Two-Step Culture System in Isolated Microspore Cultures of Hot Pepper (*Capsicum annuum* L.). *Plant Cell Tissue Organ Cult.* **2013**, *112*, 191–201. [CrossRef]
58. Nowaczyk, P.; Kisiała, A.; Olszewska, D. Induced Androgenesis of *Capsicum frutescens* L. *Acta Physiol. Plant* **2006**, *28*, 35–39. [CrossRef]
59. Escobar-Guzmán, R.E.; Hernández-Godínez, F.; Martínez de la Vega, O.; Ochoa-Alejo, N. In Vitro Embryo Formation and Plant Regeneration from Anther Culture of Different Cultivars of Mexican Husk Tomato (*Physalis ixocarpa* Brot.). *Plant Cell Tissue Organ Cult.* **2009**, *96*, 181–189. [CrossRef]
60. García-Arias, F.; Sanchez, E.; Núñez, V. Fertility Recovery of Anther-Derived Haploid Plants in Cape Gooseberry (*Physalis peruviana* L.). *Agron. Colomb.* **2018**, *36*, 201–209. [CrossRef]
61. Zagorska, N.A.; Shtereva, L.A.; Kruleva, M.M.; Sotirova, V.G.; Baralieva, D.L.; Dimitrov, B.D. Induced Androgenesis in Tomato (*Lycopersicon esculentum* Mill.). III. Characterization of the Regenerants. *Plant Cell Rep.* **2004**, *22*, 449–456. [CrossRef] [PubMed]
62. Shariatpanahi, M.E.; Niazian, M.; Ahmadi, B. Methods for Chromosome Doubling. In *Doubled Haploid Technology: Volume 1: General Topics, Alliaceae, Cereals*; Segui-Simarro, J.M., Ed.; Springer: New York, NY, USA, 2021; pp. 127–148, ISBN 978-1-07-161315-3.
63. Domblides, E.A.; Shmykova, N.A.; Shumilina, D.V.; Zayachkovskaya, T.V.; Vjurtts, T.S.; Kozar, E.V.; Kan, L.Y.; Romanov, V.S.; Domblides, A.S.; Pivovarov, V.F.; et al. Biotechnological Approaches for Breeding Programs in Vegetable Crops. In *VIII International Scientific Agriculture Symposium, "Agrosym 2017", Jahorina, Bosnia and Herzgovina, October 2017. Book of Proceedings*; Faculty of Agriculture, University of East Sarajevo: Jahorina, Bosnia and Herzegovina, 2017; pp. 452–460.
64. Joubès, J.; Chevalier, C. Endoreduplication in Higher Plants. In *The Plant Cell Cycle*; Inzé, D., Ed.; Springer: Dordrecht, The Netherlands, 2000; pp. 191–201. ISBN 978-94-010-0936-2.
65. Daghma, D.E.S.; Hensel, G.; Rutten, T.; Melzer, M.; Kumlehn, J. Cellular Dynamics during Early Barley Pollen Embryogenesis Revealed by Time-Lapse Imaging. *Front. Plant Sci.* **2014**, *5*, 118468. [CrossRef]
66. Boerman, N.A.; Frei, U.K.; Lübberstedt, T. Impact of Spontaneous Haploid Genome Doubling in Maize Breeding. *Plants* **2020**, *9*, 369. [CrossRef]
67. Vagera, J. Pepper (*Capsicum* spp.): In Vitro Induction of Haploids. In *Haploids in Crop Improvement I*; Bajaj, Y.P.S., Ed.; Biotechnology in Agriculture and Forestry; Springer: Berlin/Heidelberg, Germany, 1990; Volume 12, pp. 374–392. ISBN 978-3-642-64856-4.
68. Magyar, Z.; De Veylder, L.; Atanassova, A.; Bakó, L.; Inzé, D.; Bögre, L. The Role of the Arabidopsis E2FB Transcription Factor in Regulating Auxin-Dependent Cell Division. *Plant Cell* **2005**, *17*, 2527–2541. [CrossRef] [PubMed]
69. Lukaszewska, E.; Virden, R.; Sliwinska, E. Hormonal Control of Endoreduplication in Sugar Beet (*Beta vulgaris* L.) Seedlings Growing in Vitro. *Plant Biol.* **2012**, *14*, 216–222. [CrossRef]
70. Mishiba, K.; Okamoto, T.; Mii, M. Increasing Ploidy Level in Cell Suspension Cultures of Doritaenopsis by Exogenous Application of 2,4-Dichlorophenoxyacetic Acid. *Physiol. Plant.* **2001**, *112*, 142–148. [CrossRef]
71. Kubaláková, M.; Doležel, J.; Lebeda, A. Ploidy Instability of Embryogenic Cucumber (*Cucumis sativus* L.) Callus Culture. *Biol. Plant.* **1996**, *38*, 475. [CrossRef]
72. Bartels, P.G.; Hilton, J.L. Comparison of Trifluralin, Oryzalin, Pronamide, Propham, and Colchicine Treatments on Microtubules. *Pestic. Biochem. Physiol.* **1973**, *3*, 462–472. [CrossRef]
73. Leung, Y.Y.; Yao Hui, L.L.; Kraus, V.B. Colchicine—Update on Mechanisms of Action and Therapeutic Uses. *Semin. Arthritis Rheum.* **2015**, *45*, 341–350. [CrossRef]
74. Vasilchenko, E.N.; Zhuzhzhalova, T.P.; Vashchenko, T.G.; Kolesnikova, E.O. The Technology of Creating Restitution Sugar Beet Lines. *Vestn. Voronezh State Agrar. Univ.* **2018**, *1*, 56–64. [CrossRef]

75. PubChem N-(1,2,3,10-Tetramethoxy-9-Oxo-5,6,7,9-Tetrahydrobenzo[a]Heptalen-7-Yl)Acetamide. Available online: https://pubchem.ncbi.nlm.nih.gov/compound/2833 (accessed on 15 March 2024).
76. Sagorin, C.; Ertel, N.H.; Wallace, S.L. Photoisomerization of Colchicine. Loss of Significant Antimitotic Activity in Human Lymphocytes. *Arthritis Rheum.* **1972**, *15*, 213–217. [CrossRef]
77. PubChem Trifluralin. Available online: https://pubchem.ncbi.nlm.nih.gov/compound/5569 (accessed on 15 March 2024).
78. Zhao, J.; Simmonds, D.H. Application of Trifluralin to Embryogenic Microspore Cultures to Generate Doubled Haploid Plants in *Brassica napus*. *Physiol. Plant.* **1995**, *95*, 304–309. [CrossRef]
79. Hansen, A.L.; Gertz, A.; Joersbo, M.; Andersen, S.B. Antimicrotubule Herbicides for in Vitro Chromosome Doubling in *Beta vulgaris* L. Ovule Culture. *Euphytica* **1998**, *101*, 231–237. [CrossRef]
80. Mochalova, O.V.; Gusev, D.A. Induction of Polyploidy for Frutescent Cherry and Bessey Cherry by in Vitro Culture. *Achiev. Sci. Technol. AICis* **2016**, *30*, 36–39.
81. Grzebelus, E.; Adamus, A. Effect of Anti-Mitotic Agents on Development and Genome Doubling of Gynogenic Onion (*Allium cepa* L.) Embryos. *Plant Sci.* **2004**, *167*, 569–574. [CrossRef]
82. Hansen, N.J.P.; Andersen, S.B. In Vitro Chromosome Doubling Potential of Colchicine, Oryzalin, Trifluralin, and APM in *Brassica napus* Microspore Culture. *Euphytica* **1996**, *88*, 159–164. [CrossRef]
83. Gürel, S.; Gürel, E.; Kaya, Z. Doubled Haploid Plant Production from Unpollinated Ovules of Sugar Beet (*Beta vulgaris* L.). *Plant Cell Rep.* **2000**, *19*, 1155–1159. [CrossRef]
84. Wan, Y.; Duncan, D.R.; Rayburn, A.L.; Petolino, J.F.; Widholm, J.M. The Use of Antimicrotubule Herbicides for the Production of Doubled Haploid Plants from Anther-Derived Maize Callus. *Theor. Appl. Genet.* **1991**, *81*, 205–211. [CrossRef]
85. Thao, N.T.P.; Ureshino, K.; Miyajima, I.; Ozaki, Y.; Okubo, H. Induction of Tetraploids in Ornamental Alocasia through Colchicine and Oryzalin Treatments. *Plant Cell Tissue Organ. Cult.* **2003**, *72*, 19–25. [CrossRef]
86. Alan, A.R.; Lim, W.; Mutschler, M.A.; Earle, E.D. Complementary Strategies for Ploidy Manipulations in Gynogenic Onion (*Allium cepa* L.). *Plant Sci.* **2007**, *173*, 25–31. [CrossRef]
87. Hansen, N.J.P.; Andersen, S.B. Efficient Production of Doubled Haploid Wheat Plants by in Vitro Treatment of Microspores with Trifluralin or APM. *Plant Breed.* **1998**, *117*, 401–405. [CrossRef]
88. Melchinger, A.E.; Molenaar, W.S.; Mirdita, V.; Schipprack, W. Colchicine Alternatives for Chromosome Doubling in Maize Haploids for Doubled-Haploid Production. *Crop Sci.* **2016**, *56*, 559–569. [CrossRef]
89. Ren, X. Doubling Effect of Anti-Microtubule Herbicides on the Maize Haploid. *Emir. J. Food Agric.* **2018**, *30*, 903–908. [CrossRef]
90. Alan, A.R.; Brants, A.; Cobb, E.; Goldschmied, P.A.; Mutschler, M.A.; Earle, E.D. Fecund Gynogenic Lines from Onion (*Allium cepa* L.) Breeding Materials. *Plant Sci.* **2004**, *167*, 1055–1066. [CrossRef]
91. Campion, B.; Perri, E.; Azzimonti, M.T.; Vicini, E.; Schiavi, M. Spontaneous and Induced Chromosome Doubling in Gynogenic Lines of Onion (*Allium cepa* L.). *Plant Breed.* **1995**, *114*, 243–246. [CrossRef]
92. Foschi, M.L.; Martínez, L.E.; Ponce, M.T.; Galmarini, C.R.; Bohanec, B. Effect of Colchicine and Amiprophos-Methyl on the Production of in Vitro Doubled Haploid Onion Plants and Correlation Assessment between Ploidy Level and Stomatal Size. *Rev. Fac. Cienc. Agrar., Univ. Nac. Cuyo* **2013**, *45*, 130–140.
93. Jakše, M.; Havey, M.J.; Bohanec, B. Chromosome Doubling Procedures of Onion (*Allium cepa* L.) Gynogenic Embryos. *Plant Cell Rep.* **2003**, *21*, 905–910. [CrossRef]
94. Klutschewski, S. Methodical Improvements in Microspore Culture of *Brassica napus* L. Ph.D. Thesis, Universität Göttingen, Göttingen, Germany, 2013.
95. Fábián, A.; Földesiné Füredi, P.K.; Ambrus, H.; Jäger, K.; Szabó, L.; Barnabás, B. Effect of N-Butanol and Cold Pretreatment on the Cytoskeleton and the Ultrastructure of Maize Microspores When Cultured in Vitro. *Plant Cell Tissue Organ Cult.* **2015**, *123*, 257–271. [CrossRef]
96. Hirase, A.; Hamada, T.; Itoh, T.J.; Shimmen, T.; Sonobe, S. N -Butanol Induces Depolymerization of Microtubules in Vivo and in Vitro. *Plant Cell Physiol.* **2006**, *47*, 1004–1009. [CrossRef] [PubMed]
97. Soriano, M.; Cistué, L.; Castillo, A.M. Enhanced Induction of Microspore Embryogenesis after N-Butanol Treatment in Wheat (*Triticum aestivum* L.) Anther Culture. *Plant Cell Rep.* **2008**, *27*, 805–811. [CrossRef] [PubMed]
98. Castillo, A.M.; Nielsen, N.H.; Jensen, A.; Vallés, M.P. Effects of N-Butanol on Barley Microspore Embryogenesis. *Plant Cell Tissue Organ Cult.* **2014**, *117*, 411–418. [CrossRef]
99. Füredi, P.F.; Ambrus, H.; Barnabás, B. Development of Cultured Microspores of Maize in the Presence of N-Butanol and 2-Aminoethanol. *Acta Agron. Hung.* **2012**, *60*, 183–189. [CrossRef]
100. Füredi, P.K.F.; Ambrus, H.; Barnabás, B. The Effect of N-Butanol and 2-Aminoethanol on the in Vitro Androgenesis of Maize. *Acta Biol. Szeged.* **2011**, *55*, 77–78.
101. Pelliccione, S. Microspore Embryogenesis Induction in Hot Pepper Genotypes. Universidad de Granada. 2013. Available online: http://purl.org/dc/dcmitype/Text (accessed on 16 March 2024).
102. Hansen, A.L.; Gertz, A.; Joersbo, M.; Andersen, S.B. Short-Duration Colchicine Treatment for in Vitro Chromosome Doubling during Ovule Culture of *Beta vulgaris* L. *Plant Breed.* **1995**, *114*, 515–519. [CrossRef]
103. Hansen, A.L.; Gertz, A.; Joersbo, M.; Andersen, S.B. Chromosome Doubling in Vitro with Amiprophos-Methyl in Beta Ulgaris Ovule Culture. *Acta Agric. Scand. Sect. B—Soil Plant Sci.* **2000**, *50*, 89–95. [CrossRef]

104. Svirshchevskaya, A.M.; Dolezel, J. Production and Performance of Gynogenetic Sugarbeet Lines. *J. Sugar Beet Res.* **2000**, *37*, 117–133. [CrossRef]
105. D'Halluin, K.; Keimer, B. Production of Haploid Sugarbeets (*Beta vulgaris* L.) by Ovule Culture. In *Genetic Manipulation in Plant Breeding. Proceedings International Symposium Organized by Eucarpia, September 8–13, 1985, Berlin (West), Germany*; Walter de Gruyter: Berlin, Germany, 1986; pp. 307–309.
106. Dixit, V.; Chaudhary, B.R. Colchicine-Induced Tetraploidy in Garlic (*Allium sativum* L.) and Its Effect on Allicin Concentration. *J. Hortic. Sci. Biotechnol.* **2014**, *89*, 585–591. [CrossRef]
107. Gémes Juhasz, A.; Venczel, G.; Sagi, Z.; Gajdos, L.; Kristof, Z.; Vagi, P.; Zatyko, L. Production of Doubled Haploid Breeding Lines in Case of Paprika, Spice Paprika, Eggplant, Cucumber, Zucchini and Onion. *Acta Hortic.* **2006**, *1282*, 845–854. [CrossRef]
108. Fayos, O.; Vallés, M.P.; Garcés-Claver, A.; Mallor, C.; Castillo, A.M. Doubled Haploid Production from Spanish Onion (*Allium cepa* L.) Germplasm: Embryogenesis Induction, Plant Regeneration and Chromosome Doubling. *Front. Plant Sci.* **2015**, *6*, 384. [CrossRef] [PubMed]
109. Song, P.; Kang, W.; Peffley, E.B. Chromosome Doubling of Allium Fistulosum × A. Cepa Interspecific F1 Hybrids through Colchicine Treatment of Regenerating Callus. *Euphytica* **1997**, *93*, 257–262. [CrossRef]
110. Timin, N.I.; Pyshnaya, O.N.; Agafonov, A.F.; Mamedov, M.I. *Interspecific Hybridization of Vegetable Crops (Allium L., Daucus L., Capsicum L.)*; VNIISSOK publishing: Moscow, Russia, 2013; ISBN 978-5-901695-59-3.
111. Farhadi, N.; Panahandeh, J.; Motallebi-Azar, A.; Mokhtarzadeh, S. Production of Autotetraploid Plants by in Vitro Chromosome Engineering in *Allium hirtifolium*. *Hortic. Plant J.* **2023**, *9*, 986–998. [CrossRef]
112. Sadat Noori, S.A.; Norouzi, M.; Karimzadeh, G.; Shirkool, K.; Niazian, M. Effect of Colchicine-Induced Polyploidy on Morphological Characteristics and Essential Oil Composition of Ajowan (*Trachyspermum ammi* L.). *Plant Cell Tissue Organ Cult.* **2017**, *130*, 543–551. [CrossRef]
113. Smýkalová, I.; Horáček, J. Caraway (*Carum carvi* L.): Anther Culture and Production of DH Plants Caraway. In *Doubled Haploid Technology: Volume 2: Hot Topics, Apiaceae, Brassicaceae, Solanaceae*; Segui-Simarro, J.M., Ed.; Methods in Molecular Biology; Springer: New York, NY, USA, 2021; Volume 2, pp. 91–102. ISBN 978-1-07-161335-1.
114. Górecka, K.; Kowalska, U.; Krzyżanowska, D.; Kiszczak, W. Obtaining Carrot (*Daucus carota* L.) Plants in Isolated Microspore Cultures. *J. Appl. Genet.* **2010**, *51*, 141–147. [CrossRef]
115. Voronina, A.V.; Vishnyakova, A.V.; Monakhos, S.G.; Monakhos, G.F.; Ushanov, A.A.; Mironov, A.A. Effect of Cultivation Factors on Embryogenesis in Isolated Microspore Culture of Carrot (*Daucus carota* L.). *J. Water Land. Dev.* **2022**, *55*, 125–128. [CrossRef]
116. Ferrie, A.M.R. Doubled Haploidy for Fennel (*Foeniculum vulgare* Mill.) and Dill (*Anethum graveolens* L.). In *Doubled Haploid Technology: Volume 2: Hot Topics, Apiaceae, Brassicaceae, Solanaceae*; Segui-Simarro, J.M., Ed.; Methods in Molecular Biology; Springer: New York, NY, USA, 2021; Volume 2, pp. 103–111. ISBN 978-1-07-161335-1.
117. Nasirvand, S.; Zakaria, R.A.; Zare, N.; Esmaeilpoor, B. Polyploidy Induction in Parsley (*Petroselinum crispum* L.) by Colchicine Treatment. *Cytologia* **2018**, *83*, 393–396. [CrossRef]
118. Mwathi, M.W.; Gupta, M.; Quezada-Martinez, D.; Pradhan, A.; Batley, J.; Mason, A.S. Fertile Allohexaploid Brassica Hybrids Obtained from Crosses between B. Oleracea and B. Juncea via Ovule Rescue and Colchicine Treatment of Cuttings. *Plant Cell Tissue Organ Cult.* **2020**, *140*, 301–313. [CrossRef]
119. Pen, S.; Nath, U.K.; Song, S.; Goswami, G.; Lee, J.-H.; Jung, H.-J.; Kim, H.-T.; Park, J.-I.; Nou, I.-S. Developmental Stage and Shape of Embryo Determine the Efficacy of Embryo Rescue in Introgressing Orange/Yellow Color and Anthocyanin Genes of Brassica Species. *Plants* **2018**, *7*, 99. [CrossRef]
120. Claveria, E.; Garcia-Mas, J.; Dolcet-Sanjuan, R. Optimization of Cucumber Doubled Haploid Line Production Using In Vitro Rescue of In Vivo Induced Parthenogenic Embryos. *J. Am. Soc. Hortic. Sci.* **2005**, *130*, 555–560. [CrossRef]
121. d'Hooghvorst, I.; Torrico, O.; Nogués, S. Doubled Haploid Parthenogenetic Production of Melon 'Piel de Sapo'. In *Doubled Haploid Technology: Volume 3: Emerging Tools, Cucurbits, Trees, Other Species*; Segui-Simarro, J.M., Ed.; Methods in Molecular Biology; Springer: New York, NY, USA, 2021; Volume 3, pp. 87–95. ISBN 978-1-07-161331-3.
122. Kurtar, E.S.; Seymen, M. Anther Culture in Cucurbita Species. In *Doubled Haploid Technology: Volume 3: Emerging Tools, Cucurbits, Trees, Other Species*; Segui-Simarro, J.M., Ed.; Methods in Molecular Biology; Springer: New York, NY, USA, 2021; Volume 3, pp. 111–121. ISBN 978-1-07-161331-3.
123. Nasr, M.; Habib, H.; Ibrahim, I.; Kapiel, T. In Vitro Induction of Autotetraploid Watermelons Using Colchicine and Four Dinitroaniline Compounds. In Proceedings of the International Conference Genetic Engineering and Its Applications, Sharm Elsheik, Egypt, 8 April 2004; pp. 1–20.
124. Zhu, H.; Zhao, S.; Lu, X.; He, N.; Gao, L.; Dou, J.; Bie, Z.; Liu, W. Genome Duplication Improves the Resistance of Watermelon Root to Salt Stress. *Plant Physiol. Biochem.* **2018**, *133*, 11–21. [CrossRef]
125. Ojiewo, C.O.; Agong, S.G.; Murakami, K.; Masuda, M. Chromosome Duplication and Ploidy Level Determination in African Nightshade Solanum Villosum Miller. *J. Hortic. Sci. Biotechnol.* **2006**, *81*, 183–188. [CrossRef]
126. González-Chavira, M.M.; Guerrero-Aguilar, B.Z.; Pons-Hernández, J.L.; Escobedo-Landín, M.d.l.Á.; García-Reyna, J.F.; Mora-Avilés, M.A. Induction of Androgenic Embryos and Regeneration of Haploid Plants in Experimental Genotypes of Poblano Chili through Anthers Culture. *Rev. Mex. Cienc. Agríc.* **2023**, *14*, 277–287. [CrossRef]
127. Panda, R.C.; Kumar, O.A.; Rao, K.G.R. Cytomorphology of Induced Octoploid Chili Pepper (*Capsicum annuum* L.). *Theor. Appl. Genet.* **1984**, *68*, 567–570. [CrossRef]

128. Rotino, G.L. Anther Culture in Eggplant (*Solanum melongena* L.). In *In Vitro Embryogenesis in Higher Plants*; Germana, M.A., Lambardi, M., Eds.; Methods in Molecular Biology; Springer: New York, NY, USA, 2016; pp. 453–466. ISBN 978-1-4939-3061-6.
129. Vural, G.E.; Ari, E.; Zengin, S.; Ellialtioglu, S.S. Development of Androgenesis Studies on Eggplant (*Solanum melongena* L.) in Turkey from Past to Present. In *Sustainable Crop Production*; IntechOpen: London, UK, 2019; ISBN 978-1-78985-318-6.
130. Başay, S.; Seniz, V.; Ellialtioglu, S. Obtaining Dihaploid Lines by Using Anther Culture in the Different Eggplant Cultivars. *J. Food Agric. Environ.* **2011**, *9*, 188–190.
131. Bhattacharya, A.; Sonone, Y.; Char, B. Improvement in Tissue Culture-Assisted Induction of Double Haploidy in Brinjal (*Solanum melongena* L.). *JAH* **2019**, *21*, 178–181. [CrossRef]
132. Lantos, C.; Juhász, A.G.; Somogyi, G.; Ötvös, K.; Vági, P.; Mihály, R.; Kristóf, Z.; Somogyi, N.; Pauk, J. Improvement of Isolated Microspore Culture of Pepper (*Capsicum annuum* L.) via Co-Culture with Ovary Tissues of Pepper or Wheat. *Plant Cell Tissue Organ Cult.* **2009**, *97*, 285–293. [CrossRef]
133. Gémesné Juhász, A.; Petus, M.; Venczel, G.; Zatykó, L.; Gyulai, G.; Cséplö, M. Genetic Variability of Anther Donor versus Spontaneous Doubled Haploid Descendents and Colchicine Induced Doubled Haploid Sweet Pepper (*Capsicum annuum* l.) Lines. *Acta Hortic.* **2001**, *560*, 149–152. [CrossRef]
134. İlhan, M.; Kurtar, E.S. Doublehaploidization Efficiency of Selected Pepper Genotypes Via in Vitro Anther Culture. *Selcuk. J. Agr. Food Sci.* **2022**, *36*, 253–259. [CrossRef]
135. Ari, E.; Yildirim, T.; Mutlu, N.; Büyükalaca, S.; Gökmen, Ü.; Akman, E. Comparison of Different Androgenesis Protocols for Doubled Haploid Plant Production in Ornamental Pepper (*Capsicum annuum* L.). *Turk. J. Biol.* **2016**, *40*, 944–954. [CrossRef]
136. Olszewska, D.; Jędrzejczyk, I.; Nowaczyk, P.; Sendel, S.; Gaczkowska, B. In Vitro Colchicine Treatment of Anther-Derived Pepper Haploids. *Bulg. J. Agric. Sci.* **2015**, *21*, 806–810.
137. Mitykó, J.; Fári, M. Problems and Results of Doubled Haploid Plant Production in Pepper (*Capsicum annuum* L.) via Anther- and Microspore Culture. *Acta Hortic.* **1997**, *447*, 281–288. [CrossRef]
138. Shmykova, N.; Pyshnaya, O.; Shumilina, D.; Dzhos, E. Morphological Characteristics of Doubled Haploid Plants of Pepper Produced Using Microspore/Anther in Vitro Culture of the Interspecies Hybrids of *Capsicum annum* L. and *C. chinense* Jacq. *Russ. Agric. Sci.* **2015**, *40*, 417–421. [CrossRef]
139. Öcal, Y.; Taşkın, H.; Pınar, H.; Keleş, D.; Onsinejad, R.; Büyükalaca, S. Development of Nematode Resistant Pure Pepper Lines via Anther Culture Method. *Acta Hortic.* **2019**, *1257*, 29–36. [CrossRef]
140. Soriano, M.; Cistué, L.; Vallés, M.P.; Castillo, A.M. Effects of Colchicine on Anther and Microspore Culture of Bread Wheat (*Triticum aestivum* L.). *Plant Cell Tissue Organ Cult.* **2007**, *91*, 225–234. [CrossRef]
141. Weber, S.; ÜNker, F.; Friedt, W. Improved Doubled Haploid Production Protocol for *Brassica napus* Using Microspore Colchicine Treatment in Vitro and Ploidy Determination by Flow Cytometry. *Plant Breed.* **2005**, *124*, 511–513. [CrossRef]
142. Bürün, B.; Emiroğlu, Ü. A Comparative Study on Colchicine Application Methods in Obtaining Doubled Haploids of Tobacco (*Nicotiana tabacum* L.). *Turk. J. Biol.* **2008**, *32*, 105–111.
143. Mohammadi, P.P.; Moieni, A.; Ebrahimi, A.; Javidfar, F. Doubled Haploid Plants Following Colchicine Treatment of Microspore-Derived Embryos of Oilseed Rape (*Brassica napus* L.). *Plant Cell Tissue Organ Cult.* **2012**, *108*, 251–256. [CrossRef]
144. Węzdony, M. Protocol for Doubled Haploid Production in Hexaploid Triticale (× *Triticosecale* Wittm.) by Crosses with Maize. In *Doubled Haploid Production in Crop Plants: A Manual*; Maluszynski, M., Kasha, K.J., Forster, B.P., Szarejko, I., Eds.; Springer: Dordrecht, The Netherlands, 2003; pp. 135–140, ISBN 978-94-017-1293-4.
145. Bhatia, R.; Dey, S.S.; Sood, S.; Sharma, K.; Parkash, C.; Kumar, R. Efficient Microspore Embryogenesis in Cauliflower (*Brassica oleracea* var. *Botrytis* L.) for Development of Plants with Different Ploidy Level and Their Use in Breeding Programme. *Sci. Hortic.* **2017**, *216*, 83–92. [CrossRef]
146. Klíma, M.; Vyvadilová, M.; Kučera, V. Chromosome Doubling Effects of Selected Antimitotic Agents in *Brassica napus* Microspore Culture. *Czech J. Genet. Plant Breed.* **2008**, *44*, 30–36. [CrossRef]
147. Vanous, K.; Vanous, A.; Frei, U.K.; Lübberstedt, T. Generation of Maize (*Zea mays*) Doubled Haploids via Traditional Methods. *Curr. Protoc. Plant Biol.* **2017**, *2*, 147–157. [CrossRef]
148. Murthy, J.V.; Kim, H.H.; Hanesworth, V.R.; Hugdahl, J.D.; Morejohn, L.C. Competitive Inhibition of High Affinity Oryzalin Binding to Plant Tubulin by the Phosphoric Amide Herbicide Amiprophos-Methyl. *Plant Physiol.* **1994**, *105*, 309–320. [CrossRef] [PubMed]
149. Hamill, S.D.; Smith, M.K.; Dodd, W.A. In Vitro Induction of Banana Autotetraploids by Colchicine Treatment of Micropropagated Diploids. *Aust. J. Bot.* **1992**, *40*, 887–896. [CrossRef]
150. Eeckhaut, T.G.R.; Werbrouck, S.P.O.; Leus, L.W.H.; Van Bockstaele, E.J.; Debergh, P.C. Chemically Induced Polyploidization in Spathiphyllum Wallisii Regel through Somatic Embryogenesis. *Plant Cell Tissue Organ. Cult.* **2004**, *78*, 241–246. [CrossRef]
151. Nakasone, H.Y.; Kamemoto, H. *Artificial Induction of Polyploidy in Orchids by the Use of Colchicine*; Hawaii Agricultural Experiment Station, University of Hawaii: Honolulu, HI, USA, 1961.
152. Broughton, S.; Castello, M.; Liu, L.; Killen, J.; Hepworth, A.; O'Leary, R. The Effect of Caffeine and Trifluralin on Chromosome Doubling in Wheat Anther Culture. *Plants* **2020**, *9*, 105. [CrossRef] [PubMed]
153. Jakše, M.; Bohanec, B. *Studies of Alternative Approaches for Genome Doubling in Onion*; Office for Official Publications of the European Community: Bled, Slovenia, 2000; pp. 101–104.

154. Kurtar, E.S. The Effects of Anti-Mitotic Agents on Dihaploidization and Fertility in Winter Squash (*Cucurbita maxima* Duch.) and Pumpkin (*Cucurbita moschata* Duch.) Androgenic Haploids. *Acta Sci. Pol. Hortorum Cultus* **2018**, *17*, 3–14. [CrossRef]
155. Hwang, J.K.; Paek, K.Y.; Cho, D.H. Breeding of Resistant Pepper Lines (*Capsicum annuum* l.) to Bacterial Spot (*Xanthomonas campestris* Pv. *vesicatoria*) through Anther Culture. *Acta Hortic.* **1998**, *461*, 301–310. [CrossRef]
156. Todorova, V.; Grozeva, S.; Rodeva, V.; Masheva, S. Breeding Evaluation of Pepper Lines Obtained by in Vitro Anther Culture. *Genetika* **2013**, *45*, 601–610. [CrossRef]
157. Shrestha, S.L.; Luitel, B.P.; Kang, W.H. Agro-Morphological Characterization of Anther Derived Plants in Sweet Pepper (*Capsicum annuum* L. Cv. Boogie). *Hortic. Environ. Biotechnol.* **2011**, *52*, 196–203. [CrossRef]
158. Trajkova, F.; Gudeva, L.K. Evaluation and Agronomic Potential of Androgenic Pepper Genotypes Derived from Piran (*Capsicum annuum* L. Cv. Piran). *J. Exp. Agric. Int.* **2017**, *16*, 1–12. [CrossRef]
159. Luitel, B.; Adhikari, P.; Shrestha, S.; Kang, W. Morphological Characterization of Anther Derived Plants in Minipaprika (*Capsicum annuum* L.). *Korean J. Breed. Sci.* **2012**, *44*, 450–461. [CrossRef]
160. Akagi, T.; Jung, K.; Masuda, K.; Shimizu, K.K. Polyploidy before and after Domestication of Crop Species. *Curr. Opin. Plant Biol.* **2022**, *69*, 102255. [CrossRef]
161. Wijesinghe, S.A.E.C.; Evans, L.J.; Kirkland, L.; Rader, R. A Global Review of Watermelon Pollination Biology and Ecology: The Increasing Importance of Seedless Cultivars. *Sci. Hortic.* **2020**, *271*, 109493. [CrossRef]
162. Tammu, R.M.; Nuringtyas, T.R.; Daryono, B.S. Colchicine Effects on the Ploidy Level and Morphological Characters of Katokkon Pepper (*Capsicum annuum* L.) from North Toraja, Indonesia. *J. Genet. Eng. Biotechnol.* **2021**, *19*, 31. [CrossRef] [PubMed]
163. Mallet, J. Hybrid Speciation. *Nature* **2007**, *446*, 279–283. [CrossRef]
164. Song, P.; Kang, W.; Peffley, E.B. Chromosome Doubling of Allium Fistulosum × A. Cepa Interspecific F1 Hybrids through Colchicine Treatment of Regenerating Callus. *HortScience* **1993**, *28*, 269C-269. [CrossRef]
165. Fomicheva, M.; Domblides, E. Mastering DNA Content Estimation by Flow Cytometry as an Efficient Tool for Plant Breeding and Biodiversity Research. *Methods Protoc.* **2023**, *6*, 18. [CrossRef]
166. Domblides, E.; Ermolaev, A.; Belov, S.; Kan, L.; Skaptsov, M.; Domblides, A. Efficient Methods for Evaluation on Ploidy Level of *Cucurbita pepo* L. Regenerant Plants Obtained in Unpollinated Ovule Culture In Vitro. *Horticulturae* **2022**, *8*, 1083. [CrossRef]
167. Shumilina, D.; Kozar, E.; Chichvarina, O.; Korottseva, K.; Domblides, E. *Brassica rapa* L. ssp. Chinensis Isolated Microspore Culture Protocol. In *Doubled Haploid Technology: Volume 2: Hot Topics, Apiaceae, Brassicaceae, Solanaceae*; Segui-Simarro, J.M., Ed.; Springer: New York, NY, USA, 2021; pp. 145–162, ISBN 978-1-07-161335-1.

horticulturae

MDPI

Review

Onion Male Sterility: Genetics, Genomics and Breeding

Hela Chikh-Rouhou [1,*,†], Saurabh Singh [2,†], Srija Priyadarsini [3] and Cristina Mallor [4,5,*]

1 Regional Research Centre on Horticulture and Organic Agriculture (CRRHAB), LR21AGR03, University of Sousse, Sousse 4042, Tunisia
2 Department of Vegetable Science, College of Horticulture and Forestry, Rani Lakshmi Bai Central Agricultural University (RLBCAU), Jhansi 284003, India; horticulturesaurabh@gmail.com
3 Department of Vegetable Science, Odisha University of Agriculture and Technology (OUAT), Bhubaneswar 751003, India
4 Agrifood Research and Technology Centre of Aragon (CITA), Avda., Montañana 930, 50059 Zaragoza, Spain
5 AgriFood Institute of Aragon—IA2 (CITA-University of Zaragoza), 50013 Zaragoza, Spain
* Correspondence: hela.chikh.rouhou@gmail.com (H.C.-R.); cmallor@cita-aragon.es (C.M.)
† These authors contributed equally to this work.

Abstract: Onion, belonging to the *Allium* genus, is an essential and versatile vegetable crop that plays a pivotal role in culinary traditions worldwide. Renowned for its distinctive flavor and nutritional value, onion is an indispensable ingredient in countless dishes. As the global demand for onion continues to surge, securing a stable supply of high-quality, high-yielding onion varieties becomes ever more pressing. The onion umbel bears numerous tiny flowers that are protandrous in nature. Hybrid breeding is limited in onion due to high inbreeding depression, tedious emasculation and lack of elite inbreds. In this quest for crop improvement, the phenomenon of male sterility stands out as a key tool in modern onion breeding. Male sterility, which is recognized as the incapacity to produce viable pollen grains, inhibition of anther dehiscence and production of non-functional male gametes, has been harnessed as a mechanism to control cross-pollination and escalating hybrid development. The successful utilization of stable male sterile lines in onion holds the promise of producing uniform, high-yielding and disease-resistant hybrids. In recent decades, scientific advances have illuminated the molecular intricacies underlying male sterility systems in onion. Much progress has been made in elucidating the regulation of male sterility systems in the post-genomics era. This review highlights the current status of molecular markers linked with male sterility and provides genetic and molecular insights into its regulation. Additionally, it discusses the role of male sterility as a transformative tool in onion breeding in the genomics era.

Keywords: *Allium cepa*; male sterility systems; cytoplasmic-genic male sterility; hybrids; breeding

1. Introduction

Onion (*Allium cepa* L.) belongs to the genus *Allium*, which encompasses over 1100 species with ethno medicinal value including various other well-known members such as garlic (*A. sativum*), leeks (*A. ampeloprasum* var. *porrum*), chives (*A. cepa* var. *aggregatum*) and shallots (*A. schoenoprasum*) [1]. Onion is diploid, with a chromosome number of 2n = 2x = 16 and boasts a notably extensive genome size, approximately 16 Gbp (16,400 Mb/1C) [2]. This genome size ranks it as one of the largest cultivated diploid crops, surpassing the genome sizes of *Arabidopsis* and tomato by a factor of 100 and 18, respectively, which is also comparable to size of cultivated staple cereal, wheat [3].

Onion is a biennial cross-pollinated crop, but grown as an annual plant for bulb production. It produces a distinctive bulb composed of fleshy leaves designed to store carbohydrates and water for the plant's sustenance. Onions can be propagated through various means, including seeds, sets (small bulbs) or transplants. They are categorized into long-day or short-day cultivars based on their response to day length. A day length of 14–16 h is required to initiate bulbing in long-day type cultivars, while a shorter day-length of 10–12 h is sufficient in short-day type cultivars for bulb formation [4,5]. On the other hand, the day-neutral types exhibit flexibility in bulb formation across a wide range of day lengths. The onion inflorescence is known as umbel that bears numerous perfect and protandrous flowers [6]. Self-pollination is infrequent, as onions are entomophilous and primarily pollinated by insects. Honeybees are the principal pollinators, and it is a common practice to place beehives in the fields to enhance seed yield.

Beyond its culinary appeal, onion holds a promising position in production and export value globally, ranking second only to tomatoes (http://www.fao.org/faostat/en/, accessed on 1 December 2024). Notably, China and India collectively contribute 47.49% of the world's onion production. Onions are cherished for their versatility, being consumed in both immature and mature bulb stages, either raw or cooked. In addition to their culinary application, these plants have served as medicinal resources for over 5000 years. Their nutritional and medicinal properties make them a valuable commodity [7]. Onions are rich in bioactive compounds like quercetin and rutin [8]; hence, the regular intake is associated with reduction in different types of cancer, skin and heart ailments and control generation of ROS (reactive oxygen species) [9]. Onions are considered as low-calorie food and are abundant in mineral nutrients, vitamins and flavonoid compounds. Moreover, they contain sulfur compounds known for their anti-inflammatory and anti-cancer properties.

Wide variation is present in onions with respect to bulb shape, bulb color and flavor, bulb size and quality attributes [8,10]. This inherent diversity in onion cultivars led to extensive efforts in onion breeding to enhance desirable traits for both commercial and consumer preferences. Breeders focus on developing cultivars with improved disease resistance, higher yield and adaptability to different growing conditions. One key aspect of onion breeding involves the selection and manipulation of genetic traits to achieve desired outcomes. Traditional breeding methods often rely on controlled pollination to cross plants with specific characteristics, followed by rigorous selection of the resulting progeny. Modern breeding techniques, including MAS (marker-assisted selection), genomics assisted breeding, genome engineering via CRISPR/Cas9 and RNAi tools, have expedited the breeding process by allowing breeders to identify and introduce desired traits more efficiently [11]. Moreover, advancements in molecular biology have facilitated the elucidation of genetics of important commercial traits in onion. The molecular markers linked with desirable traits have been identified which further facilitates for more precise and accelerated breeding programs [12].

In this context, hybrid breeding has become a valuable strategy. Hybrid onions, derived from the cross-breeding of distinct parental lines, often exhibit improved vigor, uniformity and yield compared to their non-hybrid counterparts [13,14]. Male sterility, which refers to a condition wherein plants do not produce viable pollen or are devoid of anthers, have a lack of anther dehiscence, or produce malformed and shriveled anthers, is often exploited in hybrid onion seed production [13]. The advantages of male sterility include preventing self-pollination, assurance of hybrid purity and facilitating the production of F_1 hybrid onions with desirable characteristics [14–16]. The cytoplasmic genic male sterility (CGMS) system, which is the result of interaction of sterile cytoplasm and nuclear genes, has been exploited commercially in onion breeding [13]. Since the inception of the CGMS system by Jones and Emsweller [17], a substantial improvement in onion productivity has been recorded [18]. Male sterility has emerged as a pivotal factor in revolutionizing onion crop improvement. Male sterility systems play a crucial role in crop improvement, offering significant advantages in the realm of agricultural productivity and breeding programs to escalate hybrid breeding and overall crop performance [13,19,20].

Male sterility in onions opens a pathway for controlled hybrid seed production, providing breeders with a powerful tool to regulate the pollination process [15]. In onion, CMS-S, CMS-R and CMS-T types of cytoplasm have been reported for the production of elite male sterile parents and enhancing hybrid onion breeding [13,17,21]. This carefully orchestrated breeding approach, involving the strategic crossing of male sterile female parents with fertile lines, has demonstrated remarkable success in boosting onion yields and improving overall crop quality [22].

This review aims at shedding light on the current status of understanding molecular mechanisms of male sterility systems in onion, their utilization in onion breeding and future prospects. Understanding the intricacies of CGMS in onion holds great promise for improving the efficiency and effectiveness of onion breeding programs. This review also highlights the progress of genomic insights and scope of genome editing in onion breeding based on male sterility systems. As we explore the multifaceted role of male sterility in onion breeding programs, we uncover a transformative tool that holds the key to addressing the evolving challenges of modern agriculture.

2. Male Sterility in Onion

2.1. Types of Male Sterility Systems in Onion

Male sterility is widespread in angiosperms and is vital in hybrid breeding of crops like onion, cole vegetables and solanaceae vegetables where flowers are tiny and hand emasculation is cumbersome [13]. Male sterility can be defined as the condition where plants are devoid of male reproductive organs, i.e., stamens, and if anthers are formed, they are deformed or non-dehiscent in nature, the male gametes are non-viable and the female organ is viable [13,19,20,23–25]. Hence, male sterility avoids self-pollination and ensures increase in genetic diversity through cross-pollination [26]. Although male sterility rarely occurs in nature, as these plants are eliminated by natural selection forces, it has been maintained during the domestication process, because it could be a valuable tool for onion breeding. In that way, male sterility provides a natural and effective mean for genetic emasculation of plants facilitating hybrid breeding in onion [11]. The maternally inherited sterility controlled by the mitochondrial genome is designated as cytoplasmic male sterility. The sterility encoded by nuclear genes is regarded as nuclear or genic male sterility. Meanwhile, the cytoplasmic-genic male sterility (CGMS) is a type of CMS where sterility is caused by interaction of nuclear genes and sterile cytoplasm [20,27–29]. The CGMS system is widespread in onion and has been exploited commercially [13].

The effect of cytoplasmic genes can be masked by dominant fertility restorer genes (*Rf* genes) [30]. In onion, the CGMS system has been proved instrumental, where male sterility is regulated by the interaction of different sterile cytoplasms (CMS-S, CMS-T, CMS-R) and recessive nuclear genes (Table 1). The first documentation of male sterility in onion was made in the material of Italian Red cultivar by Jones during 1925 [31]. The male sterility in this material was reported to be associated with the interaction of sterile cytoplasm "S" (CMS-S) and recessive nuclear genes, which was recognized as the CGMS system [17,31]. The nuclear gene, when present in dominant form (*Ms*), leads to the fertile phenotype, while in recessive homozygous state (*msms*) with sterile cytoplasm it leads to the male sterile phenotype (Figure 1). The male sterile gene can be present in three different genotypes in diploid onion, i.e., dominant homozygous with *MsMs* genotype, heterozygous as *Msms* and recessive homozygous state with *msms* genotype (Figure 1). These genotypes may occur with either sterile "S" or normal "N" cytoplasm, and interaction with nuclear genes and cytoplasm yields respective phenotypes [26]. The male sterile (*S msms*) lines can be maintained by crossing with maintainer (*N msms*) [32].

Table 1. Characteristics of sterile cytoplasm types governing CGMS in onion.

Features	CMS-S	CMS-T	CMS-R	CMS-Y
Description	First identified in the material of Italian-Red in 1925	Discovered in the French variety "Jaune Paille Des Vertus" in 1965	Originated from the onion cultivar "Rijnsburge"	Reported in two onion accessions, PI273626 and PI236025
Male sterility mechanism	Increased genomic shift of the mitochondrial gene *orf725* along with a reduction in the copy number of *coxI* gene	Enhanced genomic shift of *orf725* gene along with an increase in the copy number of *coxI* gene	Male sterility is linked to a specific mitochondrial gene, *orf725*	Male sterility is linked to the mitochondrial gene, *orf725* with the *cox1* gene
Fertility restoration (Nuclear *Rf* gene)	Fertility is restored by a single nuclear dominant locus *Ms*	Three independent nuclear loci are involved in fertility restoration (A, C, D)	Restored by a nuclear dominant *Ms* locus	Potentially restored by an *Rf* gene (not fully characterized)
Utility	Stable and widely exploited	Not used commercially	Less common as compared to CMS-S cytoplasm	Unstable and rarerly used

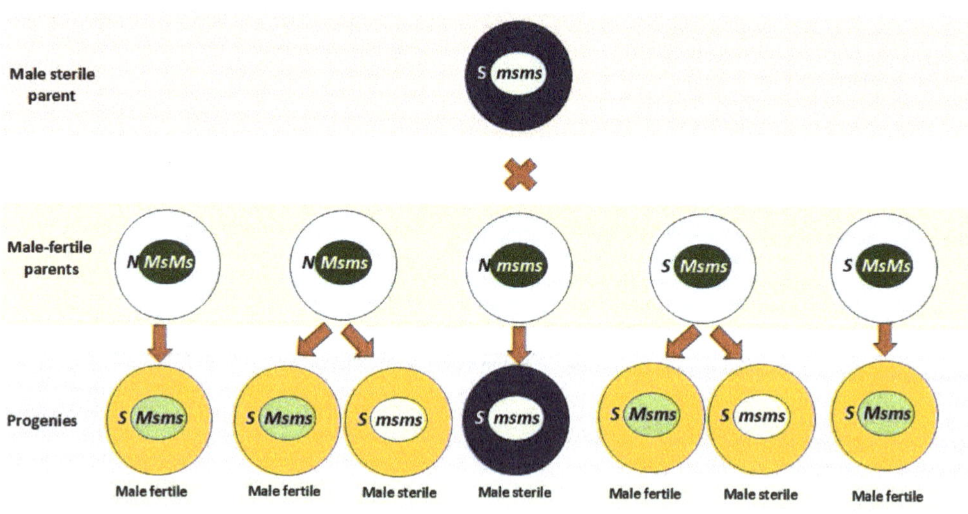

Figure 1. Different progenies generated through pollination of male sterile plants in cytoplasmic-genic male sterility (CGMS) system.

Another cytoplasmic source in the CGMS system of onion, CMS-T, was discovered in France in the material of Jaune-Paille Des-Vertus cultivar by Berninger [33]. This cytoplasm has been used in breeding hybrid cultivars of onion in Europe and Japan [34]. The male sterile phenotype in CMS-T is conditioned by the "T" cytoplasm with three male fertile loci [35]. Plants carrying the "T" cytoplasm along with dominant fertile loci "A" or two complementary dominant fertility restoring loci "C" and "D" are male fertile in nature [35,36]. Among CMS-S and CMS-T sources of sterile cytoplasms in the CGMS system of onion, the CMS-S has been widely exploited widely in onion hybrid breeding due to its greater stability under varying environments and simple inheritance [21,37,38]. Another type of cytoplasm, identified as CMS-R, has been reported in the Rijnsburger cultivar in the Netherlands [39]. The CMS-R can be reverted to male fertility by fertile locus "*Ms*" [39–42]. However, recently Kim et al. [43] reported another CMS type in onion, i.e., "CMS-Y", which was discovered in the "PI273626" accession of onion (Table 1).

2.2. Utility of Male Sterile Lines in Onion Hybrid Breeding

Onion hybrid breeding is limited by lack of stable male sterile lines particularly in the tropical environments. Onion flowers are tiny and protandrous, which favours cross-pollination [6], but this does not completely rule out chances of self-pollination, as a single onion umbel may bear numerous flowers and the anther dehiscence may coincide with stigma receptivity of adjacent flowers of same umbel. Therefore, it is necessary to use methodologies like male sterility systems to ensure hybrid seed production. Heterosis breeding in onion was initiated with the discovery of male sterility in onion [17] that was exploited for developing high yielding hybrids [31]. The CGMS system, governed by interactions between, sterile cytoplasms and nuclear genes, has been instrumental in accelerating hybrid development in onions [44–46]. The higher production and productivity of onion in the current scenario is due to the exploitation of heterosis breeding via male sterility. Hence, much progress has been made in onion with respect to generation and utilization of male sterile lines along with identification of suitable maintainers [47,48].

Realizing the value of male sterility, Kazakova and Yakovlev [49] initiated the heterosis breeding in onion by generating twenty male sterile lines. The male sterile lines like Oriental S61, Oriental S 57, Valencia S1, etc., were utilized in the development of 98 hybrid progenies, which depicted heterotic performance for yield and quality traits. Sharma [50] isolated eight different male sterile lines (MS20-MS23, MS34, MS35, MS37 and MS40) and maintainer lines in the material of Hisar-2. Based on combining ability analysis, they identified MS34 and MS40 as best general combiner lines and Pusa Red as best tester for average bulb size and weight. Male sterile plants have been recognized in the material of Nasik White Globe, IIHR-20 and Pusa Red genotypes in India [51]. Their work led to the identification of male sterile S-cytoplasm in Arka Pragti and Red Coral [51]. ICAR-IIHR released two hybrids based on the male sterility system in onion, namely Arka Kirthiman and Arka Lalima [52]; however, these hybrids did not meet much popularity among Indian onion farmers. Recently, in quest of identification of novel male sterile lines in short-day type onion genotypes in India, Manjunathagowda and Anjanappa [47] identified male sterile and maintainer lines using the black card assay (a visual method to assess pollen dispersal) and a pollen viability test employing aceto-carmine dye (used to stain the nuclei of viable pollen). The male sterility was confirmed using the molecular markers which were designed based on the *orf725* gene. Development of ideal male sterile lines in onion is of significant value for escalating hybrid breeding in onion.

Onion is highly cross-pollinated crop and suffers from high inbreeding depression in the process of development of inbred lines through continuous inbreeding. Hence, identification and isolation of stable male sterile systems is a boon for onion breeders. Some of the breeding lines of onion carrying different cytoplasms reported from different geographical locations are B1750A (CMS-S), B1750B (CMS-N), RJ70A (CMS-T), RJ70B (CMS-N), OM113, M1111 (Male sterile lines isolated from the material of Nasik White Globe), and CMS-ga614A, CMS-ga8111A and CMS-8152A (*ga*-cytoplasm from *A. galanthum*) [31,53,54]. The male sterile lines with the genotype "*S msms*" can be maintained by repeated backcrossing with maintainer (B line) with genotype, "*N msms*", and subsequent cross-pollination of male sterile line with ideal pollen parent [11] (Figure 2). Thus, the CGMS system requires maintenance of three breeding lines in onion: A line (male sterile parent), B line (male fertile maintainer) and R line (fertility restorer) [22].

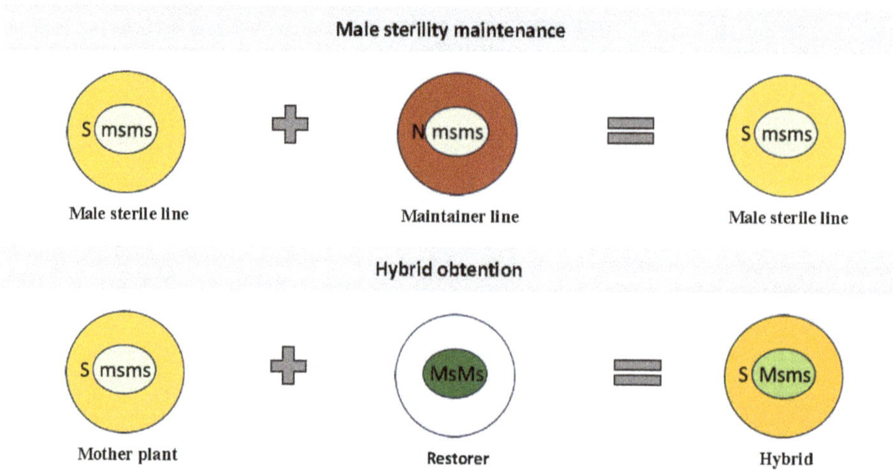

Figure 2. Maintaining male sterility through hybridization: The first lane depicts maintenance of male sterile line (*S msms*) by crossing with maintainer line (*N msms*). The second lane depicts hybrid breeding by crossing male sterile line as female parent with homozygous pollen parent.

3. Genetic and Molecular Basis of Male Sterility in Onion

3.1. Genetics of Male Fertility and Sterility in Onion

The critical understanding of genetic and molecular basis of male sterility in variable male sterile lines of onion carrying variable cytoplasms is crucial for enhancing an onion hybrid breeding program. The CMS-S cytoplasm (*S msms*) has been widely exploited in onion [13] and is investigated extensively using different populations. The inheritance of fertility restoration in onion was investigated using the F_2 mapping population derived from a cross of BYG15–13 (*N MsMs*) × AC43 (*N msms*). The results fitted well in the estimated segregation of ratio of 1 (14 *N MsMs*): 2 (28 *N Msms*): 1 (13 *N msms*) and testcross progenies segregated in 1 *N MsMs*: 1 *N msms* ratio [55]. Bang et al. [56] developed a F_2 mapping population of 188 plants from a cross of 506L (male sterile parent) × H6 (DH based male fertile parent), and the results fitted well in a segregation ratio of 3 (male fertile): 1 (male sterile), indicating the dominance of fertility locus. Huo et al. [57] derived a backcross population from a three-way cross combination [118 (*S msms*) × {118 × 12-10 (*S MsMs*)}] and studied the segregation pattern, which fitted in a ratio of 1:1 with 112 male fertile individuals and 128 male sterile individuals having a differential gene expression for the *AcPME* gene that was expressed in male fertile lines at the flower bud stage. To verify the inheritance of fertility restoration in CMS-S type male sterility in onion, two backcross populations derived from 118 (*S msms*) × 12-12 (*S MsMs*) and 110 (*S msms*) × 12-12 were investigated [58]. The genotyping of backcross populations with SCAR markers linked with dominant locus "*Ms*" (DNF-567) and recessive locus "*ms*" (RNS-375) indicated the segregation ratio of 1:1. These results revealed that fertility restoration in CMS-S types is under the genetic control of the dominant gene [58]. The degree of fertility restoration varies with the source of fertile *Ms* locus in onion germplasm. The testcross progenies based on three different sources of *Ms* locus, viz. Sapporo-Ki (Ski), Ailsa Craig (AC) and B2354B, were screened for male fertility through visual phenotyping and acetocarmine staining technique [59]. The results revealed variable degrees of fertility restoration in these three different sources of *Ms* locus. The segregation ratio fitted in 1:1 proportion for male fertility and male sterility in the testcross progenies derived from AC and Ski sources [59]. The genotyping with molecular marker *"AcPms1"* confirmed the segregation ratio of 1:1 [59].

The CMS-T cytoplasm exhibits maternal inheritance and fertility restoration in CMS-T cytoplasm depicts a complex inheritance pattern governed by interaction of three *Rf* genes, while only one *Rf* gene is required in case of CMS-S cytoplasm [33,35]. The findings revealed by Kim [60], based on the analysis of four F_2 populations developed using CMS-T-like male sterile lines and male fertile lines, suggested perfect co-segregation of the *"jnurf13"* marker for male fertile phenotypes and genotypes. The segregation ratio based on analysis of these F_2 populations fitted in 3:1 ratio, indicating single dominant inheritance of fertility restoration. These results contradicted earlier findings of inheritance of fertility restoration in the CMS-T cytoplasm by three *Rf* genes.

CMS-Y, also referred as "cytotype-Y", is a novel maternally inherited CMS type in onion (PI273626 and PI236025) containing unique combination of mitochondrial genes, *"coxI"* and *"orf725"* [43,61]. The inheritance studies based on segregation generation of a single plant from the accession "PI273626" indicated a single dominant gene, *"Ms"*, for fertility restoration in CMS-Y type male sterility.

3.2. Utility of Molecular Markers in Onion Male Sterility

The cytoplasmic male sterility (CMS) phenomenon is triggered in the crop plants if there is duplication or increase in the copy number of specific regions within the mitochondrial genome containing CMS-linked genes. The higher proportion of copy number of mitochondrial genes is associated with more chances of occurrence of male sterility [62–64]. The majority of CMS-linked genes are chimeric in nature [20,28,65–68] and the ATP synthase encoding genes are mainly responsible for the generation of CMS causing chimeric genes [28]. The frequent recombinations in plant mitochondrial genomes are mediated by short repeat sequences of less than 100 bp, that ultimately leads to the development of chimeric genes inducing CMS [69–73]. However, the CMS-causing genes may be masked by fertility restorer genes (*Rf*) [27]. These *"Rf"* genes mainly encode PPR (pentatrico peptide repeat) proteins [20,74], although some of the *"Rf"* genes encode other proteins like aldehyde dehydrogenase proteins, acyl-carrier proteins, glycine-rich proteins and peptidases [75–78].

The three main CMS types, CMS-S, CMS-T and CMS-R, are mainly involved in onion hybrid breeding programs. The comparative mitochondrial genome sequence analysis of these CMS types recognized *"orf725"* as a common CMS-causing gene in onion [61,79–81]. The restoration of fertility in CMS-R type is conditioned by *Ms* and *Ms2* loci [41,60,82]. The dominant *Ms2* locus with CMS-R confers restoration of male fertility despite *Ms* locus being in homozygous recessive state [41]. On the other hand, a genomic region located at the 3′ end of the *"orf725"* gene that exhibits high homology with *"orfA501"* has been detected in the CMS-T cytoplasm [83]. The genome walking PCR analysis using homologous sequences of *"orfA501"* followed by sequencing of 5′ PCR products indicated the association of 128-bp of *atp1* exon1 sequences with *orfA501*-homolog sequences. Furthermore, the close linkage of 5′ partial sequences of the *nad7* gene with *orf725* was revealed via sequencing of 3′ genome walking PCR products. Thus, a novel candidate gene, *"orf219"*, was suggested to be involved in conferring male sterility by the CMS-T cytoplasm [83].

To facilitate the fast identification of male sterile and fertile lines for enhancing onion hybrid breeding, molecular markers are playing a crucial role. Molecular markers have been designed for discriminating mitotypes in onions [60,79,84]. The development of PCR-based markers such as RFLP, CAPS, SCAR and SNP markers has facilitated the marker-assisted selection of *Ms* locus in onion [56,85,86]. The other markers reported for identification of *Ms* locus in onion germplasm of India are *cob* (PCR marker), *MKFR* (PCR marker), *accD* (InDel marker), *Jnurf13* (InDel marker) and *AcPMS1* (PCR marker) [85,86]. Initially, a molecular marker linked with mitochondrial "*cob*" gene was developed, which was capable of distinguishing CMS-S and -N cytoplasms in onion [87]. However, this marker was not able to differentiate between CMS-T and -N cytoplasms. A remarkable success was achieved by Engelke et al. [88] who developed a molecular marker, that was a combination of *cob* and *orfA501*, to distinguish CMS-S, -T and -N cytoplasms in onion breeding material. Later on, Kim et al. [79] reported another mitochondrial marker "*orf725*" based on sequences of the "*cox1*" gene and chimeric "*orf725*" gene to distinguish CMS-S, -T and -N cytoplasms in onion. However, mitochondrial "*orf725*" marker was not able to distinguish bona fide T cytoplasm (RJ70A) and N cytoplasm (B1750B), both producing amplicon of 833 bp. The cytoplasm-producing amplicon of both 833 bp and 628 bp was referred to as "R" cytoplasm, tracing its origin from the "Rijnsburger" onion cultivar [79].

Havey [21] stated that the dominant "*Ms*" locus does not restore male fertility of CMS-T, while Kim [60] documented that fertility in CMS-T is restored by dominant "*Ms*" locus. This contradiction can be explained by the fact that CMS-R and CMS-S cytoplasms generate 628 bp amplicon of mitochondrial "*orf725*" marker, and here the dominant "*Ms*" locus restores the male fertility in both cytoplasms. On the other hand, a 628 bp amplicon of "*orf725*" is not generated by the CMS-T cytoplasm, and hence the "*Ms*" locus fails to restore the male fertility [21,39]. A dominant marker "*orfA501*" was reported to distinguish CMS-T cytoplasm type by Havey and Kim [39]. However, being dominant in nature, "*orfA501*" is not ideal for screening large numbers of male sterile lines in onion. Hence, a novel molecular marker was developed based on the *orf219* candidate gene [83]. In addition, three HRM (high-resolution melting) markers (accD-HRM, orf725-HRM, orf219-HRM) were generated to screen the large breeding material [83,89]. Recently, a two-step technique was devised to distinguish CMS-N, CMS-T, CMS-S and CMS-R cytoplasms based on HRM markers (AcM-HRM-F1, AcM-HRM-R1 and AcM-HRM-R2) [90]. Firstly, a forward primer and two reverse primers based on both *cox1* and *orf725* gene sequences was designed to differentiate CMS-N, CMS-R and CMS-S cytoplasms. In the second step, to distinguish CMS-N and CMS-T cytoplasms, screening of breeding lines carrying -N cytoplasm was performed using molecular markers developed based on the sequence of the *orfA501* gene [90]. The genetic linkage mapping of *Ms* locus using F_2 mapping population derived from a cross of male sterile and doubled haploid line (506L × H6) located the *Ms* locus on chromosome 2 linked with CAPS markers (jnurf05 and jnurf17) [86]. Previously, Martin et al. [67] also located the *Ms* locus on chromosome 2 using genetic linkage mapping of F_2 population (BYG15−23 × AC43) and depicted the role of SNPs, InDel and SSR markers [67].

Onion hybrid breeding using CGMS requires A (male sterile), B (maintainer) and R lines (restorer), where *Ms*/*ms* alleles are used in restorer and maintainer lines [15]. Thus, development and identification of R (*N MsMs*) and B (*N msms*) lines is crucial in onion hybridization. In line with this, the molecular markers linked with the *Ms* locus are instrumental in accelerating onion hybrid breeding [15]. Gökçe et al. [55] identified RFLP-based allelic diversity in the *Ms* locus, which facilitated the determination of maintainer lines in onion germplasm [55]. Furthermore, PCR-based co-dominant oligopeptide-transporter (OPT) markers specific to the *Ms* locus were reported by Gökçe and Havey [38]. Another PCR-based marker "WHR240" linked with the "*AcPME*" gene was reported to identify fertility restorer lines [57]. Yang et al. [58] reported SCAR markers (DNF-566 and RNS-357) linked with *Ms* locus to differentiate between homozygous dominant (*MsMs*), homozygous recessive (*msms*) and heterozygous (*Msms*) phenotypes among varieties, hybrids and OP populations.

4. Phenotypic Features

Male sterility is a crucial trait in onion breeding programs due to its pivotal role in hybrid seed production, ensuring uniformity, vigor and yield stability. Phenotypic observation of stamen morphology and pollen viability remains a traditional yet effective method for identifying male sterility. Indeed, the overall appearance of the inflorescence can be indicative of fertility or sterility (Figure 3).

Figure 3. Onion inflorescence and breeding. (**a**) Onion umbel in the initial stage of dehiscence; (**b**) male fertile onion umbel; (**c**) bagging with microperforated bags of onion umbels of individual plants for selfing to maintain male fertile parent or pollen parent in the Mediterranean region; (**d**) bagging of individual plants for selfing to maintain male fertile parent or pollen parent in the temperate Region of Himachal Pradesh, India.

The phenotype of male sterile plants can be distinguished from their male fertile counterparts based on floral, morphological, reproductive and seed-related traits [24]. The male sterile plants often exhibit deformed flowers, malformed or shriveled anthers (Figure 4), or failure to produce functional pollen grains, anther indehiscence or failure of male reproductive success [13,73,91]. In onion, the plants with the male sterile cytoplasm may exhibits light green, dark green or yellow anthers [92]. In contrast to that, Saini et al. [93] did not record any association between male sterility and color of anthers in onion breeding material; instead, they noticed erratic production of pollen grains in male sterile plants. This could be explained by the influence of environmental conditions, particularly high temperature. Recently, Singh and Khar [94] documented significant phenotypic differences between male sterile and male fertile counterparts in short-day Indian onion lines. The phenotypic observations were made by examining male sterile and fertile flowers, anthers and pollen grains.

Figure 4. Onion inflorescence: (**A**) fertile and (**B**) sterile inflorescences exhibiting differences in their overall architecture (malformed or shriveled anthers, the arrangement of flowers, the presence or absence of certain structures). The overall appearance of the inflorescence can be indicative of fertility or sterility.

However, advancements in molecular techniques have enabled the development of molecular markers linked with cytoplasm types in onion, that revolutionized an onion breeding program [18,21]. As discussed, these markers enable precise identification and selection of sterile lines, facilitating the creation of superior hybrid varieties with desired traits. By integrating both phenotypic and molecular approaches, onion breeders can accelerate the development of high-performing cultivars, ultimately enhancing crop productivity and achieving the goals of food security.

5. Genomic and Transcriptomic Insights into Onion Male Sterility

In the post-genomics era, the increase in availability of genomic resources such as reference genomes, molecular markers, transcriptome data and genomic databases across the crop plants has accelerated plant breeding [95,96]. In the case of onion, the first draft genome assembly of a doubled haploid line "DHCU066619" was of 14.9 Gb genome size [97]. The gene prediction analysis based on *ab initio* estimated 540,925 genes. The availability of onion genome assembly has significantly facilitated onion breeding. The male sterility in onion is vital for escalating hybrid breeding.

The CMS is governed by mitochondrial genome and, in this context, Kim et al. [80] provided the complete mitochondrial genome sequence of a breeding line carrying S- cytoplasm. Furthermore, the structure of mitochondrial genome and analysis of DNA rearrangements linked with male sterility in onion has been studied extensively [61,81,98]. The next-generation sequencing (NGS) platform was used to sequence the mitochondrial genome of CMS-S type onion cultivar "Momiji-3" [81]. The results indicated that the mitochondrial genome of "Momiji-3" is represented by three circles due to high recombination frequency via repeated sequences. The transcriptome data analysis of mitochondrial genome revealed the presence of 635 RNA-editing positions in the coding regions of the gene. The RNA editing positions indicated the start and stop codons in six genes, viz. *nad1*, *nad4L*, *atp6*, *atp9*, *ccmFC* and *orf725* [81]. The study did not report any presence of novel *orf* transcripts and indicated "*orf725*" as candidate gene for CMS in "Momiji-3" variety [81]. The comparative mitochondrial genome sequence of CMS-T and male fertile counterpart in onion was provided by Kim et al. [61]. They also reported high genomic similarity between CMS-T and normal cytoplasm except for the presence of "*orf725*" along with "*cox1*" sequences in CMS-T mitochondrial genome.

Recently, Bishnoi et al. [98] reported mitochondrial genome sequence of a short-day tropical onion CMS line "97A" and its maintainer line "97B". The mitochondrial genome of CMS line "97A" was of 3,16,321 bp, while the mitochondrial genome of "97B" was comparatively 15 scaffolds due to repetitive genomic regions. Both male sterile and fertile genomes contained 13 and 20 chloroplast-derived fragments, respectively. Further genome analysis depicted 24 protein coding genes in the mitochondrial genome. The comparative genome analysis revealed high genome similarity between male sterile and fertile mitochondrial genomes, except for the presence of chimeric *orf725* gene in CMS 97A line [98].

Various studies employing RNA sequencing and transcriptomic analysis have contributed significantly to the identification of genes exhibiting differential expression between male sterile lines and their male fertile counterparts, thereby shedding light on the intricate physiological and molecular pathways responsible for male sterility. Although limited in number, research specifically conducted on onion has played a crucial role in this regard. In this context, Yuan et al. [99] analyzed the microscopic structure of the anthers in CMS plants (SA2) and their male-fertile maintainers (SB2) in onion. They discovered that, in male sterile plants, pollen production was disrupted at a specific stage called the tetrad stage, which is different from what happens in male-fertile plants. To understand the genetic basis of this differential expression, comparative RNA sequencing of anthers collected from both SA2 and SB2 plants was performed using Illumina HiSeq platform and a large number of 146,413 unique genetic sequences, termed all-unigenes, were identified. Further genomic analysis revealed the role of two cytoplasmic genes, *atp9* and *cox1*, in controlling male sterility. Three nuclear genes, viz. *SERK1*, *AG* and *AMS*, exhibited differential expression between CMS and male-fertile plants. They confirmed these findings using a method called fluorescence quantitative PCR.

In another study, Liu et al. [100] used RNA-seq analysis to determine the differential expression of genes between CMS line "64-2" and its maintainer line "64-1", in Welsh onion. The study revealed significant differences between two lines demonstrating differential expression of 1504 unigenes in 2013 and 2928 unigenes in 2014. The validation of CMS genes (F-type ATPase, NADH dehydrogenase and cytochrome c oxidase) was performed with qPCR analysis. The study revealed the role of both mitochondrial and nuclear genes in regulating CMS in Welsh onion. Despite differences in species, onion (*A. cepa* L.) and Welsh onion (*A. fistulosum* L.) share genetic similarities and physiological traits, suggesting that conclusions drawn from Welsh onion studies could be relevant and informative for understanding aspects including cytoplasmic male sterility.

The transcriptome analysis and the investigation of MADS-box genes in onion male sterility revealed the downregulation of class-B and class-C MADS-box genes, leading to stamen developmental failure and male sterility in onion [101].

Figure 5 integrates genomic and transcriptomic insights to illustrate the complex regulatory network governing male sterility in onion. It provides an overview of the molecular mechanisms underlying this process, highlighting the role of mitochondrial genes (*orf725*) in mitochondrial dysfunction. Additionally, it emphasizes the differential expression of nuclear genes associated with floral organ identity, particularly the downregulation of MADS-box genes, which leads to stamen developmental failure in male sterile anthers [101]. It also illustrates the interactions between mitochondrial and nuclear genes, detailing functional consequences such as tapetal cell degeneration and pollen abortion.

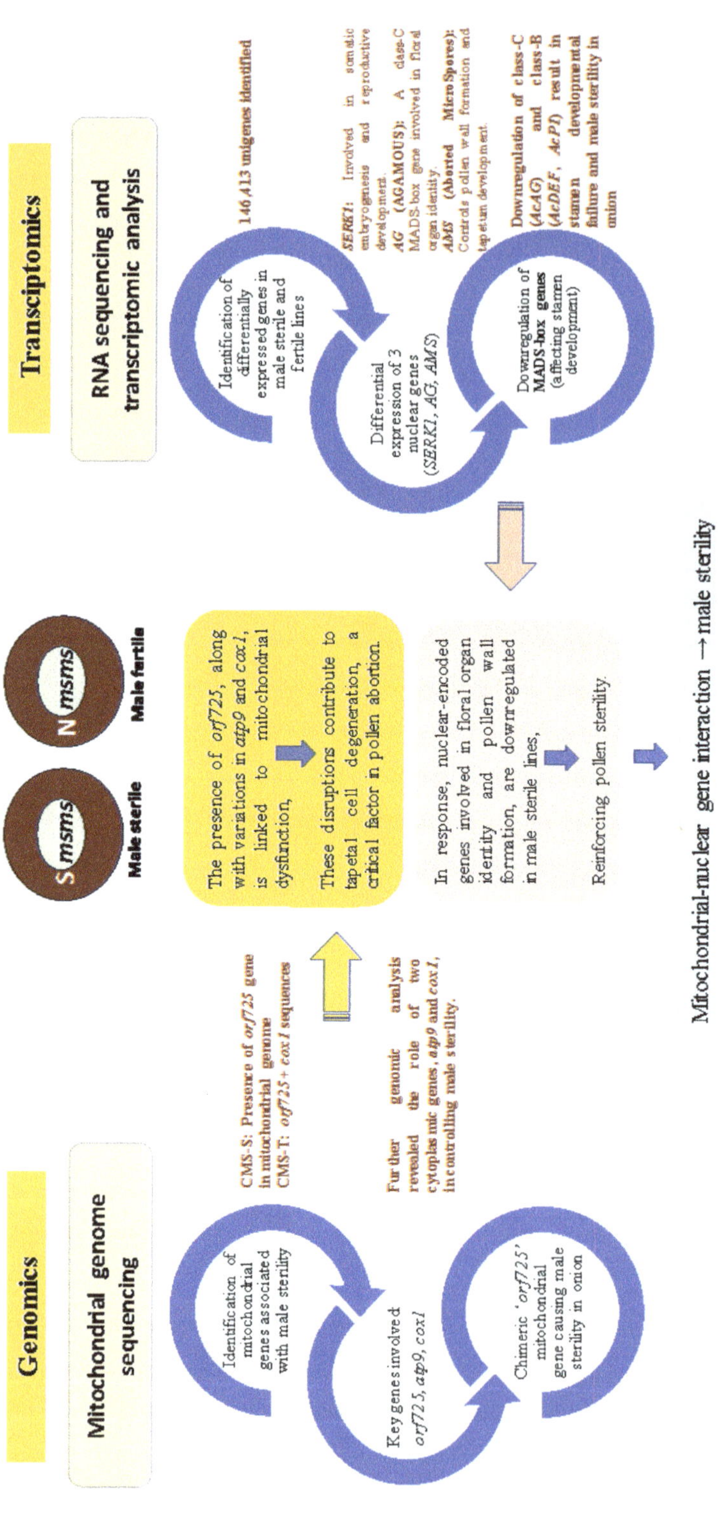

Figure 5. Key genomic and transcriptomic insights into the molecular mechanisms of male sterility in onion (*A. cepa* L.).

6. Conclusions

Research on male sterility in onion holds great promise for enhancing crop improvement efforts in the future. Male sterility systems serve as a powerful tool for hybrid seed production, enabling the development of high-performing onion hybrids. Male sterility, primarily governed by mitochondrial genes, plays a crucial role in hybrid breeding. With the advances in molecular tools and genomics, the complete mitochondrial genome has been sequenced, shedding light on the genetic basis of male sterility in onion. This knowledge has opened new opportunities for harnessing male sterility in onion breeding programs, allowing breeders to exploit male sterility mechanisms to develop hybrid onion varieties with enhanced yield potential and improved agronomic traits. Genomic studies have facilitated the development and identification of molecular markers linked to male sterility, particularly the mitochondrial genes *orf725*, *orf219* and *cox1*, enabling marker-assisted selection of male-sterile lines in onion. Furthermore, transcriptomic studies have revealed the differential expression of nuclear genes, which plays a role in the regulation of male sterility in onion. These findings highlight the complex mitochondrial–nuclear interactions underlying this trait.

By exploiting male sterility mechanisms, breeders can overcome the limitations of intensive manual emasculation, facilitating efficient hybrid seed production. This approach also helps mitigate inbreeding depression and achieve hybrid vigor, resulting in increased yield potential and improved uniformity in onion crops. Continued research on male sterility will be essential for advancing onion hybrid breeding strategies and addressing challenges in sustainable agriculture.

Author Contributions: H.C.-R. and C.M., Writing—original draft preparation; H.C.-R., S.S. and S.P. Writing—review and editing. All authors have read and agreed to the published version of the manuscript.

Funding: This research received no external funding.

Informed Consent Statement: Not applicable.

Data Availability Statement: Not applicable.

Conflicts of Interest: The authors declare no conflicts of interest.

References

1. Teotia, D.; Agrawal, A.; Goyal, H.; Jain, P.; Singh, V.; Verma, Y.; Perveen, K.; Bukhari, N.A., Chandra, A., Malik, V. Pharmacophylogeny of genus *Allium* L. *J. King Saud Univ. Sci.* **2024**, *36*, 103330. [CrossRef]
2. Ricroch, A.; Yockteng, R.; Brown, S.C.; Nadot, S. Evolution of genome size across some cultivated *Allium* species. *Genome* **2005**, *48*, 511–520. [CrossRef] [PubMed]
3. Marcussen, T.; Sandve, S.R.; Heier, L.; Spannagl, M.; Pfeifer, M.; Jakobsen, K.S.; Wulff, B.B.H.; Steuernagel, B.; Mayer, K.F.X., Olsen, O.-A. Ancient hybridizations among the ancestral genomes of bread wheat. *Science* **2014**, *345*, 1250092. [CrossRef]
4. Rashid, H.A.; Cheng, W.; Thomas, B. Temporal and spatial expression of arabidopsis gene homologs control daylength adaptation and bulb formation in onion (*Allium cepa* L.). *Sci. Rep.* **2019**, *9*, 14629. [CrossRef]
5. Atif, M.J.; Ahanger, M.A.; Amin, B.; Ghani, M.I.; Ali, M.; Cheng, Z. Mechanism of allium crops bulb enlargement in response to photoperiod: A review. *Int. J. Mol. Sci.* **2020**, *21*, 1325. [CrossRef] [PubMed]
6. Currah, L.; Ockendon, D.J. Protandry and the sequence of flower opening in the onion (*Allium cepa* L.). *New Phytol.* **1978**, *81*, 419–429. [CrossRef]
7. Ricciardi, L.; Mazzeo, R.; Marcotrigiano, A.R.; Rainaldi, G.; Iovieno, P.; Zonno, V.; Pavan, S.; Lotti, C. Assessment of genetic diversity of the "Acquaviva Red Onion" (*Allium cepa* L.) apulian landrace. *Plants* **2020**, *9*, 260. [CrossRef]
8. Chalbi, A.; Chikh-Rouhou, H.; Tlahig, S.; Mallor, C.; Garcés-Claver, A.; Haddad, M.; Sta-Baba, R.; Bel-Kadhi, M.S. Biochemical characterization of local onion genotypes (*Allium cepa* L.) in the arid regions of Tunisia. *Pol. J. Environ. Stud.* **2022**, *32*, 15–26. [CrossRef]

9. Jimenez, L.; Alarcón, E.; Trevithick-Sutton, C.; Gandhi, N.; Scaiano, J.C. Effect of γ-radiation on green onion DNA integrity: Role of ascorbic acid and polyphenols against nucleic acid damage. *Food Chem.* **2011**, *128*, 735–741. [CrossRef]
10. Chalbi, A.; Chikh-Rouhou, H.; Mezghani, N.; Slim, A.; Fayos, O.; Bel-Kadhi, M.S.; Garcés-Claver, A. Genetic diversity analysis of Onion (*Allium cepa* L.) from the arid region of Tunisia using phenotypic traits and SSR markers. *Horticulturae* **2023**, *9*, 1098. [CrossRef]
11. Sharma, P.; Nair, S.A.; Sharma, P. Male sterility and its commercial exploitation in hybrid seed production of vegetable crops: A review. *Agric. Rev.* **2019**, *40*, 261–270. [CrossRef]
12. Jo, J.; Purushotham, P.M.; Han, K.; Lee, H.-R.; Nah, G.; Kang, B.-C. Development of a genetic map for onion (*Allium cepa* L.) using reference-free genotyping-by-sequencing and SNP assays. *Front. Plant Sci.* **2017**, *8*, 1606. [CrossRef] [PubMed]
13. Priyadarsini, S.; Singh, S.; Nandi, A. Male sterility systems in the genomics era for expediting vegetable breeding. *Sci. Hortic.* **2024**, *338*, 113774. [CrossRef]
14. Mallor, C.; Arnedo, M.S.; Garcés, A. Assessing the genetic diversity of Spanish *Allium cepa* landraces for onion breeding using microsatellite markers. *Sci. Hortic.* **2014**, *170*, 24–31. [CrossRef]
15. Manjunathagowda, D.C. Perspective and application of molecular markers linked to the cytoplasm types and male-fertility restorer locus in onion (*Allium cepa*). *Plant Breed.* **2021**, *140*, 732–744. [CrossRef]
16. Serra, A.D.B.; Currah, L. Agronomy of Onions. Allium Crop Science: Recent Advances. CABI. 2002, pp. 187–232. Available online: https://www.cabi.org/VetMedResource/ebook/20023117383 (accessed on 1 December 2024).
17. Jones, H.A.; Emsweller, S.L. Development of the flower and macrogametophyte of *Allium cepa*. *Hilgardia* **1936**, *10*, 415–428. [CrossRef]
18. McCallum, J. Onion. In *Genome Mapping and Molecular Breeding in Plants*; Kole, C., Ed.; Springer: Berlin/Heidelberg, Germany; New York, NY, USA, 2008; Volume 5, p. 331.
19. Singh, S.; Bhatia, R.; Kumar, R.; Sharma, K.; Dash, S.; Dey, S.S. Cytoplasmic male sterile and doubled haploid lines with desirable combining ability enhances the concentration of important antioxidant attributes in Brassica oleracea. *Euphytica* **2018**, *214*, 207. [CrossRef]
20. Singh, S.; Dey, S.S.; Bhatia, R.; Kumar, R.; Behera, T.K. Current understanding of male sterility systems in vegetable Brassicas and their exploitation in hybrid breeding. *Plant Reprod.* **2019**, *32*, 231–256. [CrossRef]
21. Havey, M.J. Diversity among male-sterility-inducing and male-fertile cytoplasms of onion. *Theor. Appl. Genet.* **2000**, *101*, 778–782. [CrossRef]
22. Soto, V.C.; Caselles, A.; Silva, M.F.; Galmarini, C.R. Onion hybrid seed production: Relation with nectar composition and flower traits. *J. Econ. Entomol.* **2018**, *111*, 1023–1029. [CrossRef] [PubMed]
23. Singh, S.; Dey, S.S.; Bhatia, R.; Kumar, R.; Sharma, K.; Behera, T.K. Heterosis and combining ability in cytoplasmic male sterile and doubled haploid based Brassica oleracea progenies and prediction of heterosis using microsatellites. *PLoS ONE* **2019**, *14*, e0210772. [CrossRef] [PubMed]
24. Singh, S.; Bhatia, R.; Kumar, R.; Behera, T.K.; Kumari, K.; Pramanik, A.; Ghemeray, H.; Sharma, K.; Bhattacharya, R.C.; Dey, S.S. Elucidating mitochondrial DNA markers of ogura-based cms lines in indian cauliflowers (*Brassica oleracea var. botrytis* L.) and their floral abnormalities due to diversity in cytonuclear interactions. *Front. Plant Sci.* **2021**, *12*, 631489. [CrossRef]
25. Singh, S.; Bhatia, R.; Kumar, R.; Das, A.; Ghemeray, H.; Behera, T.K.; Dey, S.S. Characterization and genetic analysis of OguCMS and doubled haploid based large genetic arsenal of Indian cauliflowers (*Brassica oleracea var. botrytis* L.) for morphological, reproductive and seed yield traits revealed their breeding potential. *Genet. Resour. Crop. Evol.* **2021**, *68*, 1603–1623. [CrossRef]
26. Block, E. *Garlic and Other Alliums: The Lore and the Science*; Royal Society of Chemistry: Cambridge, UK, 2010; 480p.
27. Chen, L.; Liu, Y.-G. Male sterility and fertility restoration in crops. *Annu. Rev. Plant Biol.* **2013**, *65*, 579–606. [CrossRef]
28. Hanson, M.R.; Bentolila, S. Interactions of mitochondrial and nuclear genes that affect male gametophyte development. *Plant Cell* **2004**, *16*, S154–S169. [CrossRef]
29. Bohra, A.; Jha, U.C.; Adhimoolam, P.; Bisht, D.; Singh, N.P. Cytoplasmic male sterility (CMS) in hybrid breeding in field crops. *Plant Cell Rep.* **2016**, *35*, 967–993. [CrossRef]
30. Chase, C.D. Cytoplasmic male sterility: A window to the world of plant mitochondrial–nuclear interactions. *Trends Genet.* **2007**, *23*, 81–90. [CrossRef] [PubMed]
31. Jones, H.A.; Clarke, A.E. Inheritance of male sterility in the onion and the production of hybrid seed. *Proc. Am. Soc. Hortic. Sci.* **1943**, *43*, 189–194.
32. Jones, H.A.; Davis, G. *Inbreeding and Heterosis and Their Relation to the Development of New Varieties of Onions*; Technical Bulletin 874; USDA: Washington, DC, USA, 1944.
33. Berninger, E. Contribution à l'étude de la stérilité male de l'oignon (*Allium cepa* L.). *Ann. Amelior. Plant* **1965**, *15*, 183–199.

34. Brewster, J.L. *Onions and Other Vegetable Alliums*, 2nd ed.; Crop Production Science in Horticulture, 15; CABI Publishing: Wallingford, UK, 2008; 432p, ISBN 13:978-1845933999.
35. Schweisguth, B. Etude d'un nouveau type de stérilité male chez l'oignon, *Allium cepa*. *Annu. Amélior Plant* **1973**, *23*, 221–233.
36. Manjunathagowda, D.C.; Muthukumar, P.; Gopal, J.; Prakash, M.; Bommesh, J.C.; Nagesh, G.C.; Megharaj, K.C.; Manjesh, G.N.; Anjanappa, M. Male sterility in onion (*Allium cepa* L.): Origin, evolutionary status, and their prospectus. *Genet. Resour. Crop. Evol.* **2021**, *68*, 421–439. [CrossRef]
37. Havey, M.J. A putative donor of S-cytoplasm and its distribution among open-pollinated populations of onion. *Theor. Appl. Genet.* **1993**, *86*, 128–134. [CrossRef] [PubMed]
38. Gokçe, A.F.; Havey, M.J. Linkage equilibrium among tightly linked RFLPs and the Ms locus in open-pollinated onion populations. *J. Am. Soc. Hortic. Sci.* **2002**, *127*, 944–946. [CrossRef]
39. Havey, M.J.; Kim, S. Molecular marker characterization of commercially used cytoplasmic male sterilities in onion. *J. Am. Soc. Hortic. Sci.* **2021**, *146*, 351–355. [CrossRef]
40. Kim, T.; Kim, S. Identification of a novel haplotype of the *Ms* locus controlling restoration of male-fertility and its implication in origination of cytoplasmic male-sterility in onion (*Allium cepa* L.). *J. Hortic. Sci. Biotechnol.* **2021**, *96*, 750–758. [CrossRef]
41. Yu, N.; Kim, S. Identification of *Ms2*, a novel locus controlling male-fertility restoration of cytoplasmic male-sterility in onion (*Allium cepa* L), and development of tightly linked molecular markers. *Euphytica* **2021**, *217*, 191. [CrossRef]
42. Khar, A.; Galván, G.A.; Singh, H. Allium Breeding Against Biotic Stresses. In *Genomic Designing for Biotic Stress Resistant Vegetable Crops*; Kole, C., Ed.; Springer: Cham, Switzerland, 2022. [CrossRef]
43. Kim, B.; Kim, C.-W.; Kim, S. Inheritance of fertility restoration of male-sterility conferred by cytotype Y and identification of instability of male fertility phenotypes in onion (*Allium cepa* L.). *J. Hortic. Sci. Biotechnol.* **2019**, *94*, 341–348. [CrossRef]
44. Jones, H.A.; Perry, B.A. Vegetative propagation of short-day varieties of onions as an aid in a breeding program. *Proc. Am. Soc. Hortic. Sci.* **1949**, *53*, 367–370.
45. Fujieda, K.; Matsuoka, N.; Fujita, Y. Vegetative multiplication of onion, *Allium cepa* L., through tissue culture. *J. Jpn. Soc. Hortic. Sci.* **1979**, *48*, 186–194. [CrossRef]
46. Pike, L.M.; Yoo, K.S. A tissue culture technique for the clonal propagation of onion using immature flower buds. *Sci. Hortic.* **1990**, *45*, 31–36. [CrossRef]
47. Manjunathagowda, D.C.; Anjanappa, M. Identification and development of male sterile and their maintainer lines in short-day onion (*Allium cepa* L.) genotypes. *Genet. Resour. Crop. Evol.* **2020**, *67*, 357–365. [CrossRef]
48. Ahmad, R.; Hassan, M.; Akhtar, G.B.; Saeed, S.; Khan, S.A.; Shah, M.K.N.; Khan, N. Identification and characterization of important sterile and maintainer lines from various genotypes for advanced breeding programmes of onion (*Allium cepa*). *Plant Breed.* **2020**, *139*, 988–995. [CrossRef]
49. Kazakova, A.A.; Yakovlev, G.V. Expression of heterosis in onion hybrid produced on male sterility basis. *Proc. Bot. Genet.* **1973**, *49*, 268–280.
50. Sharma, P.K. Studies on general and specific combining ability effects in onion using male sterile, maintainer and restorer lines and hybrids. *Ekin J. Crop Breed. Genet.* **2022**, *8*, 77–85.
51. Pathak, C.S.; Singh, D.P.; Deshpande, A.A. *Annual Report*; Indian Institute Horticultral Research (IIHR): Bangalore, India, 1980; pp. 34–36.
52. Pathak, C. A possible new source of male sterility in onion. *Acta Hortic.* **1997**, *4*, 313–316. [CrossRef]
53. Havey, M.J. Seed yield, floral morphology, and lack of male-fertility restoration of male-sterile onion (*Allium cepa* L.) populations possessing the cytoplasm of Allium galanthum Kir. et Kar. *J. Am. Soc. Hortic. Sci.* **1999**, *124*, 626–629. [CrossRef]
54. Pathak, C.; Gowda, R.V. Breeding for the development of onion hybrids in India: Problems and prospects. *Acta Hortic.* **1993**, *358*, 239–242. [CrossRef]
55. Gökçe, A.F.; McCallum, J.; Sato, Y.; Havey, M.J. Molecular tagging of the Ms locus in onion. *J. Am. Soc. Hortic. Sci.* **2002**, *127*, 576–582. [CrossRef]
56. Bang, H.; Cho, D.Y.; Yoo, K.S.; Yoon, M.K.; Patil, B.S.; Kim, S. Development of simple PCR–based markers linked to the Ms locus, a restorer–of– fertility gene in onion (*Allium cepa* L.). *Euphytica* **2011**, *179*, 439–449. [CrossRef]
57. Huo, Y.; Miao, J.; Liu, B.; Yang, Y.; Zhang, Y.; Tahara, Y.; Meng, Q.; He, Q.; Kitano, H.; Wu, X. The expression of pectin methylesterase in onion flower buds is associated with the dominant male-fertility restoration allele. *Plant Breed.* **2012**, *131*, 211–216. [CrossRef]
58. Yang, Y.Y.; Huo, Y.M.; Miao, J.; Liu, B.J.; Kong, S.P.; Gao, L.M.; Liu, C.; Wang, Z.B.; Tahara, Y.; Kitano, H.; et al. Identification of two SCAR markers co-segregated with the dominant Ms and recessive ms alleles in onion (*Allium cepa* L.). *Euphytica* **2013**, *190*, 267–277. [CrossRef]

59. Lee, H.-I.; Havey, M.J. Variable penetrance among different sources of the male fertility restoration allele of onion. *HortScience* **2020**, *55*, 543–546. [CrossRef]
60. Kim, S. A codominant molecular marker in linkage disequilibrium with a restorer–of–Fertility gene (Ms) and its applica-tion in reevaluation of inheritance of fertility restoration in onions. *Mol. Breed.* **2014**, *34*, 769–778. [CrossRef]
61. Kim, B.; Yang, T.-J.; Kim, S. Identification of a gene responsible for cytoplasmic male-sterility in onions (*Allium cepa* L.) using comparative analysis of mitochondrial genome sequences of two recently diverged cytoplasms. *Theor. Appl. Genet.* **2019**, *132*, 313–322. [CrossRef] [PubMed]
62. Bellaoui, M.; Martin-Canadell, A.; Pelletier, G.; Budar, F. Low-copy-number molecules are produced by recombination, actively maintained and can be amplified in the mitochondrial genome of Brassicaceae: Relationship to reversion of the male sterile phenotype in some cybrids. *Mol. Genet. Genom.* **1998**, *257*, 177–185. [CrossRef]
63. Janska, H.; Sarria, R.; Woloszynska, M.; Arrieta-Montiel, M.; Mackenzie, S.A. Stoichiometric shifts in the common bean mitochondrial genome leading to male sterility and spontaneous reversion to fertility. *Plant Cell* **1998**, *10*, 1163–1180. [CrossRef]
64. Sandhu, A.P.; Abdelnoor, R.V.; Mackenzie, S.A. Transgenic induction of mitochondrial rearrangements for cytoplasmic male sterility in crop plants. *Proc. Natl. Acad. Sci. USA* **2007**, *104*, 1766–1770. [CrossRef]
65. Schnable, P. The molecular basis of cytoplasmic male sterility and fertility restoration. *Trends Plant Sci.* **1998**, *3*, 175–180. [CrossRef]
66. Budar, F.; Touzet, P.; De Paepe, R. The nucleo-mitochondrial conflict in cytoplasmic male sterilities revised. *Genetica* **2003**, *117*, 3–16. [CrossRef]
67. Martin, W.J.; McCallum, J.; Shigyo, M.; Jakse, J.; Kuhl, J.C.; Yamane, N.; Pither-Joyce, M.; Gokce, A.F.; Sink, K.C.; Town, C.D.; et al. Genetic mapping of expressed sequences in onion and in silico comparisons with rice show scant colinearity. *Mol. Genet. Genom.* **2005**, *274*, 197–204. [CrossRef]
68. Kim, Y.; Zhang, D. Molecular control of male fertility for crop hybrid breeding. *Trends Plant Sci.* **2018**, *23*, 53–65. [CrossRef] [PubMed]
69. Small, I.; Suffolk, R.; Leaver, C.J. Evolution of plant mitochondrial genomes via substoichiometric intermediates. *Cell* **1989**, *58*, 69–76. [CrossRef]
70. Albert, B.; Godelle, B.; Gouyon, P.H. Evolution of the plant mitochondrial genome: Dynamics of duplication and deletion of sequences. *J. Mol. Evol.* **1998**, *46*, 155–158. [CrossRef] [PubMed]
71. Kmiec, B.; Woloszynska, M.; Janska, H. Heteroplasmy as a common state of mitochondrial genetic information in plants and animals. *Curr. Genet.* **2006**, *50*, 149–159. [CrossRef]
72. Woloszynska, M.; Trojanowski, D. Counting mtDNA molecules in Phaseolus vulgaris: Sublimons are constantly produced by recombination via short repeats and undergo rigorous selection during substoichiometric shifting. *Plant Mol. Biol.* **2009**, *70*, 511–521. [CrossRef]
73. Priyadarsini, S.; Singh, S.; Nandi, A. Molecular advances in research and applications of male sterility systems in tomato. *Plant Physiol. Biochem.* **2025**, *220*, 109503. [CrossRef]
74. Gaborieau, L.; Brown, G.G.; Mireau, H. The propensity of pentatricopeptide repeat genes to evolve into restorers of cytoplasmic male sterility. *Front. Plant Sci.* **2016**, *7*, 1816. [CrossRef] [PubMed]
75. Cui, X.; Wise, R.P.; Schnable, P.S. The rf2 nuclear restorer gene of male-sterile T-cytoplasm maize. *Science* **1996**, *272*, 1334–1336. [CrossRef]
76. Fujii, S.; Toriyama, K. Suppressed expression of retrograde-regulated male sterility restores pollen fertility in cytoplasmic male sterile rice plants. *Proc. Natl. Acad. Sci. USA* **2009**, *106*, 9513–9518. [CrossRef]
77. Hu, J.; Wang, K.; Huang, W.; Liu, G.; Gao, Y.; Wang, J.; Huang, Q.; Ji, Y.; Qin, X.; Wan, L.; et al. The rice pentatricopeptide repeat protein RF5 restores fertility in Hong-Lian cytoplasmic male-sterile lines via a complex with the glycine-rich protein GRP162. *Plant Cell* **2012**, *24*, 109–122. [CrossRef]
78. Kitazaki, K.; Arakawa, T.; Matsunaga, M.; Yui-Kurino, R.; Matsuhira, H.; Mikami, T.; Kubo, T. Post-translational mechanisms are associated with fertility restoration of cytoplasmic male sterility in sugar beet (*Beta vulgaris*). *Plant J.* **2015**, *83*, 290–299. [CrossRef]
79. Kim, S.; Lee, E.T.; Cho, D.Y.; Han, T.; Bang, H.; Patil, B.S.; Ahn, Y.K.; Yoon, M.-K. Identification of a novel chimeric gene, orf725, and itsuse in development of a molecular marker for distinguishing among three cytoplasm typesinonion (*Allium cepa* L.). *Theor. Appl. Genet.* **2009**, *118*, 433–441. [CrossRef] [PubMed]
80. Kim, B.; Kim, K.; Yang, T.; Kim, S. Completion of the mitochondrial genome sequence of onion (*Allium cepa* L.) containing the CMS-S male-sterile cytoplasm and identification of an independent event of the ccm FN gene split. *Curr. Genet.* **2016**, *62*, 873–885. [CrossRef] [PubMed]
81. Tsujimura, M.; Kaneko, T.; Sakamoto, T.; Kimura, S.; Shigyo, M.; Yamagishi, H.; Terachi, T. Multichromosomal structure of the onion mitochondrial genome and a transcript analysis. *Mitochondrion* **2019**, *46*, 179–186. [CrossRef] [PubMed]

82. Kim, S.; Kim, C.; Park, M.; Choi, D. Identification of candidate genes associated with fertility restoration of cytoplasmic male sterility in onion (*Allium cepa* L.) using a combination of bulked segregant analysis and RNA-seq. *Theor. Appl. Genet.* **2015**, *128*, 2289–2299. [CrossRef]
83. Ahn, W.; Kim, S. Identification of a candidate gene responsible for male sterility conferred by CMS-T cytoplasm in onion (*Allium cepa* L.) and development of molecular markers for detection of CMS-T cytoplasm. *Euphytica* **2023**, *219*, 28. [CrossRef]
84. Engelke, T.; Tatlioglu, T. APCR-marker for the CMS1 inducing cytoplasm in chives derived from recombination events affecting the mitochondrial gene *atp9*. *Theor. Appl. Genet.* **2002**, *104*, 698–702. [CrossRef]
85. Khar, A.; Zimik, M.; Verma, P.; Singh, H.; Mangal, M.; Singh, M.C.; Gupta, A.J. Molecular marker-based characterization of cytoplasm and restorer of male sterility (Ms) locus in commercially grown onions in India. *Mol. Biol. Rep.* **2022**, *49*, 5535–5545. [CrossRef]
86. Park, J.; Bang, H.; Cho, D.Y.; Yoon, M.-K.; Patil, B.S.; Kim, S. Construction of high-resolution linkage map of the Ms locus, a restorer-of-fertility gene in onion (*Allium cepa* L.). *Euphytica* **2013**, *192*, 267–278. [CrossRef]
87. Sato, Y. PCR amplification of CMS-specific mitochondrial nucleotide sequences to identify cytoplasmic genotypes of onion (*Allium cepa* L.). *Theor. Appl. Genet.* **1998**, *96*, 367–370. [CrossRef]
88. Engelke, T.; Terefe, D.; Tatlioglu, T. APCR-based marker system monitoring CMS-(S), CMS-(T) and (N)-cytoplasm in the onion (*Allium cepa* L.). *Theor. Appl. Genet.* **2003**, *107*, 162–167. [CrossRef] [PubMed]
89. Kim, B.; Kim, S. Identification of a variant of CMS-T cytoplasm and development of high resolution melting markers for distinguishing cytoplasm types and genotypinga restorer-of-fertility locus in onion (*Allium cepa* L.). *Euphytica* **2019**, *215*, 164. [CrossRef]
90. Khrustaleva, L.; Nzeha, M.; Ermolaev, A.; Nikitina, E.; Romanov, V. Two-step identification of N-, S-, R- and T-cytoplasm types in onion breeding lines using high-resolution melting (HRM)-based markers. *Int. J. Mol. Sci.* **2023**, *24*, 1605. [CrossRef] [PubMed]
91. Holford, P.; Croft, J.H.; Newbury, H.J. Differences between, and possible origins of the cytoplasms found in fertile and male-sterile onions (*Allium cepa* L.). *Theor. Appl. Genet.* **1991**, *82*, 737–744. [CrossRef]
92. Khar, A.; Saini, N. Limitations of PCR based molecular markers to identify male sterile and maintainer plants from Indian onion (*Allium cepa* L.) populations. *Plant Breed.* **2016**, *135*, 519–524. [CrossRef]
93. Saini, N.; Hedau, N.K.; Khar, A.; Yadav, S.; Bhatt, J.C.; Agrawal, P.K. Successful deployment of marker assisted selection (MAS) for inbred and hybrid development in long-day onion (*Allium cepa* L.). *Indian J. Genet. Plant Breed.* **2015**, *75*, 93. [CrossRef]
94. Singh, H.; Khar, A. Perspectives of onion hybrid breeding in India: An overview. *Indian J. Agric. Sci.* **2021**, *91*, 1426–1432. [CrossRef]
95. Singh, S.; Singh, R.; Priyadarsini, S.; Ola, A.L. Genomics empowering conservation action and improvement of celery in the face of climate change. *Planta* **2024**, *259*, 42. [CrossRef]
96. Chikh-Rouhou, H.; Abdedayem, W.; Solmaz, I.; Sari, N.; Garcés-Claver, A. Melon (*Cucumis melo* L.): Genomics and Breeding. In *Smart Plant Breeding for Vegetable Crops in Post-Genomics Era*; Singh, S., Sharma, D., Sharma, S.K., Singh, R., Eds.; Springer: Singapore, 2023; pp. 25–52. [CrossRef]
97. Finkers, R.; van Kaauwen, M.; Ament, K.; Burger-Meijer, K.; Egging, R.; Huits, H.; Kodde, L.; Kroon, L.; Shigyo, M.; Sato, S.; et al. Insights from the first genome assembly of Onion (*Allium cepa*). *G3 Genes | Genomes | Genet.* **2021**, *11*, jkab243. [CrossRef]
98. Bishnoi, R.; Solanki, R.; Singla, D.; Mittal, A.; Chhuneja, P.; Meena, O.P.; Dhatt, A.S. Comparative mitochondrial genome analysis reveals a candidate ORF for cytoplasmic male sterility in tropical onion. *3 Biotech* **2024**, *14*, 6. [CrossRef]
99. Yuan, Q.; Song, C.; Gao, L.; Zhang, H.; Yang, C.; Sheng, J.; Ren, J.; Chen, D.; Wang, Y. Transcriptome de novo assembly and analysis of differentially expressed genes related to cytoplasmic male sterility in onion. *Plant Physiol. Biochem.* **2018**, *125*, 35–44. [CrossRef] [PubMed]
100. Liu, Q.; Lan, Y.; Wen, C.; Zhao, H.; Wang, J.; Wang, Y. Transcriptome sequencing analyses between the cytoplasmic male sterile line and its maintainer line in welsh onion (*Allium fistulosum* L.). *Int. J. Mol. Sci.* **2016**, *17*, 1058. [CrossRef] [PubMed]
101. Lou, H.; Huang, Y.; Zhu, Z.; Xu, Q. Cloning and expression analysis of onion (*Allium cepa* L.) Mads-Box genes and regulation mechanism of cytoplasmic male sterility. *Biochem. Genet.* **2023**, *61*, 2116–2134. [CrossRef] [PubMed]

horticulturae

MDPI

Article

Pan-Genome Analysis of *TRM* Gene Family and Their Expression Pattern under Abiotic and Biotic Stresses in Cucumber

Lili Zhao †, Ke Wang †, Zimo Wang, Shunpeng Chu, Chunhua Chen, Lina Wang and Zhonghai Ren *

Shandong Collaborative Innovation Center of Fruit & Vegetable Quality and Efficient Production, Key Laboratory of Biology and Genetic Improvement of Horticultural Crops in Huang-Huai Region, Ministry of Agriculture, College of Horticultural Science and Engineering, Shandong Agricultural University, Tai'an 271018, China; 2020010080@sdau.edu.cn (L.Z.); wangke_0524@outlook.com (K.W.); 2021110304@sdau.edu.cn (Z.W.); 2022010058@sdau.edu.cn (S.C.); chenchunhua@sdau.edu.cn (C.C.)

* Correspondence: zhren@sdau.edu.cn

† These authors contributed equally to this work.

Abstract: Cucumber (*Cucumis sativus* L.) is a vital economic vegetable crop, and the TONNEAU1 Recruiting Motif (TRM) gene plays a key role in cucumber organ growth. However, the pan-genomic characteristics of the *TRM* gene family and their expression patterns under different stresses have not been reported in cucumber. In this study, we identified 29 *CsTRMs* from the pan-genomes of 13 cucumber accessions, with *CsTRM29* existing only in PI183967. Most CsTRM proteins exhibited differences in sequence length, except five CsTRMs having consistent protein sequence lengths among the 13 accessions. All CsTRM proteins showed amino acid variations. An analysis of *CsTRM* gene expression patterns revealed that six *CsTRM* genes strongly changed in short-fruited lines compared with long-fruited lines. And four *CsTRM* genes strongly responded to salt and heat stress, while *CsTRM14* showed responses to salt stress, powdery mildew, gray mold, and downy mildew. Some *CsTRM* genes were induced or suppressed at different treatment timepoints, suggesting that cucumber *TRM* genes may play different roles in responses to different stresses, with expression patterns varying with stress changes. Remarkably, the expression of *CsTRM21* showed considerable change between long and short fruits and in responses to abiotic stresses (salt stress and heat stress), as well as biotic stresses (powdery mildew and gray mold), suggesting a dual role of *CsTRM21* in both fruit shape determination and stress resistance. Collectively, this study provided a base for the further functional identification of *CsTRM* genes in cucumber plant growth and stress resistance.

Keywords: cucumber; pan-genome; *TRM* gene; fruit shape; abiotic stress response; biotic stress response

1. Introduction

TONNEAU1 Recruiting Motif (TRM) family genes are crucial for the growth and development of plants, exerting significant functions in various plant species. In Arabidopsis, 34 TRM proteins were identified, and half of them are putative microtubule-associated proteins [1]. *AtTRM1* and *AtTRM2* regulate leaf morphology by positively promoting longitudinal polar cell elongation [2]. The *Attrm5* mutant causes slow leaf growth, delayed flowering, and shortened root length [3]. *AtTRM61* has a conserved functional structure and possesses conserved binding motifs for cofactor S-adenosyl-L-methionine (AdoMet), affecting embryo arrest and seed abortion [4]. Additionally, TRMs can interact with TONNEAU1 (TON1) and Protein Phosphatase 2A (PP2A) through their M2 and M3 domains, respectively, forming the TTP (TON1-TRM-PP2A) protein complex. This complex is targeted to microtubules (MTs) [1], regulating microtubule organization and preprophase band (PPB) formation, thus influencing cell division and/or growth. This regulation ultimately affects the size and shape of plant organs [1,5–9]. In tomatoes, TRMs can interact with OVATE FAMILY PROTEINS (OFP) through their M8 domain. The OFP-TRM protein complex undergoes relocalization between the cytoplasm and microtubules, maintaining a dynamic balance to regulate cell division and organ growth, ultimately affecting fruit shape [10,11]. *SlTRM5* enhances fruit elongation by influencing cell division [12]. In the LA1589 genetic background, although SlTRM3/4 minimally influenced the fruit shape, the absence of SlTRM5 led to a slight flattening of the fruit [13]. The fruit shape of the double mutant lacking both *SlTRM3/4* and *SlTRM5* closely resembles that of the single mutant lacking only *SlTRM5* [13]. Introducing the non-functional versions of either *SlTRM3/4* or *SlTRM5* into *ovate/sov1* near-isogenic lines (NILs) partially restored the pear shape of the fruit. Moreover, when both non-functional alleles of *SlTRM3/4* and *SlTRM5* were combined in *ovate/sov1* NILs, the fruit shape index (FSI) was similar to wild-type (WT) fruits [11,13], suggesting that *SlTRM3/4* and *SlTRM5* have additive effects in regulating fruit elongation. The fruit shape analyses of the *SlTRM17/20a*, *SlTRM19*, or *SlTRM26a* null mutants in the LA1589 genetic background, generated using CRISPR/Cas9, revealed an interesting finding, indicating that SlTRM17/20a and SlTRM19 act synergistically to regulate fruit elongation, whereas SlTRM26a has a limited impact on fruit shape. The null alleles of SlTRM5 and SlTRM19, in both the LA1589 and ovate/sov1 backgrounds, were observed to mutually balance their effects on fruit elongation. This suggests that *SlTRM5* and *SlTRM19* have opposing effects on fruit elongation [13]. In rice, the *TRM* homologous genes *OsGW7/GL7/SLG7* interact with TON1 and PP2A through their M2 and M3 domains, respectively, and target them to the cortical microtubules. By influencing the cell length and width, they regulate the grain size and quality [14–16]. In cucumber, *CsTRM5* plays a role in shaping fruits by influencing the direction of cell division and expansion. Additionally, ABA was involved in regulating cucumber fruit elongation through *CsTRM5*-mediated cell expansion [17].

TRM gene family members are often localized to microtubules [1,2,12]. Microtubules are crucial components of the plant cell skeleton, and they play vital roles in maintaining cell shape, adapting to growth, development, and environmental changes, as well as in processes such as cell division, intracellular transport, immune responses, and stress tolerance [18–27]. MICROTUBULE-DESTABILIZING PROTEIN 25 (MDP25) is a hydrophilic cation-binding protein of the plant-specific developmentally regulated plasma membrane polypeptide (DREPP) family [28]. It is postulated that *AtMDP25* similarly modulates stomatal closure, root hydrotropic response, and immune responses by influencing microtubule dynamics [29–31]. *OsDREPP2* exhibits an affinity for microtubules and, in vitro, it inhibits microtubule polymerization [32], and *MtDREPP* induces the fragmentation of microtubules within membrane nanodomains during rhizobial infections [33]. Ethylene signaling regulates microtubule reassembly by upregulating microtubule-stabilizing protein WAVE-DAMPENED2-LIKE5 (WDL5) expression in response to salt stress [34]. Katanin1 (KTN1) acts as a microtubule-severing protein. It aids in maintaining the organized microtubule structure. Under hypersalinity, the microtubule-associated protein KTN1 regulates hypersalinity-induced microtubule disassembly/assembly, thereby enhancing salinity tolerance [35]. Microtubules under high temperature stress undergo depolymerization [36]. High temperature stress (35–37 °C) disrupts the formation of microtubule-organizing centers, leading to changes in microtubule dynamics, including their elongation and the shortening of microtubule arrays [37]. The changes in microtubule dynamics impact vesicular transport, protein trafficking, and cell wall deposition [38–41]. Currently, there are no reports on the involvement of cucumber *TRM* family genes in biotic or abiotic stress.

Pan-genomics aims to encompass the entire range of genetic diversity within a species by assembling and comparing genome sequences from multiple individuals and displayed a powerful potential in discovering novel genes or gene novel function [42,43]. Pan-genomes have been constructed for major crops like maize, rice, wheat, and soybean, utilizing high-quality genomes from various samples, leading to significant advances in the study of plant genome evolution and the identification of key genes linked to important agronomic traits [44–48]. In 2014, a pan-genome for wild soybean was developed, offering a valuable potential resource for enhancing the genetic diversity of cultivated soybean, which was reduced during domestication. This effort also identified numerous variations associated with agronomic traits, including seed composition, flowering, maturity time, and biotic resistance [49]. The most comprehensive rice graph-based pan-genome constructed was based on the high-quality genomes of 33 genetically diverse rice accessions [47], which not only provide detailed insights into genomic variations and their mechanisms, but also systematically assessed their effects on genome evolution, gene expression, crop domestication, and environmental adaptability for the first time. Pan-genomes have also been developed for several key vegetable crops, including tomato, cucumber, eggplant, and rapeseed, and have also been released, providing a valuable foundation for future biological studies and breeding programs [50–54]. The first cucumber pan-genome was constructed using genome data from 12 representative accessions, employing PacBio sequencing technology and a graph-based assembly strategy. This study elucidated the karyotype evolution of cucumber during domestication and identified several potentially significant genes associated with agronomic traits. These findings provide a foundation for key gene discovery, breeding, and the improvement of cucumber.

In this study, we identified 29 *CsTRM* genes in the cucumber pan-genome and discovered that the majority exhibit variations in protein length across the 13 accessions, and all CsTRM proteins showed amino acid variations. Additionally, we analyzed the expression patterns of the *CsTRM* genes with RNA sequencing (RNA-seq) data in fruit and under different stresses, which may have distinct roles in response to these stresses. Consequently, our study offers a basis for exploring the potential role of *TRMs* for fruit shape and stress tolerance in cucumber.

2. Materials and Method

2.1. Identification of TRM Genes in Cucumber

The cucumber pan-genome assembly and annotation files were download from https://www.ncbi.nlm.nih.gov/ (accessed on 3 March 2024), and the 'PI183967' genome assembly from http://www.cucurbitgenomics.org/ (accessed on 3 March 2024). CDS sequences were extracted and we translated them into protein sequences using TBtools software (v2.031). The *AtTRM* family members were retrieved from https://www.arabidopsis.org/ (accessed on 20 March 2024), and these sequences were employed as queries in TBtools to predicted *TRM* family members in cucumber. A conserved motif analysis was performed using MEME (https://www.omicsclass.com/article/432) (accessed on 20 March 2024). The results were visualized with TBtools, and the final *CsTRM* family members were screened based on the conserved M2 motif.

2.2. Protein Length, Motif Composition, and Gene Structure Analysis Amino Acid Variations

The protein sequences of CsTRMs from various cucumber accessions were obtained, and their lengths were measured with TBtools. The amino acid variations were assessed with the DNAman6.0 program. The conserved motifs were characterized with TBtools. The locations of CDSs and UTRs were retrieved from the genomic annotation database and visualized with TBtools [55].

2.3. Gene Duplication and Synteny Analysis

The genomic databases for cucumber, maize, rice, Arabidopsis, and tomato were obtained from http://cucurbitgenomics.org/organism/20 (accessed on 25 March 2024) and http://plants.ensembl.org/index.html (accessed on 25 March 2024). The gene duplication events and the syntenic relationships were identified using the Multiple Collinearity Scan toolkit (MCScan X) [56] employing the standard settings. The results were visualized and constructed used by TBtools [55].

2.4. Transcriptome Profiling of CsTRMs in Fruit

The publicly available transcriptomic data of cucumber fruit at anthesis from inbred lines 32X and Gui Fei Cui carpel numbers (SPR182933) [57], long fruit 408 and short fruit 409 (SPR045470) [58], and fruit of WT and *CsFUL1^{A}*-OX-29 (SPR117025) [59] were acquired from NCBI (https://www.ncbi.nlm.nih.gov/geo/browse) (accessed on 25 May 2024) to analyze the expression patterns of *CsTRMs* in fruit. The genome-wide expression patterns of the *CsTRM* gene were displayed on a heatmap generated with TBtools [55]. For analyzing the transcriptome of the *CsTRMs*, a threshold of p-value (or FDR) ≤ 0.05, and a value of log2 (fold-change) ≤ -1 or log2 (fold-change) ≥ 1 were employed to define DEGs.

2.5. Transcriptome Profiling of CsTRMs in Response to Abiotic and Biotic Stresses

We used the publicly accessible transcriptomic data of cucumber seedlings exposed to salt under 75 mM NaCl and 0.3 mM Si (GSE116265) [60], heated at 42 °C in Chinese Long 9930 (GSE151055) [61], and cucumber leaves were inoculated by powdery mildew (PM), Pm5.1 (GSE81234) [62], and gray mold (GM) (SRP062592) [63]; and powdery mildew (DM) (SRP009350) [64] were acquired from https://www.ncbi.nlm.nih.gov/ (accessed on 25 May 2024) to investigate *CsTRM* expression profiles in response to different stresses. After aligning the gene IDs to the cucumber genome, the genome-wide expression of the *CsTRM* gene was displayed on a heatmap generated with TBtools [55]. For analyzing the transcriptome of the *CsTRMs*, a level of p-value (or FDR) ≤ 0.05, and a value of log2 (fold-change) ≤ -1 or log2 (fold-change) ≥ 1 were employed to define DEGs.

2.6. RNA Isolation and Quantitative Real-Time PCR (qRT-PCR) Analysis

To acquire samples for the purpose of expression analysis, the ovary was collected from cucumber plants 4 days before anthesis (4 DBA) and 0 day after anthesis (0 DAA) of the long fruit line CSSL2-7 and the round fruit line RNS7 [65], and the cotyledons of cucumber seedlings inoculated with GM 0 h, 6 h, 24 h, and 72 h for RNA extraction. Total RNA was extracted using TRIzol (Accurate Biotechnology, AG21102, Changsha, China), and then we verified quality and measured concentration. And, for cDNA Synthesis with the Evo M-MLV RT Mix Kit with gDNA Clean for qPCR (Accurate Biotechnology, AG11728, Changsha, China), qRT-PCR was performed using the 2X SYBR Green Pro Taq HS Premix (Accurate Biotechnology, AG11718) with CFX Opus 96 real-time PCR system (BIO-RAD). The gene of cucumber Actin served as reference gene. Three biological replicates were included in the expression analysis. The relative expression levels of *CsTRMs* were calculated using the $2^{-\Delta\Delta Ct}$ approach. Primers were listed in Table S5.

3. Result

3.1. Identification of CsTRM Genes Based on the Cucumber Pan-Genome

To investigate the variation of the *TRM* genes across cucumber accessions, we identified *CsTRM* genes from the pan-genome, which includes 13 cucumber accessions [13]. A total of 29 putative *TRM* genes were identified among the genomes of the 13 cucumber accessions (Table 1, Table S1). We renamed them *CsTRM01*–*CsTRM29* according to their chromosomal order to avoid confusion in this study (Table 1). Additionally, *CsTRM04* exhibits multiple copies in W4. There were 28 *CsTRM* genes identified from '9930', being consistent with the previous study [54], and from PI183967, lacking *CsTRM03* and possessing a unique *CsTRM29* (Table 1); 27 from 'Cu2', 'Cuc64', 'W4', 'Hx14', 'Hx17' 'Cuc37', 'Gy14', and '9110gt'; 26 from 'XTMC'; and 25 from 'Cuc80' and 'W8' (Table 1). For *CsTRM01, 02, 05, 07, 09, 11, 12, 13, 14, 15, 17, 18, 19, 21, 23, 24, 25, 26, 27,* and *28*, all are present in the 13 cucumber accessions. *CsTRM3* is absent in Cuc80 and PI; *CsTRM04* is absent in Cuc80 and W8; *CsTRM05* is absent in *XTMC* and W8; *CsTRM08* is absent in *XTMC; CsTRM10* is absent in Cu2 and Cuc80; *CsTRM16* is absent in Cuc64, W4, W8, Hx14, Hx17, and Cuc37; *CsTRM20* is absent in Gy14; *CsTRM22* is absent in 9110gt. *CsTRM29* only existing in PI183967 is identified as a new member of the *CsTRM* gene family in the 13 cucumber accessions.

Table 1. Identification of *TRM* genes in the 13 cucumber accessions.

Gene Name	9930	XTMC	Cu2	Cuc80	PI	Cuc64	W4	W8	Hx14	Hx117	Cuc37	Gy14	9110gt
CsTRM01	1G003000	1G002960	1G003190	1G003020	1G02950	1G003020	1G003020	1G003050	1G008150	1G003070	1G003050	1G002970	1G003150
CsTRM02	1G006080	1G006040	1G006230	1G006090	1G06220	1G006110	1G006160	1G006130	1G012280	1G009330	1G006130	1G005930	1G006390
CsTRM03	1G024390	1G023100	1G019380			1G034690	1G022250	1G022640	1G033380	1G035450	1G025870	1G016710	1G021460
CsTRM04	1G036240	1G038300	1G031670		1G23930	1G045200	1G039460 1G050980		1G048780	1G052690	1G036240	1G023410	1G034790
CsTRM05	2G013800	2G013420	2G014430	2G016150	2G11310	2G012220	2G015170	2G022290	2G022190	2G018160	2G012230	2G011350	2G015370
CsTRM06	2G002210		2G001130	2G002170	2G01120	2G001150	2G001150		2G002120	2G002150	2G001120	2G001120	2G002200
CsTRM07	2G006910	2G004780	2G004780	2G005680	2G04550	2G004690	2G004760	2G005690	2G006680	2G005730	2G004660	2G004680	2G005790
CsTRM08	3G000320		3G000290	3G000270	3G00310	3G000310	3G000300	3G000290	3G000310	3G000300	3G000310	3G000260	3G000300
CsTRM09	3G008900	3G014120	3G011330	3G009320	3G08770	3G009130	3G011320	3G009390	3G013030	3G018440	3G009280	3G008870	3G011230
CsTRM10	3G009320	3G014570			3G09200	3G009570	3G011790	3G009840	3G013470	3G018890	3G009740	3G009280	3G011660
CsTRM11	3G016640	3G023990	3G019120	3G016980	3G16440	3G027380	3G019160	3G017460	3G023810	3G029290	3G017050	3G016550	3G018960
CsTRM12	3G020250	3G028160	3G024300	3G021230	3G20290	3G031530	3G023320	3G021650	3G030910	3G038450	3G021080	3G020040	3G025120
CsTRM13	3G028590	3G044970	3G034490	3G039640	3G27110	3G050790	3G034230	3G032590	3G049730	3G053340	3G043310	3G025270	3G034790
CsTRM14	3G033690	3G052230	3G039760	3G045880	3G31210	3G055180	3G039400	3G038880	3G057040	3G059810	3G049600	3G029050	3G041170
CsTRM15	3G035160	3G053700	3G041160	3G047320	3G32570	3G056620	3G040870	3G040290	3G058490	3G061290	3G050990	3G030380	3G042680
CsTRM16	3G036950	3G056500	3G043950	3G049050	3G34290							3G032070	3G044470
CsTRM17	3G045060	3G067760	3G055880	3G057270	3G42630	3G066680	3G051990	3G050610	3G070610	3G071560	3G061110	3G040150	3G052920
CsTRM18	4G024630	4G030170	4G024030	4G078840	4G14290	4G027840	4G018900	4G021930	4G026900	4G030090	4G084710	4G013800	4G026010
CsTRM19	4G031780	4G042910	4G034540	4G090510	4G21450	4G044030	4G027440	4G033570	4G040430	4G039790	4G095540	4G020000	4G035410
CsTRM20	5G002760	5G003630	5G003640	5G002610	5G05360	5G003650	5G002590	5G005490	5G006540	5G003570	5G002660		5G003770
CsTRM21	5G003260	5G004130	5G004140	5G003110	5G05880	5G003020	5G003100	5G004990	5G007040	5G004090	5G003170	5G003160	5G004310
CsTRM22	5G005590	5G007580	5G007530	5G005560	5G08200	5G000650	5G006510	5G001620	5G009400	5G009680	5G005630	5G005580	
CsTRM23	5G026190	5G042130	5G041190	5G050680	5G17200	5G021470	5G028290	5G024190	5G042910	5G054980	5G043620	5G016900	5G034200
CsTRM24	5G038540	5G060730	5G054610	5G063070	5G29400	5G043920	5G040890	5G049340	5G063180	5G067580	5G056820	5G028860	5G046740
CsTRM25	6G016870	6G024320	6G018060	6G025320	6G14470	6G015450	6G015250	6G019390	6G025200	6G018200	6G015300	6G014340	6G017280
CsTRM26	6G022550	6G035270	6G023850	6G053560	6G17180	6G022100	6G019950	6G023190	6G032860	6G030040	6G019540	6G016650	6G023980
CsTRM27	6G040450	6G052040	6G035430	6G079450	6G25260	6G035100	6G032700	6G033070	6G052400	6G045570	6G036810	6G024690	6G035970
CsTRM28	7G025430	7G031600	7G024250	7G035470	7G13640	7G025890	7G021920	7G034950	7G031600	7G031470	7G037050	7G012470	7G023340
CsTRM29					UnG00530								

Note: PI represent cucumber accession PI183967.

3.2. Analysis of Protein Length and Amino Acid Variations in the CsTRM Proteins

To further understand the protein length variation of CsTRMs among the cucumber accessions, the lengths of the identified CsTRM proteins are presented in Table 2. Five CsTRMs showed the same protein lengths across 13 cucumber accessions, CsTRM04, 11, 14, 15, and 21, respectively. The length of CsTRM01, 02, 05, 06, 13, 18, 22, 24, and 26 varied by just one of the accessions. And others exhibited variations in protein length across different accessions. The data highlighting these length differences are shown in red in Table 2.

Table 2. The predicted lengths of TRM proteins (amino acid residues) in the 13 cucumber accessions.

Protein Number	9930	XTMC	Cu2	Cuc80	PI	Cuc64	W4	W8	Hx14	Hx117	Cuc37	Gy14	9110gt
CsTRM01	1048	1048	1048	1048	1048	1048	1043	1048	1048	1048	1048	1063	1048
CsTRM02	1040	1067	1067	1067	1067	1067	1067	1067	1067	1067	1067	1067	1067
CsTRM03	780	788	788			788	781	781	781	781	781	788	781
CsTRM04	402	402	402		402	402	402/402		402	402	402	402	402
CsTRM05	776	803	803	803	803	803	803	803	803	803	803	803	803
CsTRM06	722		722	722	722	722	722		722	750	722	722	722
CsTRM07	893	478	893	891	891	893	893	899	893	891	891	922	891
CsTRM08	893		879	879	893	879	879	879	893	879	879	904	879
CsTRM09	930	933	933	933	932	933	933	933	933	933	933	932	933
CsTRM10	346	346			344	344	344	344	346	344	346	305	346
CsTRM11	616	616	616	616	616	616	616	616	616	616	616	616	616
CsTRM12	953	953	954	954	953	953	953	953	953	952	953	954	953
CsTRM13	963	963	963	963	963	963	963	963	963	963	963	927	963
CsTRM14	353	353	353	353	353	353	353	353	353	353	353	353	353
CsTRM15	888	888	888	888	888	888	888	888	888	888	888	888	888
CsTRM16	472	472	550	550	472							472	550
CsTRM17	1091	1038	1091	1091	1091	1091	210	353	440	600	1091	1058	357
CsTRM18	961	961	961	961	961	922	961	961	961	961	961	961	961
CsTRM19	903	903	987	987	906	987	906	906	906	906	906	906	906
CsTRM20	785	785	785	785	781	781	781	781	785	785	745		785
CsTRM21	476	476	476	476	476	476	476	476	476	476	476	476	476
CsTRM22	495	495	495	495	495	495	495	495	495	495	495	449	
CsTRM23	794	794	794	848	795	795	794	795	794	794	795	794	736
CsTRM24	1049	1049	1049	1049	1049	1049	1049	1049	1049	1049	1049	1049	958
CsTRM25	1011	1011	1011	1011	902	959	1009	1011	1022	1009	1011	940	1011
CsTRM26	936	936	936	936	936	936	938	936	936	936	936	936	936
CsTRM27	505	505	505	505	505	505	505	736	505	473	505	505	505
CsTRM28	960	960	960	959	959	994	995	1047	895	995	995	976	978
CsTRM29					788								

Note: PI represent cucumber accession PI183967.

Among the proteins differing in lengths, CsTRM01 in 'W4'; CsTRM02 in '9930'; CsTRM03 in '9930'; CsTRM05 in '9930'; CsTRM07 in 'XTMC'; CsTRM09 in '9930'; CsTRM13 in 'Gy14'; CsTRM18 in 'Cuc64'; CsTRM19 in '9930' and 'XTMC'; CsTRM20 in 'Cuc37'; CsTRM22 in 'Gy14'; CsTRM23 and CsTRM24 in '9110gt'; CsTRM25 in 'PI183967'; CsTRM27 in 'Hx117'; and CsTRM28 in 'Hx14' were shorter in length compared to those in other accessions, whereas CsTRM16 in 'Cu2', 'Cuc80', and 'Gy14'; CsTRM19 in 'Cu2', 'Cuc80', and 'Cuc64'; and CsTRM26 in 'W4' were longer compared to those in other accessions. Furthermore, the lengths of certain proteins exhibited multiple instances of polymorphism. For example, the protein length of CsTRM17 showed no difference in '9930', 'Cu2', 'Cuc80', 'PI', 'Cuc64', and 'Cuc37', but was totally different in other accessions, furthermore, dramatically shortened in 'W4', 'W8', 'Hx14', 'Hx117', and '9110gt' (Table 2).

Aside from changes in protein length, amino acid substitutions can also impact a protein's function [66]. The analysis focused on the amino acid variations in CsTRMs across various cucumber accessions (Table S1). Amino acid variations were annotated using the CsTRM protein sequence of 9930 as reference, and all CsTRM proteins exhibit amino acid variations. CsTRM04, 11, 14, 15, and 21 have six, five, three, seven, and two amino acid variations, respectively, but these do not lead to changes in protein length (Table 2). Some CsTRMs exhibit amino acid insertions leading to an increase in protein length. For example, CsTRM02 has 27 amino acid insertions in accessions other than 9930. In CsTRM06, 17,

26, 27, and 28, there are frame shifts leading to amino acid variations. Some amino acid variations are quite significant, such as CsTRM07, 17, and 24.

Some amino acid variations are quite significant, such as CsTRM07, 17, and 24 (Table 2). Further comparisons of the CsTRM07, 17, and 24 gene structures and gene conservative motifs are conducted (Figure 1). CsTRM07 in XTMC has only 478 amino acids, which is significantly shorter than that in the other 12 accessions (Table 2), and its gene structure underwent changes along with alterations in some conserved motifs, experiencing an increase in gene length, but not leading to the loss of conserved motifs in Gy14 and PI183967 (Figure 1A). For CsTRM17, the protein length varied from 210 amino acids to 1091 amino acids across the 13 accessions (Table 2), with corresponding changes in the gene structure and some conserved motifs; especially in W4, there are only two conserved motifs (Figure 1B). In CsTRM24 of 9110gt, alterations in the gene structure caused the decreased protein length, but without a reduction in conserved motifs.

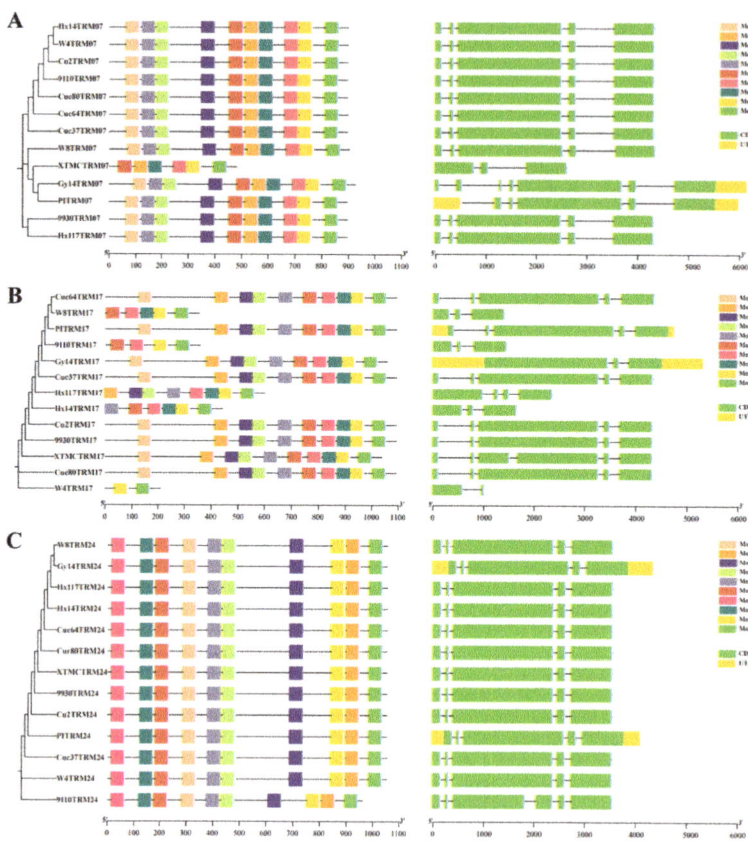

Figure 1. Comparison of the conserved motifs and gene structures of *CsTRM07* (**A**), *CsTRM17* (**B**), and *CsTRM24* (**C**) in the 13 cucumber accessions.

3.3. Synteny Analysis of CsTRM Genes

The phylogenetic relationship of the cucumber *TRM* family was further examined by creating comparative syntenic maps of cucumber and comparing them with four representative species, containing two dicots (Arabidopsis and tomato) and two monocots (rice and maize) (Figure 2). One, three, eight, and nineteen *CsTRM* genes showed syntenic relationships with those in the other four species: maize, rice, Arabidopsis, and tomato, respectively (Figure 2). Only one *TRM* collinear gene pair was found between cucumber and maize, followed by cucumber and rice (four), cucumber and Arabidopsis (nine), and cucumber and tomato (twenty) (Table S2). It is evident that dicotyledonous plants exhibit a notably higher number of homologous genes compared to those shared between dicotyledonous and monocotyledonous plants. This observation aligns with the patterns expected in biological evolution. *CsTRM18* and its collinear gene pairs with maize are observed in rice and tomato, but not in Arabidopsis, indicating differences in the evolutionary process of *CsTRM18*. Additionally, collinear gene pairs between cucumber and rice, maize, and Arabidopsis are observed in cucumber and tomato, suggesting that cucumber and tomato may have undergone a common evolutionary history.

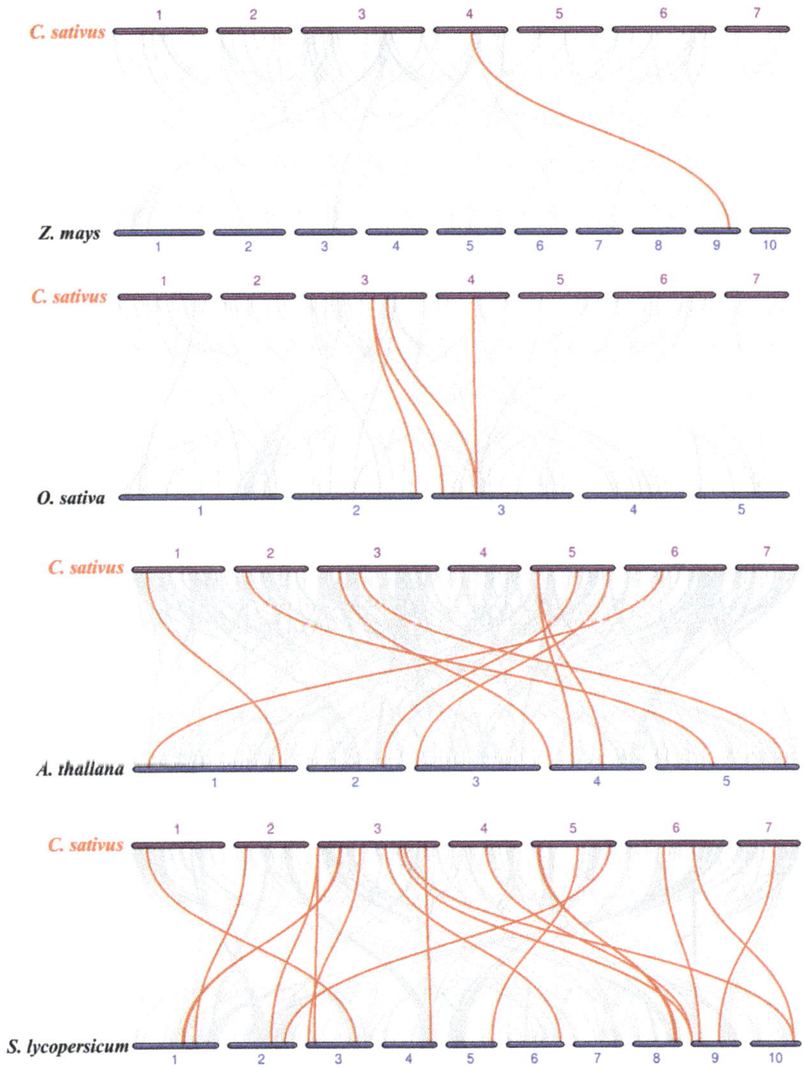

Figure 2. Synteny analysis of *TRMs* among cucumber and other plant species: Gray lines indicate the collinear blocks, while red lines highlight the collinear gene pairs involving TRM genes. '*C. sativus*', '*Z. mays*', '*O. sativa*', '*A. thaliana*', and '*S. lycopersicum*' indicate *Cucumis sativus*, *Zea mays*, *Oryza sativa*, *Arabidopsis thaliana*, and *Solanum lycopersicum*, respectively.

3.4. Expression Profiles of CsTRM Genes in the Fruit

In cucumber and tomato, some *TRM* genes can regulate the fruit shape [11,13,18]. To investigate the function of *CsTRMs* in fruit shape, we conducted the expression analysis of *CsTRMs* using published RNA-seq data on fruits with different carpel numbers and lengths [57–59]. Relative to the South China-type cucumber 32X (carpel number = 3), the transcription levels of *CsTRMs* in the mutant Gui Fei Cui (GFC, carpel number = 5) from 32X showed no significant changes (Figure 3A, Table S3), indicating that *CsTRMs* might not play a crucial role in regulating the cucumber fruit carpel number. Compared to long fruit 408, there were eight genes downregulated in short fruit 409, namely, *CsTRM5*, *6*, *10*, *11*, *14*, *21*, *26*, and *27* (Figure 3B). Compared to empty vector/control transgenic plants WT, *CsFUL1^A*-OX-29 had a total of 12 genes downregulated, namely, *CsTRM1*, *2*, *5*, *6*, *7*, *10*, *12*, *13*, *16*, *20*, *21*, and *27*; and 4 genes upregulated, namely, *CsTRM8*, *17*, *25*, and *26* (Figure 3C). In *CsFUL1^A*-OX-29 versus empty vector/control plants and 409 versus 408, *CsTRM5*, *6*, *10*, *21*, and *27* were significantly downregulated (Figure 3B,C), indicating that these genes play a crucial role in regulating the fruit shape. However, the expression trend of *CsTRM26* in the two groups of long and short fruit materials is opposite (Figure 3B,C), which may be due to the different genetic backgrounds of the materials.

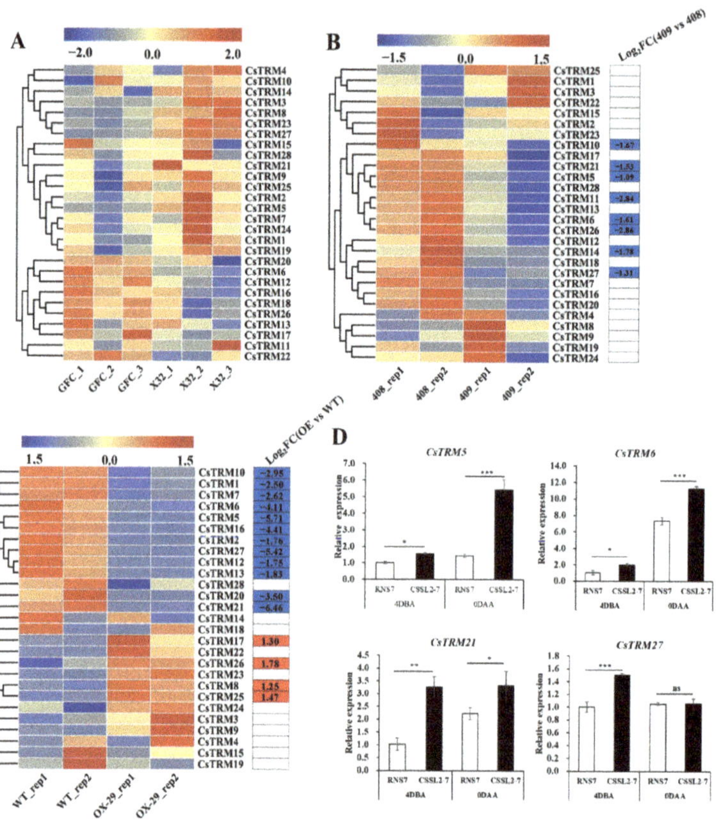

Figure 3. Expression analysis of *CsTRMs* in the fruit: The transcriptional levels of *CsTRM* genes in GFC (carpel number = 5) and 32X (carpel number = 3) (**A**), 408 (long fruit) and 409 (short fruit) (**B**), and WT and *CsFUL1^A*-OX (**C**) are shown on the heatmaps. A color scale range of −2.0 to 2.0 and −1.5 to 1.5 was applied, based on the normalized values. The color gradient, from blue to red, represents

increasing expression levels. GFC, mutant Gui Fei Cui (GFC) from South China-type cucumber 32X. The carpel number changed from 3 in 32X to 5 in GFC, despite the number of other floral organs, such as sepal, petal, and stamen, remaining unchanged. WT, empty vector/control transgenic plants. FC, fold-change. (**D**) qRT-PCR analysis of *CsTRM* expression of the cucumber ovary at 4 days before anthesis (4 DBA) and 0 days after anthesis (0 DAA) at the long fruit CSSL2-7 and the round fruit RNS7. The gene of cucumber Actin served as reference gene. The standard error of the mean is represented by the error bars ($n = 3$). Significance analysis was performed with the two-tailed Student's t-test (ns $p > 0.05$, * $p < 0.05$, ** $p < 0.01$, *** $p < 0.001$).

To further verify the reliability of the RNA-seq results, we conducted qRT-PCR analyses of *CsTRM* gene expression in the ovary of the long-fruited line CSSL2-7 and the round-fruited line RNS7 at 4 days before anthesis (4DBA) and on the day of anthesis (0DAA). Collectively, the qRT-PCR results corresponded well with the transcriptomic data, confirming the accuracy of the datasets (Figure 3D). In RNS7, the expression levels of *CsTRM5*, *CsTRM6*, and *CsTRM21* were consistently lower than those in CSSL2-7 at both 4DBA and 0DAA. The *CsTRM27* expression in RNS7 was lower than in CSSL2-7 at 4DBA but showed no significant difference at 0DAA, suggesting that gene expression levels change during fruit development.

3.5. Expression Patterns of CsTRM Genes under Abiotic and Biotic Stresses

TRM gene family members are often localized to microtubules; microtubules are involved in immune responses and stress tolerance. To further investigate the roles of *CsTRM* genes under various stresses, we analyzed their comprehensive expression patterns in response to various stresses, including salt, heat, downy mildew (DM, *Pseudoperonospora cubensis*), gray mold (GM, *Botrytis cinerea*), and powdery mildew (PM, *Podosphaera fusca*) based on public RNA-seq data [60–64].

Initially, we examined the roles of *CsTRM* genes in response to salt stress (Table S4). The transcriptomic data were visualized using a heatmap (Figure 4A). We observed that the expression levels of *CsTRM4*, *8*, and *14* considerably increased in response to NaCl stress, and four genes exhibited the opposite trend with exposure to NaCl stress: they are *CsTRM5*, *11*, *21*, and *24* (Figure 4A). Under the conditions treated with Silicon (Si) only, the expression of *CsTRM3* and *CsTRM14* was upregulated, whereas the expression of *CsTRM11*, *21*, and *24* was downregulated. The expression of *CsTRM14* was upregulated under both individual NaCl treatment and individual Si treatment, while the expression of *CsTRM11*, *21*, and *24* was downregulated. Previous research has demonstrated that the application of Silicon (Si) can enhance plant growth when subjected to salt stress. After treatment with Si, the gene expression levels of *CsTRM11*, *14*, and *24*, which exhibited significant changes under salt stress, returned to normal levels; *CsTRM5*, *8*, and *21* showed only slight regression, while the expression level of upregulated *CsTRM4* showed a slight increase. We also investigated the responses of *CsTRM* genes to heat tolerance (Figure 4B, Table S4). At three hours after high-temperature treatment, *CsTRM1*, *11*, *16*, *18*, *21*, *22*, and *26* were downregulated, while *CsTRM3*, *8*, and *20* were upregulated. The expression of *CsTRM16* and *CsTRM22* at six hours after heat stress showed no significant difference compared to the 0 h heat treatment, while the changes in other differentially expressed genes were consistent with the 3 h heat treatment. Specifically, at three and six hours after heat stress, the genes upregulated were nearly identical (Figure 4B), indicating their potential significant roles in conferring thermostolerance. Additionally, *CsTRM3*, *8*, *11*, and *21* were differentially expressed in responding to heat and NaCl treatments, with consistent trends.

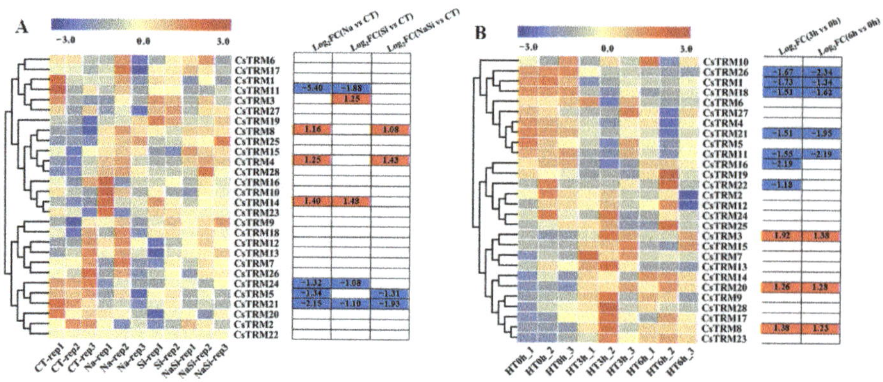

Figure 4. Expression patterns of *CsTRM* genes in response to abiotic stress: The heatmap displays the gene expression levels of *CsTRM* genes in response to salt (**A**) and heat (**B**) tolerance. A color scale range of –3.0 to 3.0 was applied, based on the normalized values. The color gradient, from blue to red, represents increasing expression levels. Abbreviations include CT for control treatment; HT for heat treatment; HT0h for HT at 0 h; HT3h for HT at 3 h; and HT6h for HT at 6 h. FC, fold-change.

To investigate the possible roles of *CsTRMs* in resisting biotic stresses, we analyzed the expression of *CsTRMs* using RNA-Seq data from cucumber seedlings infected with PM for 48 h, GM for 96 h, and DM for 8 days [62–64]. Four genes were found to be differentially expressed after PM inoculation in the susceptible cucumber line D8 leaves compared with the control. The expression of *CsTRM14*, *21*, and *27* were upregulated, while *CsTRM20* were downregulated; and four genes were differentially expressed in the resistant cucumber line SSL508-28 leaves compared with the control. The expression of *CsTRM4*, *14*, and *27* were upregulated, while *CsTRM21* were downregulated (Figure 5A). In the susceptible and the resistant cucumber line affected by PM, *CsTRM14*, and *CsTRM27* had similar expression trends, while *CsTRM21* had opposite expression trends (Figure 5A). After 96 h of GM inoculation, cucumber seedlings showed a significant downregulation of 14 *CsTRM* genes compared to the uninoculated control, namely, *CsTRM1*, *2*, *5*, *6*, *7*, *10*, *11*, *13*, *14*, *16*, *20*, *21*, *27*, and *28*, and significant upregulation of three genes, namely, *CsTRM3*, *18*, and *26* (Figure 5B). In the transcriptomic analysis of cucumber seedlings infected with DM, only five *TRMs* genes exhibited significant changes in expression (Figure 5C). *CsTRM1*, *7*, *14*, and *28* were upregulated at least at one time point during treatment, while *CsTRM8* was downregulated at 6 days post inoculation (dpi) and 8 dpi. *CsTRM28* was upregulated at 2 dpi, 3 dpi, 4 dpi, 6 dpi, and 8 dpi (Figure 5C), indicating its significant role in responding to the DM. In summary, the expression of *CsTRM14* was significantly upregulated in cucumber seedlings inoculated with PM, BC, and DM, indicating its broad-spectrum role in responding to biotic stress.

To further confirm the reliability of the RNA-seq results, we conducted qRT-PCR analyses of the *CsTRM* gene expression of the cotyledons of cucumber seedlings inoculated with gray mold (GM) at 0 h, 6 h, 24 h, and 72 h, and maintaining environmental humidity after inoculation was necessary. Collectively, the qRT-PCR results corresponded well with the RNA-seq data, confirming the accuracy of the datasets (Figure 5D). The expression of *CsTRM5* gradually decreased after inoculation. In contrast, the expression of *CsTRM18* and *CsTRM26* increased after inoculation at 6 h and remained at elevated levels. The expression of *CsTRM21* showed a rapid increased after inoculation at 6 h, but the expression decreased at 24 h and 72 h to below that observed at 0 h post inoculation. This indicates that *CsTRMs* exhibit different response mechanisms to gray mold infection.

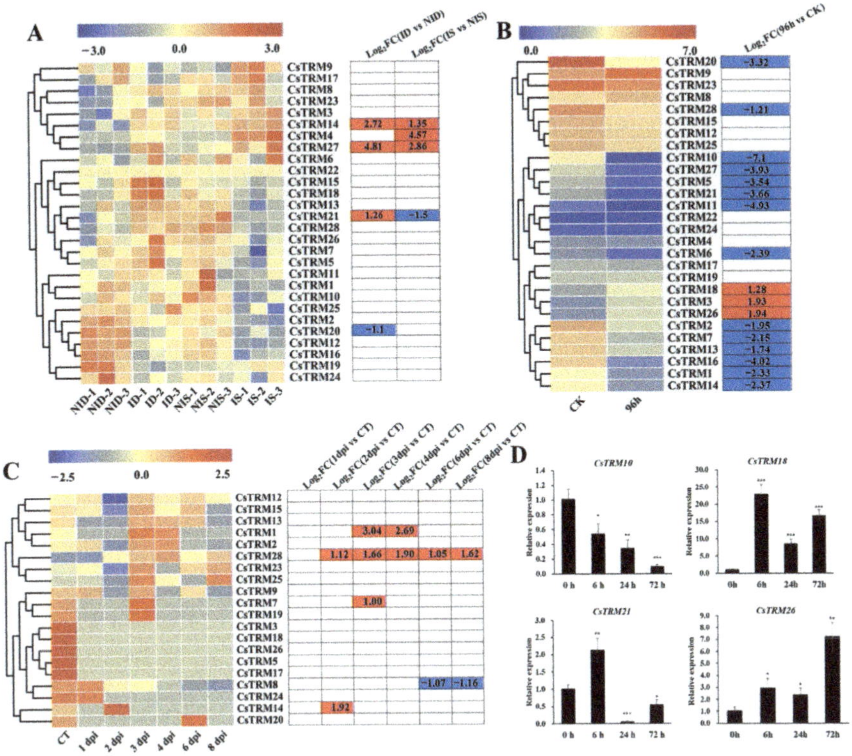

Figure 5. Expression analysis of *CsTRMs* under biotic stresses: The heatmaps displays the transcriptional levels of *CsTRM* genes in response to powdery mildew (PM) for 48 h (**A**), gray mold (GM) for 96 h (**B**), and downy mildew (DM) for 1–8 days post inoculation (**C**). A color scale range of −3.0 to 3.0 was applied, based on the normalized values. The color gradient, from blue to red, represents increasing expression levels. Abbreviations include ID for PM-inoculated susceptible cucumber line D8 leaves; NID for non-inoculated D8 leaves; IS for PM-inoculated resistant cucumber line SSL508-28 leaves; NIS for non-inoculated SSL508-28 leaves; CT for without inoculation; DPI for days post inoculation; and FC for fold-change. (**D**) qRT-PCR analysis of *CsTRM* expression of the cotyledons of cucumber seedlings inoculated with gray mold (GM) at 0 h, 6 h, 24 h, and 72 h, and maintaining environmental humidity after inoculation was necessary. The gene of cucumber Actin served as reference gene. The standard error of the mean is represented by the error bars ($n = 3$). Significance analysis was performed with the two-tailed Student's t test (* $p < 0.05$, ** $p < 0.01$, *** $p < 0.001$).

4. Discussion

Studies have shown that a single reference genome is insufficient for capturing the full diversity within a species [67]. Therefore, we carried out a thorough analysis to identify and characterize the *TRM* family in 13 different cucumber varieties. Although, in the previous study, 28 members of the *TRM* family were identified [54], in this study, a novel member, *CsTRM29*, which is only present in PI183967, was discovered (Table 1). Moreover, only five *CsTRMs* show a uniform protein length in all 13 cucumber accessions, and all the identified TRM proteins have amino acid variations including insertions, deletions, single amino acid changes, and frame shifts (Table S1). Some *CsTRMs* underwent changes not only in gene structure but also in conserved motifs (Figure 1). Therefore, in this study, we found rich variations occurred in *CsTRM*s from the pan-genomes of 13 cucumber accessions, and these variations will provide a base for discovering *TRM* genes with novel functions, which will accelerate the breeding of new cucumber varieties [44].

It is widely recognized that there exists a correlation between gene expression and gene function. The cucumber fruit typically have three fused carpels [68]; the carpel number is an important fruit trait that affects the fruit shape, size, and internal quality [57]. In the lines with different carpel numbers, there were no significant differences observed in the expression of *CsTRMs* (Figure 3A), suggesting that *CsTRMs* might not play a critical role in regulating the number of carpels in cucumber fruits. However, in the short-fruited lines (409 and *CsFUL1^{A}*-OX-29), *CsTRM5, 6, 10, 21,* and *27* were significantly downregulated (Figure 3B,C), indicating that these genes might play crucial roles in regulating cucumber fruit length. Interestingly, the expression of *CsTRM26* is lower in the short-fruited line 409 than in the long-fruited line 408, but higher in the short-fruited line CsFUL1A-OX-29 than in the wild type. This could be due to differing genetic backgrounds or the possibility that *CsTRM26* does not regulate cucumber fruit length.

Thus far, *TRMs* have been reported to be functional in plant organ growth, but not in plant response to stresses. But an increasing number of works of research suggested that, apart from their crucial roles in mechanical architecture and cell division, microtubules are also implicated in plants' adaptation to severe environmental conditions [69]. Since some TRMs are microtubule-binding proteins, they might participate in stress responses. Therefore, we analyzed the expression patterns of *CsTRMs* under certain stress conditions in this study. Many *CsTRM* genes showed expression changes at varying degrees under different stress conditions (Figures 4 and 5). Under salt and heat stress conditions, the expression of *CsTRM3* and *CsTRM8* was significantly upregulated, while *CsTRM11* and *CsTRM21* were significantly downregulated (Figure 4); however, under inoculation with PM, BC, or DM, the expression of *CsTRM14* was significantly increased, while the expression of *CsTRM21* showed significant changes after inoculation with PM and BC (Figure 5). These results might indicate that different *CsTRMs* respond to abiotic or biotic stresses. Remarkably, *CsTRM21* plays a crucial role in regulating the fruit shape (Figure 3B,C) and in responding to biotic stresses (Figures 4 and 5). Therefore, this study provided not only a base for the function of *CsTRMs* in stress tolerance, but also a cross-talk point between organ growth and biotic stresses.

5. Conclusions

In this study, we conducted a pan-genome analysis to identify the TRM gene family in cucumber, identifying a total of 29 members, including a novel member, *CsTRM29*, which is exclusively present in PI183967. Only five of the CsTRMs have consistent protein lengths among the 13 accessions. All CsTRM proteins showed amino acid variations. Furthermore, the transcriptomic data of fruits with different shapes indicate that *CsTRMs* play a significant role in regulating the fruit length but not in controlling the carpel number. And transcriptomic data from various stress conditions highlighted the comparative analysis of the *CsTRM* expression in response to abiotic and biotic stressors, and *CsTRM14* was identified as responding to salt stress, powdery mildew, gray mold, and downy mildew. Notably, *CsTRM21* plays a role in regulating both the fruit shape and resistance. In summary, this study offers a reference for exploring the potential role of *TRMs* for fruit shape and stress resistance in cucumber.

Supplementary Materials: The following supporting information can be downloaded at: https://www.mdpi.com/article/10.3390/horticulturae10090908/s1. Table S1. The variation of amino acids in different cucumber accessions. Table S2. Orthologous relationships between cucumber and Arabidopsis, tomato, rice, maize. Table S3. RNA-seq data (FPKM values) of CsTRM genes in the fruit. Table S4. RNA-seq data (FPKM values) of CsTRM under abiotic and biotic stresses. Table S5. The primer sequences used for the qRT-PCR.

Author Contributions: Conceptualization, Z.R., L.W. and C.C.; methodology, L.Z. and K.W.; validation, L.Z.; software, L.Z., K.W., Z.W. and S.C.; formal analysis, L.Z. and K.W.; investigation, L.Z. and K.W.; data curation, L.Z. and K.W.; writing—original draft preparation, L.Z.; writing—review and editing, Z.R.; supervision, L.W. and Z.R.; project administration, Z.R., L.W. and C.C.;

funding acquisition, Z.R. and L.W. All authors have read and agreed to the published version of the manuscript.

Funding: This work was supported by funding from the National Natural Science Foundation of China (32172605 and 31972419) and the 'Taishan Scholar' Foundation of the People's Government of Shandong Province (ts20130932).

Data Availability Statement: Data are contained within the article and Supplementary Materials.

Acknowledgments: We extend our appreciation to the anonymous reviewers for their valuable suggestions to help improve this article.

Conflicts of Interest: The authors declare no conflicts of interest.

References

1. Drevensek, S.; Goussot, M.; Duroc, Y.; Christodoulidou, A.; Steyaert, S.; Schaefer, E.; Duvernois, E.; Grandjean, O.; Vantard, M.; Bouchez, D.; et al. The *Arabidopsis* TRM1-TON1 interaction reveals a recruitment network common to plant cortical microtubule arrays and eukaryotic centrosomes. *Plant Cell* **2012**, *24*, 178–191. [CrossRef]
2. Lee, Y.K.; Kim, G.T.; Kim, I.J.; Park, J.; Kwak, S.S.; Choi, G.; Chung, W.I. LONGIFOLIA1 and LONGIFOLIA2, two homologous genes, regulate longitudinal cell elongation in *Arabidopsis*. *Development* **2006**, *133*, 4305–4314. [CrossRef]
3. Guo, Q.Q.; Ng, P.Q.; Shi, S.S.; Fan, D.; Li, J.; Zhao, J.; Wang, H.; David, R.; Mittal, P.; Do, T.; et al. *Arabidopsis* TRM5 encodes a nuclear-localised bifunctional tRNA guanine and inosine-N1- methyltransferase that is important for growth. *PLoS ONE* **2019**, *14*, e0225064. [CrossRef]
4. Tang, J.; Jia, P.; Xin, P.; Chu, J.; Shi, D.Q.; Yang, W.C. The *Arabidopsis* TRM61/TRM6 complex is a bona fide tRNA N1-methyladenosine methyltransferase. *J. Exp. Bot.* **2020**, *71*, 3024–3036. [CrossRef] [PubMed]
5. Camilleri, C.; Azimzadeh, J.; Pastuglia, M.; Bellini, C.; Grandjean, O.; Bouchez, D. The *Arabidopsis* TONNEAU2 gene encodes a putative novel protein phosphatase 2A regulatory subunit essential for the control of the cortical cytoskeleton. *Plant Cell* **2002**, *14*, 833–845. [CrossRef] [PubMed]
6. Azimzadeh, J.; Nacry, P.; Christodoulidou, A.; Drevensek, S.; Camilleri, C.; Amiour, N.; Parcy, F.; Pastuglia, M.; Bouchez, D. Arabidopsis TONNEAU1 proteins are essential for preprophase band formation and interact with centrin. *Plant Cell* **2008**, *20*, 2146–2159. [CrossRef]
7. Spinner, L.; Gadeyne, A.; Belcram, K.; Goussot, M.; Moison, M.; Duroc, Y.; Eeckhout, D.; De Winne, N.; Schaefer, E.; Van De Slijke, E.; et al. A protein phosphatase 2A complex spatially controls plant cell division. *Nat. Commun.* **2013**, *4*, 1863. [CrossRef]
8. Schaefer, E.; Belcram, K.; Uyttewaal, M.; Duroc, Y.; Goussot, M.; Legland, D.; Laruelle, E.; Tauzia-Moreau, M.D.; Pastuglia, M.; Bouchez, D. The preprophase band of microtubules controls the robustness of division orientation in plants. *Science* **2017**, *356*, 186–189. [CrossRef]
9. Yang, Y.Q.; Chen, B.Q.; Dang, X.; Zhu, L.L.; Rao, J.Q.; Ren, H.B.; Lin, C.T.; Qin, Y.; Lin, D.S. Arabidopsis IPGA1 is a microtubule-associated protein essential for cell expansion during petal morphogenesis. *J. Exp. Bot.* **2019**, *70*, 5231–5243. [CrossRef] [PubMed]
10. Van der Knaap, E.; Chakrabarti, M.; Chu, Y.H.; Clevenger, J.P.; Illa-Berenguer, E.; Huang, Z.J.; Keyhaninejad, N.; Mu, Q.; Sun, L.; Wang, Y.P.; et al. What lies beyond the eye: The molecular mechanisms regulating tomato fruit weight and shape. *Front. Plant Sci.* **2014**, *5*, 227. [CrossRef]
11. Wu, S.; Zhang, B.Y.; Keyhaninejad, N.; Rodríguez, G.R.; Kim, H.J.; Chakrabarti, M.; Illa-Berenguer, E.; Taitano, N.K.; Gonzalo, M.J.; Díaz, A.; et al. A common genetic mechanism underlies morphological diversity in fruits and other plant organs. *Nat. Commun.* **2018**, *9*, 4734. [CrossRef] [PubMed]
12. Lazzaro, M.D.; Wu, S.; Snouffer, A.; Wang, Y.P.; Van der Knaap, E. Plant organ shapes are regulated by protein interactions and associations with microtubules. *Front. Plant Sci.* **2018**, *9*, 1766. [CrossRef]
13. Zhang, B.; Li, Q.; Keyhaninejad, N.; Taitano, N.; Sapkota, M.; Snouffer, A.; van der Knaap, E. A combinatorial TRM-OFP module bilaterally fine-tunes tomato fruit shape. *New Phytol.* **2023**, *238*, 2393–2409. [CrossRef] [PubMed]
14. Wang, S.K.; Li, S.; Liu, Q.; Wu, K.; Zhang, J.Q.; Wang, S.S.; Wang, Y.; Chen, X.B.; Zhang, Y.; Gao, C.X.; et al. The OsSPL16-GW7 regulatory module determines grain shape and simultaneously improves rice yield and grain quality. *Nat. Genet.* **2015**, *47*, 949–954. [CrossRef]
15. Wang, Y.X.; Xiong, G.S.; Hu, J.; Jiang, L.; Yu, H.; Xu, J.; Fang, Y.X.; Zeng, L.J.; Xu, E.; Xu, J.; et al. Copy number variation at the GL7 locus contributes to grain size diversity in rice. *Nat. Genet.* **2015**, *47*, 944–948. [CrossRef]
16. Zhou, Y.; Miao, J.; Gu, H.Y.; Peng, X.R.; Leburu, M.; Yuan, F.H.; Gu, H.W.; Gao, Y.; Tao, Y.J.; Zhu, J.Y.; et al. Natural Variations in SLG7 Regulate Grain Shape in Rice. *Genetics* **2015**, *201*, 1591–1599. [CrossRef] [PubMed]
17. Xie, Y.; Liu, X.F.; Sun, C.Z.; Song, X.F.; Li, X.L.; Cui, H.N.; Guo, J.Y.; Liu, L.; Ying, A.; Zhang, Z.Q.; et al. *CsTRM5* regulates fruit shape via mediating cell division direction and cell expansion in cucumber. *Hortic. Res.* **2023**, *10*, uhad007. [CrossRef] [PubMed]
18. Wade, R.H. On and around microtubules: An overview. *Mol. Biotechnol.* **2009**, *43*, 177–191. [CrossRef]
19. Landrein, B.; Hamant, O. How mechanical stress controls microtubule behavior and morphogenesis in plants: History; experiments and revisited theories. *Plant J.* **2013**, *75*, 324–338. [CrossRef] [PubMed]

20. Nick, P. Microtubules, signalling and abiotic stress. *Plant J.* **2013**, *75*, 309–323. [CrossRef]
21. Sampathkumar, A.; Yan, A.; Krupinski, P.; Meyerowitz, E.M. Physical forces regulate plant development and morphogenesis. *Curr. Biol.* **2014**, *24*, R475–R483. [CrossRef]
22. Zhang, Q.; Zhang, W. Regulation of developmental and environmental signaling by interaction between microtubules and membranes in plant cells. *Protein Cell* **2016**, *7*, 81–88. [CrossRef] [PubMed]
23. Zhou, S.; Chen, Q.; Sun, Y.; Li, Y. Histone H2B monoubiquitination regulates salt stress-induced microtubule depolymerization in *Arabidopsis*. *Plant Cell Environ.* **2017**, *40*, 1512–1530. [CrossRef] [PubMed]
24. Yang, P.Z.; Jin, J.W.; Zhang, J.R.; Wang, D.; Bai, X.C.; Xie, W.F.; Hu, T.M.; Zhao, X.; Mao, T.L.; Qin, T. MDP25 mediates the fine-tuning of microtubule organization in response to salt stress. *J. Integr. Plant Biol.* **2022**, *64*, 1181–1195. [CrossRef]
25. McNally, F.J.; Roll-Mecak, A. Microtubule-severing enzymes: From cellular functions to molecular mechanism. *J. Cell Biol.* **2018**, *217*, 4057–4069. [CrossRef]
26. Bao, Z.R.; Xu, Z.J.; Zang, J.Z.; Bürstenbinder, K.; Wang, P.W. The morphological diversity of plant organs: Manipulating the organization of microtubules may do the trick. *Front. Cell Dev. Biol.* **2021**, *9*, 691. [CrossRef] [PubMed]
27. Bao, Z.R.; Guo, Y.; Deng, Y.L.; Zang, J.Z.; Zhang, J.H.; Deng, Y.T.; Ouyang, B.; Qu, X.L.; Bürstenbinder, K.; Wang, P.W. Microtubule-associated protein SlMAP70 interacts with IQ67-domain protein SlIQD21a to regulate fruit shape in tomato. *Plant Cell* **2023**, *35*, 4266–4283. [CrossRef] [PubMed]
28. Gantet, P.; Masson, F.; Domergue, O.; Marquis-Mention, M.; Bauw, G.; Inze, D.; Rossignol, M.; de la Serve, B.T. Cloning of a cDNA encoding a developmentally regulated 22 kDa polypeptide from tobacco leaf plasma membrane. *Biochem. Mol. Biol. Int.* **1996**, *40*, 469–477. [CrossRef] [PubMed]
29. Nagasaki-Takeuchi, N.; Miyano, M.; Maeshima, M. A plasma membrane-associated protein of *Arabidopsis thaliana* AtPCaP1 binds copper ions and changes its higher order structure. *J. Biochem.* **2008**, *144*, 487–497. [CrossRef]
30. Tanaka-Takada, N.; Kobayashi, A.; Takahashi, H.; Kamiya, T.; Kinoshita, T.; Maeshima, M. Plasma Membrane-Associated Ca2+-Binding Protein PCaP1 is Involved in Root Hydrotropism of *Arabidopsis thaliana*. *Plant Cell Physiol.* **2019**, *60*, 1331–1341. [CrossRef]
31. Giovannoni, M.; Marti, L.; Ferrari, S.; Tanaka-Takada, N.; Maeshima, M.; Ott, T.; De Lorenzo, G.; Mattei, B. The plasma membrane-associated Ca^{2+} -binding protein, PCaP1, is required for oligogalacturonide and flagellin-induced priming and immunity. *Plant Cell Environ.* **2021**, *44*, 3078–3093. [CrossRef] [PubMed]
32. Yamada, N.; Theerawitaya, C.; Kageyama, H.; Cha-Um, S.; Takabe, T. Expression of developmentally regulated plasma membrane polypeptide (DREPP2) in rice root tip and interaction with $Ca^{(2+)}$/CaM complex and microtubule. *Protoplasma* **2015**, *252*, 1519–1527. [CrossRef] [PubMed]
33. Su, C.; Klein, M.L.; Hernández-Reyes, C.; Batzenschlager, M.; Ditengou, F.A.; Lace, B.; Keller, J.; Delaux, P.M.; Ott, T. The Medicago truncatula DREPP Protein Triggers Microtubule Fragmentation in Membrane Nanodomains during Symbiotic Infections. *Plant Cell* **2020**, *32*, 1689–1702. [CrossRef]
34. Dou, L.; He, K.; Higaki, T.; Wang, X.; Mao, T. Ethylene Signaling Modulates Cortical Microtubule Reassembly in Response to Salt Stress. *Plant Physiol.* **2018**, *176*, 2071–2081. [CrossRef]
35. Yang, J.; An, B.; Luo, H.; He, C.; Wang, Q. AtKATANIN1 Modulates Microtubule Depolymerization and Reorganization in Response to Salt Stress in *Arabidopsis*. *Int. J. Mol. Sci.* **2019**, *21*, 138. [CrossRef]
36. Kumar, S.; Jeevaraj, T.; Yunus, M.H.; Chakraborty, S.; Chakraborty, N. The plant cytoskeleton takes center stage in abiotic stress responses and resilience. *Plant Cell Environ.* **2023**, *46*, 5–22. [CrossRef] [PubMed]
37. Smirnova, E.A.; Bajer, A.S. Microtubule converging centers and reorganization of the interphase cytoskeleton and the mitotic spindle in higher plant Haemanthus. *Cell Motil. Cytoskelet.* **1994**, *27*, 219–233. [CrossRef] [PubMed]
38. Parrotta, L.; Faleri, C.; Cresti, M.; Cai, G. Heat stress affects the cytoskeleton and the delivery of sucrose synthase in tobacco pollen tubes. *Planta* **2016**, *243*, 43–63. [CrossRef]
39. Parveen, S.; Rahman, A. Actin isovariant ACT7 modulates root thermomor-phogenesis by altering intracellular auxin homeostasis. *Int. J. Mol. Sci.* **2021**, *22*, 7749. [CrossRef]
40. Pressman, E.; Harel, D.; Zamski, E.; Shaked, R.; Althan, L.; Rosenfeld, K.; Firon, N. The effect of high temperatures on the expression and activity of sucrose-cleaving enzymes during tomato (*Lycopersicon esculentum*) anther development. *J. Hortic. Sci. Biotechnol.* **2006**, *81*, 341–348. [CrossRef]
41. Zheng, Y.; Anderson, S.; Zhang, Y.; Garavito, R.M. The structure of sucrose synthase-1 from *Arabidopsis thaliana* and its functional implications. *J. Biol. Chem.* **2011**, *286*, 36108–36118. [CrossRef] [PubMed]
42. Schreiber, M.; Jayakodi, M.; Stein, N.; Mascher, M. Plant pangenomes for crop improvement; biodiversity and evolution. *Nat. Rev. Genet.* **2024**, *10*, 1038. [CrossRef] [PubMed]
43. Wang, C.; Han, J.; Wang, T.; Chen, C.; Liu, J.; Xu, Z.; Zhang, Q.; Wang, L.; Ren, Z. Pan-Genome-Wide Identification and Transcriptome-Wide Analysis of *DREB* Genes That Respond to Biotic and Abiotic Stresses in Cucumber. *Agriculture* **2022**, *12*, 1879. [CrossRef]
44. Zhao, Q.; Feng, Q.; Lu, H.; Li, Y.; Wang, A.; Tian, Q.; Zhan, Q.; Lu, Y.; Zhang, L.; Huang, T.; et al. Pan-genome analysis highlights the extent of genomic variation in cultivated and wild rice. *Nat. Genet.* **2018**, *50*, 278–284. [CrossRef]
45. Liu, Y.; Du, H.; Li, P.; Shen, Y.; Peng, H.; Liu, S.; Zhou, G.; Zhang, H.; Liu, Z.; Shi, M.; et al. Pan-Genome of Wild and Cultivated Soybeans. *Cell* **2020**, *182*, 162–176. [CrossRef]

46. Hufford, M.B.; Seetharam, A.S.; Woodhouse, M.R.; Chougule, K.M.; Ou, S.; Liu, J.; Ricci, W.A.; Guo, T.; Olson, A.; Qiu, Y.; et al. De novo assembly, annotation, and comparative analysis of 26 diverse maize genomes. *Science* **2021**, *373*, 655–662. [CrossRef] [PubMed]
47. Qin, P.; Lu, H.; Du, H.; Wang, H.; Chen, W.; Chen, Z.; He, Q.; Ou, S.; Zhang, H.; Li, X.; et al. Pan-genome analysis of 33 genetically diverse rice accessions reveals hidden genomic variations. *Cell* **2021**, *184*, 3542–3558. [CrossRef]
48. Li, W.; Liu, J.; Zhang, H.; Liu, Z.; Wang, Y.; Xing, L.; He, Q.; Du, H. Plant pan-genomics: Recent advances, new challenges, and roads ahead. *J. Genet. Genom.* **2022**, *49*, 833–846. [CrossRef]
49. Li, Y.; Zhou, G.; Ma, J.; Jiang, W.; Jin, L.; Zhang, Z.; Guo, Y.; Zhang, J.; Sui, Y.; Zheng, L.; et al. De novo assembly of soybean wild relatives for pan-genome analysis of diversity and agronomic traits. *Nat. Biotechnol.* **2014**, *32*, 1045–1052. [CrossRef]
50. Golicz, A.A.; Bayer, P.E.; Barker, G.C.; Edger, P.P.; Kim, H.; Martinez, P.A.; Chan, C.K.; Severn-Ellis, A.; McCombie, W.R.; Parkin, I.A.; et al. The pangenome of an agronomically important crop plant Brassica oleracea. *Nat. Commun.* **2016**, *7*, 13390. [CrossRef]
51. Gao, L.; Gonda, I.; Sun, H.; Ma, Q.; Bao, K.; Tieman, D.M.; Burzynski-Chang, E.A.; Fish, T.L.; Stromberg, K.A.; Sacks, G.L.; et al. The tomato pan-genome uncovers new genes and a rare allele regulating fruit flavor. *Nat. Genet.* **2019**, *51*, 1044–1051. [CrossRef]
52. Alonge, M.; Wang, X.; Benoit, M.; Soyk, S.; Pereira, L.; Zhang, L.; Suresh, H.; Ramakrishnan, S.; Maumus, F.; Ciren, D.; et al. Major Impacts of Widespread Structural Variation on Gene Expression and Crop Improvement in Tomato. *Cell* **2020**, *182*, 145–161. [CrossRef] [PubMed]
53. Song, J.; Guan, Z.; Hu, J.; Guo, C.; Yang, Z.; Wang, S.; Liu, D.; Wang, B.; Lu, S.; Zhou, R.; et al. Eight high-quality genomes reveal pan-genome architecture and ecotype differentiation of Brassica napus. *Nat. Plants* **2020**, *6*, 34–45. [CrossRef]
54. Li, H.; Wang, S.; Chai, S.; Yang, Z.; Zhang, Q.; Xin, H.; Xu, Y.; Lin, S.; Chen, X.; Yao, Z.; et al. Graph-based pan-genome reveals structural and sequence variations related to agronomic traits and domestication in cucumber. *Nat. Commun.* **2022**, *13*, 682. [CrossRef] [PubMed]
55. Chen, C.; Chen, H.; Zhang, Y.; Thomas, H.R.; Frank, M.H.; He, Y.; Xia, R. TBtools: An Integrative Toolkit Developed for Interactive Analyses of Big Biological Data. *Mol. Plant.* **2020**, *13*, 1194–1202. [CrossRef]
56. Wang, Y.; Tang, H.; Debarry, J.D.; Tan, X.; Li, J.; Wang, X.; Lee, T.H.; Jin, H.; Marler, B.; Guo, H.; et al. MCScanX: A toolkit for detection and evolutionary analysis of gene synteny and collinearity. *Nucleic Acids Res.* **2012**, *40*, e49. [CrossRef]
57. Che, G.; Gu, R.; Zhao, J.; Liu, X.; Song, X.; Zi, H.; Cheng, Z.; Shen, J.; Wang, Z.; Liu, R.; et al. Gene regulatory network controlling carpel number variation in cucumber. *Development* **2020**, *147*, dev184788. [CrossRef] [PubMed]
58. Jing, L.; Yan, S.; Yang, W.; Li, Y.; Xia, M.; Chen, Z.; Wang, Q.; Yan, L.; Song, X.; Liu, R.; et al. Transcriptomic analysis reveals the roles of microtubule-related genes and transcription factors in fruit length regulation in cucumber (*Cucumis sativus* L.). *Sci. Rep.* **2015**, *5*, 8031. [CrossRef]
59. Zhao, J.; Jiang, L.; Che, G.; Pan, Y.; Li, Y.; Hou, Y.; Zhao, W.; Zhong, Y.; Ding, L.; Yan, S.; et al. A Functional Allele of *CsFUL1* Regulates Fruit Length through Repressing *CsSUP* and Inhibiting Auxin Transport in Cucumber. *Plant Cell* **2019**, *31*, 1289–1307. [CrossRef]
60. Zhu, Y.; Yin, J.; Liang, Y.; Liu, J.; Jia, J. Transcriptomic dynamics provide an insight into the mechanism for silicon-mediated alleviation of salt stress in cucumber plants. *Ecotoxicol. Environ. Saf.* **2019**, *174*, 245–254. [CrossRef]
61. Chen, X.; Wang, Z.; Tang, R.; Wang, L.; Chen, C.; Ren, Z. Genome-Wide identification and expression analysis of Hsf and Hsp gene families in cucumber (*Cucumis sativus* L.). *Plant Growth Regul.* **2021**, *95*, 223–239. [CrossRef]
62. Xu, Q.; Xu, X.; Shi, Y.; Qi, X.; Chen, X. Elucidation of the molecular responses of a cucumber segment substitution line carrying *Pm5.1* and its recurrent parent triggered by powdery mildew by comparative transcriptome profiling. *BMC Genom.* **2017**, *18*, 21. [CrossRef]
63. Kong, W.; Chen, N.; Liu, T.; Zhu, J.; Wang, J.; He, X.; Jin, Y. Large-Scale Transcriptome Analysis of Cucumber and Botrytis cinerea during Infection. *PLoS ONE* **2015**, *10*, e0142221. [CrossRef] [PubMed]
64. Adhikari, B.N.; Savory, E.A.; Vaillancourt, B.; Childs, K.L.; Hamilton, J.P.; Day, B.; Buell, C.R. Expression Profiling of Cucumis sativus in Response to Infection by *Pseudoperonospora cubensis*. *PLoS ONE* **2012**, *7*, e34954. [CrossRef] [PubMed]
65. Wang, X.; Li, H.; Gao, Z.; Wang, L.; Ren, Z. Localization of quantitative trait loci for cucumber fruit shape by a population of chromosome segment substitution lines. *Sci. Rep.* **2020**, *10*, 11030. [CrossRef]
66. Li, H.; Hu, Y.J.; Zhang, Q.X.; Wang, L.N.; Ren, Z.H. Identification and Analysis on TRM Family in Cucumber. *J. Shandong Agric. Univ.* **2021**, *52*, 358–363.
67. Yin, S.; Zhao, L.; Liu, J.; Sun, Y.; Li, B.; Wang, L.; Ren, Z.; Chen, C. Pan-genome Analysis of *WOX* Gene Family and Function Exploration of *CsWOX9* in Cucumber. *Int. J. Mol. Sci.* **2023**, *24*, 17568. [CrossRef] [PubMed]
68. Bai, S.L.; Peng, Y.-B.; Cui, J.X.; Gu, H.T.; Xu, L.Y.; Li, Y.Q.; Xu, Z.H.; Bai, S.N. Developmental analyses reveal early arrests of the spore-bearing parts of reproductive organs in unisexual flowers of cucumber (*Cucumis sativus* L.). *Planta* **2004**, *220*, 230–240. [CrossRef]
69. Ma, H.; Liu, M. The microtubule cytoskeleton acts as a sensor for stress response signaling in plants. *Mol. Biol. Rep.* **2019**, *46*, 5603–5608. [CrossRef]

horticulturae

MDPI

Article

Integrated Phenotypic Physiology and Transcriptome Analysis Revealed the Molecular Genetic Basis of Anthocyanin Accumulation in Purple Pak-Choi

Qinyu Yang [1], Tao Huang [1], Li Zhang [1], Xiao Yang [1], Wenqi Zhang [1], Longzheng Chen [2], Zange Jing [3], Yuejian Li [1], Qichang Yang [1], Hai Xu [2,*] and Bo Song [1,*]

[1] Institute of Urban Agriculture, Chinese Academy of Agricultural Sciences, Chengdu 610213, China; yang675398120@163.com (Q.Y.); anybody119@sina.com (T.H.); zhangli10@caas.cn (L.Z.); yangxiao@caas.cn (X.Y.); zhangwenqi202408@163.com (W.Z.); yuejian_li@163.com (Y.L.); yangqichang@caas.cn (Q.Y.)

[2] Jiangsu Key Laboratory for Horticultural Crop Genetic Improvement, Institute of Vegetable Crops, Jiangsu Academy of Agricultural Sciences, Nanjing 210014, China; chenglong4510@sohu.com

[3] College of Agriculture and Life Science, Kunming University, Kunming 650214, China; jingzange@aliyun.com

* Correspondence: xuhai407@163.com (H.X.); songbo@caas.cn (B.S.)

Abstract: Purple Pak-choi is rich in anthocyanins, which have both ornamental and edible health functions, and has been used more and more widely in facility cultivation. In order to further clarify the molecular mechanism of purple Pak-choi, two Pak-choi inbred lines ('PQC' and 'HYYTC') were selected for the determination of pigment content and transcriptome analysis, and the key genes controlling the formation of purple character in leaves of Pak-choi were discovered. The results of pigment determination showed that the anthocyanin content of 'PQC' was 0.29 mg/g, which was about 100 times than 'HYYTC'; The chlorophyll content of 'HYYTC' was 2.25 mg/g, while 'PQC' only contained 1.05 mg/g. A total of 20 structural genes related to anthocyanin biosynthesis and 28 transcriptional regulatory genes were identified by transcriptome analysis. Weighted gene co-expression network analysis (WGCNA) was used to construct the weight network analysis map of 14 genes. The results showed that the cinnamate hydroxylase gene (*BraA04002213*, *BrC4H3*), flavanone-3- hydroxylase (*BraA09004531*, *BrF3H1*), and chalcone synthetase (*BraA10002265*, *BrCHS1*) were the core genes involved in the anthocyanin synthesis pathway of purple Pak-choi. The results identified the key genes controlling the formation of purple leaf traits, which laid a foundation for further analysis of the molecular mechanism of anthocyanin accumulation in purple Pak-choi and provided a theoretical basis for leaf color regulation.

Keywords: Pak-choi; purple; phenotypic physiology; transcriptome; anthocyanin accumulation

1. Introduction

Pak-choi (*Brassica campestris* ssp. *Chinensis*) is an important cruciferous leaf vegetable crop, which is widely cultivated in the middle and lower reaches of the Yangtze River in China, because of its rich germplasm resources, short growth cycle, and easy cultivation. As a main type of Pak-choi, purple Pak-choi is rich in anthocyanins and has the functions of both ornamental and edible health care, it is increasingly favored by people. However, in the production and cultivation, the purple character is often affected by external environment such as temperature and light, resulting in uneven leaf coloring and dull color, which is mainly due to the influence of environmental conditions on the expression of key genes of anthocyanin metabolism pathway [1–4]. Effective leaf color regulation was performed on the *DFR* gene of sweet potato and the *PAP1* gene of tobacco by means of RNAi and gene overexpression, respectively, indicating that leaf color metabolic engineering improvement through the regulation of key anthocyanin genes is feasible [5].

Anthocyanin is an important category of secondary metabolic substances in plants, belonging to a kind of flavonoid [6]. Anthocyanin biosynthesis originated from the branch of the flavonoid pathway. These enzymes are encoded by biosynthetic structural genes which include early biosynthetic genes (EBGs) and late biosynthetic genes (LBGs). The EDGs including *CHS* (encoding chalcone synthase), *CHI* (encoding chalcone isomerase), and *F3H* (encoding flavanone 3-hydroxylase) are involved in the production of precursors. The LBGs are involved in the production of colored anthocyanins, such as *DFR* (encoding dihydroflavonol 4-reductase) and *UFGT* (encoding UDP-glucose: flavonoid-3-O-glucosyltransferase). Although the anthocyanin biosynthesis pathway is relatively conserved and well-studied in model plants such as Arabidopsis, and many genes related to the synthesis pathway have been cloned, there are few tudies on the mining and expression of related genes in Chinese cabbage vegetables. Previous research was conducted for the transcriptomic analysis of green and purple Chinese cabbage in which 36 genes were screened for up-regulated expression [7]. Late anthocyanin synthesis genes *DFR, ANS,* and *UFGT* may play a more critical role in the anthocyanin biosynthesis of purple Pak-choi concluded by transcriptome analysis [8].

Despite some beneficial attempts made by predecessors, the regulatory mechanism and regulatory network of anthocyanin synthesis related to the formation of purple leaf traits in Pak-choi are still unclear, and the key genes controlling anthocyanin synthesis and accumulation in purple Pak-choi need to be further studied and confirmed. In this study, the purple leaf line 'PQC' and green leaf line 'HYYTC' were selected to analyze gene differential expression at the transcriptional level. Weighted gene co-expression network analysis (WGCNA) and weighted network analysis were used to identify gene modules closely related to anthocyanin synthesis. The aim was to dig out the key genes involved in anthocyanin biosynthesis and regulation of Pak-choi, and initially construct the anthocyanin molecular regulatory network related to purple leaf character formation, which laid a foundation for in-depth analysis of the molecular mechanism of anthocyanin accumulation in purple Pak-choi and provided a theoretical basis for the color regulation in facility cultivation.

2. Materials and Methods

2.1. Experimental Material

After seeding, the two materials 'PQC' and 'HYYTC' were placed in an incubator at 25 °C for germination promotion. After the germination, the seedlings were cultivated in plate substrate using Pindstrup Mosebrug A/S, Denmark. After sowing, they were placed in a light incubator. The following procedure was carried out: light culture at 22 °C for 12 h and dark culture at 18 °C for 12 h. The cotyledon extension stage (T1), cotyledon flattening stage (T2), and two true leaf stages (T3) were sampled, respectively. Cotyledon was selected at T1 and T2, and the second true leaf was selected as the sample at T3. Each sample was randomly collected from five seedlings, and six samples were collected. All samples were repeated three times, totaling eighteen samples. All samples were frozen in liquid nitrogen at −80 °C for RNA extraction (Figure 1).

2.2. Determination of Pigment

The two materials were sown separately in the plastic greenhouse and transplanted into the insect-proof net room at the seedling age of 30 d. After the transplantation for 45 d, the largest leaf free from disease and insect infestation was selected and placed in the refrigerator at 4 °C for subsequent tests.

Anthocyanin content determination: 0.1 g of leaf tissue was weighted and placed in Erlenmeyer flasks then 10 mL, 0.1 mol/L hydrochloric alcohol solution was added and soaked in a water bath at 60 °C for 30 min then poured into a 25 mL volumetric bottle, and the volume was fixed to 25 mL. With 0.1 mol/L hydrochloric ethanol solution as the control, the optical density of the extracted solution at 530 nm, 620 nm, and 650 nm was determined. The optical density of anthocyanins ODλ = (OD530 − OD620) − 0.1 (OD650 − OD620); Anthocyanin content (nmol/g) = ODλ/ε × V/m × 106, anthocyanin content (mg/g) = anthocyanin content (nmol/g) × M × 10^{-3} [9].

A B C D E F

Figure 1. Three growth periods of two lines for transcriptome analysis. (**A**) The cotyledon expanded stage (PQC-T1); (**B**) the cotyledon flattening stage of 'PQC' (PQC-T2); (**C**) the two true leaf stage of 'PQC' (PQC-T3); (**D**) the cotyledon expanded stage of 'HYYTC' (HYYTC-T1); (**E**) the cotyledon flattening stage of 'HYYTC' (HYYTC-T2); (**F**) the two true leaf stage of 'HYYTC' (HYYTC-T3).

Determination of chlorophyll content: 0.2 g of leaf tissue was weighted into a mortar, 5 mL 95% ethanol was added into the mortar, the mixture was ground into a homogenate, filtered into a 25 mL volumetric bottle, and 95% ethanol was used to set the volume to 25 mL. The absorbance was measured by spectrophotometer at 649 nm and 665 nm with 95% ethanol as the control. Chlorophyll a content Ca = 13.95A665 − 6.88A649, chlorophyll b content Cb = 24.96A649 − 7.32A665. Chlorophyll content (mg/g) = (C × V × N)/(W × 1000) [10].

2.3. RNA Extraction, cDNA Library Construction and Quality Control

Trizol kit (Invitrogen Company, Waltham, MA, USA) was used to extract total RNA from samples, and Qiaquick kit (Qiagen, Shanghai, China) was used to purify cDNA. RNA concentration and integrity were evaluated using a NanoDrop 2000 (Thermo Fisher Scientific, Waltham, MA, USA) spectrophotometer and polyacrylamide gel electrophoresis, respectively. mRNA was divided into fragments by fragmentation buffer and amplified by PCR to build a fragment sequencing library. The library was sequenced using Illumina Hi-Seq™ 2000. The original image data files obtained by Illumina Hi-Seq™ 2000 were converted into raw reads by base recognition analysis. Clean reads were obtained from the sequenced raw reads after the removal of splices and low-quality duplicate reads.

2.4. Screening Differentially Expressed Genes

Gene expression was obtained by calculating FPKM [11]. In the screening process, FC (log2 fold change) > 2 and FDR (false discovery rate) < 0.05 were used as the selection criteria for differential genes. Differential expressions between samples were analyzed by DEG-Seq [12].

2.5. Enrichment Analysis of Differentially Expressed Genes GO and KEGG

The GO annotation was carried out by a specific perl script using the TopGo R 2.5 software package to perform Fisher significance enrichment tests with the assembled transcript as the background. Kobas 3.0 software was used for KEGG enrichment analysis, and the R 2.10 software package ggplot was used for visualization of the enrichment analysis point plot.

2.6. *Functional Annotation of Differentially Expressed Genes*

The obtained clean reads were compared to the Chinese cabbage reference genome (http://brassicadb.cn/#/Download/, accessed on 6 May 2023). Differentially expressed genes were identified by BLAST (BLAST: Basic Local Alignment Search Tool nih.gov, accessed on 6 May 2023). The sequence was compared with the protein database to obtain annotation information. TMM was used to standardize read count data [13].

2.7. *Weighted Gene Co-Expression Network Analysis*

The association network of all differentially expressed genes was constructed using WGCNA in R language. The adjacency matrix is generated by calculating the correlation between all the genes, and the soft threshold β is chosen according to the scale-free topological criteria, using the adjacency matrix to calculate the topological overlap matrix (TOM). The dynamic cutting algorithm was used for gene clustering and module division, and the truncation height of 0.25 was used to merge the branches into the final module. The value of the module characteristic gene (ME) was calculated, and the correlation degree between the module and anthocyanin was estimated by ME value.

2.8. *Quantitative RT-qPCR Verification*

Nine differentially expressed anthocyanin structural genes and transcription factor genes (*BrPAL*, *BrC4H3*, *BrF3H*, *BrCHS*, *BrCHI*, *BrDFR*, *BrUFGT*, *BrANS*, *BrMYB44*, and *BrTT8*) were selected for RT-qPCR verification. The sequence of primers used for the above gene expression is shown in Table 1. The qPCR amplification reaction system was based on the SYBR Premix Ex Taq kit with single-strand cDNA as the template. The system was 20.0 μL, including Taq 10.0 μL, template 2.0 μL, positive and negative primes 1.0 μL and ddH_2O 6.0 μL. The qPCR reaction condition was set at 95 °C for 5 min; 95 °C 10 s; 60 °C for 30 s; and 72 °C for 15 s, a total of 40 cycles. The $2^{-\Delta\Delta Ct}$ method was used for relative quantitative analysis of the data [14].

Table 1. The primer sequences used for RT qPCR.

No	Gene ID	Gene Name	Forward Primer (5′–3′)	Reverse Primer (5′–3′)
1	*BraA04000661*	*BrPAL*	AGCAACATAACCAAGATG	TCTCAGATTCTCCTCAAG
2	*BraA04002213*	*BrC4H3*	TGAGGAAACGCTTGCAGT	GGCCTGAGGATAGGGATG
3	*BraA05002651*	*Br4CL4*	ATCTTTCCTCGCCGTCGTTT	CTCCGGCCGAAATCTTAGGCT
4	*BraA09004531*	*BrF3H*	ATTCATTGTCTCTAGTCATCTTC	CCGTGAGTAGTCTCTGTT
5	*BraA10002265*	*BrCHS*	TATCCTGACTACTACTTC	CTCCTTTAGAAACTCTTC
6	*BraA03005399*	*BrANS*	TCCTGATTCCATTGTGAT	TCCTAACCTTCTCCTTATTC
7	*BraA06000554*	*BrUFGT*	GTAATGTATCCGTGGTTAG	GGTAGAGGTTAAGAGGTT
8	*BraA08002374*	*BrMYB44*	TTATGAGACGGAGAATGT	TACCTCTTCCTTCCTAAC
9	*BraA09002835*	*BrTT8*	AGACGAAGAAGAAGTAGA	CCTCCATTAGATTCATCAT
10	-	*BrActin*	GTTGCTATCCAGGCTGTTC	AGCGTGAGGAAGAGCATAAC

3. Results

3.1. *Observation of Leaf Color*

In the cotyledon flattening stage, the color of the cotyledon of 'PQC' began to appear, and the distribution was uneven. The purple pigment mainly accumulated on the true leaves at the stage of two true leaves. In the adult stage, the adult plants of 'PQC' are more colorful, with dark purple on the front of the leaves and slightly purple on the back, while the leaves of 'HYYTC' are green on the front and back of the leaves at all stages (Figure 2).

Figure 2. Individual plant and leaf traits of two lines (**A**,**F**): The cotyledon flattening stage (T2); (**B**,**G**): The two true leaf stage (T3); (**C**,**H**): The adult stage; (**D**,**I**): The front side of leaves; (**E**,**J**): The opposite side of the leave.

3.2. Determination of Pigment Content in Leaves

The results showed that the anthocyanin content of the two materials was significantly different. The anthocyanin content of 'PQC' was 0.29 mg/g, and that of 'HYYTC' was only 0.002 mg/g. The chlorophyll content in 'HYYTC' was as high as 2.25 mg/g, while the chlorophyll content in 'PQC' was significantly lower than that in 'HYYTC' (Figure 3).

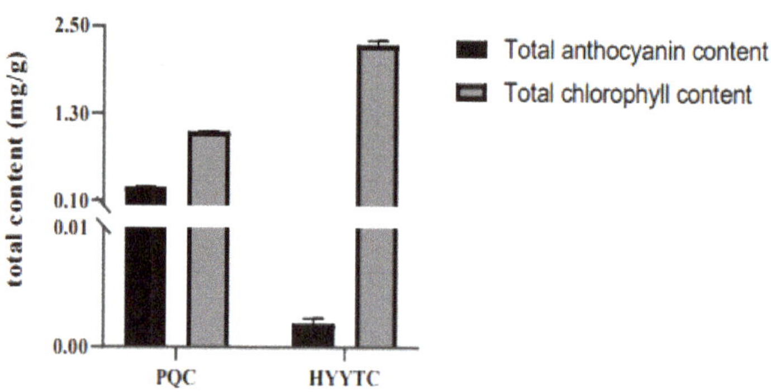

Figure 3. The content of anthocyanin and chlorophyll for two lines.

3.3. Sequencing Results and Comparison Statistics

Total RNA from leaves was extracted and used to construct 18 cDNA libraries. A total of 155.04 Gb of Clean Data were obtained from 18 samples sequenced. After removing low-quality reads and connectors, 337,561,131 pairs of clean reads were obtained, and between 22 million and 33 million pairs of clean reads were obtained for each library. The Q30 of each sample used for assembly was higher than 92.71%, indicating that the sequencing results met the quality requirements for subsequent assembly analysis (Table S1). Using the V2.5 version of the Chinese cabbage genome as the reference genome, the comparison rate between the reads of each sample and the reference genome ranged from 87.48 to 90.29%, among which the ratio of 'PQC' was 89.22–90.04%, and that of 'HYYTC' was 87.48–90.29% (Table 2).

Table 2. Overview of the transcriptome sequencing.

Samples	Clean Reads	Clean Bases	GC Content (%)	≥Q30 (%)
PQC-T1-1	26,549,215	7,928,202,428	48.92%	92.83%
PQC-T1-2	26,180,713	7,829,647,066	48.76%	93.19%
PQC-T1-3	34,780,519	10,381,289,642	48.64%	93.28%
PQC-T2-1	27,200,992	8,115,863,242	48.72%	93.43%
PQC-T2-2	26,021,139	7,764,834,504	48.67%	93.38%
PQC-T2-3	22,706,919	6,782,398,766	48.59%	93.38%
PQC-T3-1	22,221,865	6,636,130,982	48.23%	92.71%
PQC-T3-2	27,585,040	8,220,152,954	48.90%	93.62%
PQC-T3-3	28,884,210	8,615,366,610	48.63%	93.46%
HYYTC-T1-1	33,792,919	10,064,723,406	48.75%	93.10%
HYYTC-T1-2	27,771,004	8,285,974,048	48.88%	93.08%
HYYTC-T1-3	32,973,684	9,829,969,670	49.10%	92.72%
HYYTC-T2-1	27,844,329	8,323,582,204	48.66%	93.43%
HYYTC-T2-2	31,916,361	9,526,775,132	48.57%	93.13%
HYYTC-T2-3	32,197,152	9,602,692,572	49.28%	93.10%
HYYTC-T3-1	29,935,983	8,932,854,782	48.45%	93.06%
HYYTC-T3-2	28,258,874	8,433,086,514	48.74%	93.16%
HYYTC-T3-3	32,788,267	9,761,637,592	48.49%	93.07%

3.4. Identification of Differentially Expressed Genes

By comparing the FPKM values of 'PQC' and 'HYYTC' in three different leaf growth stages, the expression patterns of differential genes (DEGs) were determined. 2632 (up-regulation 1098, down-regulation 1534), 3147 (up-regulation 1427, down-regulation 1720), and 1920 (up-regulation 697, down-regulation 1223) DEGs were identified in the three comparison groups of HYYTC-T1 vs. PQC-T1 (G1), HYYTC-T2 vs. PQC-T2 (G2), and HYYTC-T3 vs. PQC-T3 (G3), respectively (Figure 4A). Through Wayne plot analysis, out of a total of 5070 DEGs, a total of 738 (308 up-regulation, 427 down-regulation) were identified. Three of the genes had inconsistent expression patterns in each of the three comparison groups, which may be the result of specific expression at different reproductive periods (Figure 4B–D).

3.5. Clustering Analysis

A cluster analysis of 738 gene expression profiles from purple material 'PQC' and green material 'HYYTC' at three different periods showed that compared with green material, the up-regulated gene expression of purple material was less than that of down-regulated gene expression (Figure 5A). According to trend expression analysis, DEG was divided into nine gene clusters, with significant differences in gene expression between cluster 0 and cluster 8 (Figure 5B). In cluster 0, there are 264 genes in total. With the growth process, the expression of related genes in purple material 'PQC' is higher than that in green material 'HYYTC', and these genes are mainly concentrated in transcription, ribosomal structure, and biotransformation (*BraA03005834*, *BraA04002715*, and *BraA05001180*). Amino acid transport and metabolism (*BraA09005771*, *BraA05000655*, and *BraA06001513*) and post-translational modification, protein turnover, companion (*BraA01001595*, *BraA01003552* and *BraA05001325*). Cluster 8 contains a total of 400 genes. With the development of the growth period, the green material 'HYYTC' showed a higher transcription level than the purple material 'PQC'. These genes are mainly involved in post-translational modification, protein turnover, chaperones (*BraA01002075*, *BraA03000073*, and *BraA08002394*), and signal transduction mechanisms (*BraA05004531*, *BraA07002138* and *BraA09003227*). In addition, carbohydrate transport and metabolism (*BraA02003506*, *BraA03000585*, and *BraA03000676*) were found to be highly expressed in 'HYYTC'. Other DEGs are grouped into clusters 1–7 according to their expression patterns.

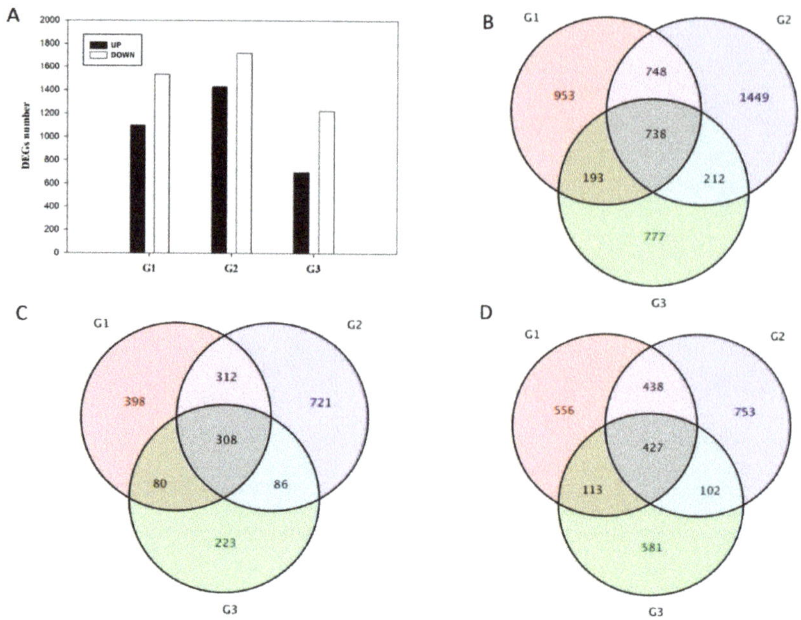

Figure 4. Number of DEGs between 'PQC' and 'HYYTC' at the cotyledon expanded stage, cotyledon flattening stage, and two true leaf stages, respectively. (**A**) The total number of upregulated and down-regulated DEGs. (**B**) Venn diagram of all DEGs. (**C**) Venn diagram of up-regulated genes. (**D**) Venn diagram of down-regulated genes. G1, G2, and G3 represent the DEGs between 'PQC' and 'HYYTC' lines at the Cotyledon expanded stage, cotyledon flattening stage, and two true leaf stages, respectively.

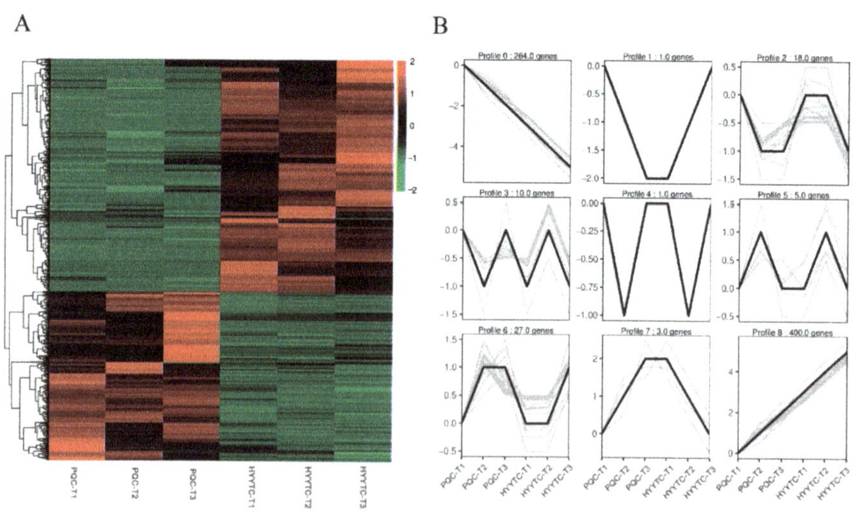

Figure 5. Clustering analysis of 738 DEGs. (**A**) Hierarchical clustering of the 738 DEGs. (**B**) Expression patterns of the 738 DEGs in the nine clusters.

3.6. GO Analysis and KEGG Enrichment Pathway Analysis

738 common DEGs were enriched in the single-organism biosynthesis process (GO:0044711), organonitrogen compound metabolic process (GO:1901564), and phosphorus metabolic process (GO: 0006793) (Figure 6). In addition, we found that some genes were also enriched in the anthocyanin-containing compound biosynthetic process (GO:009718, GO:0046283), flavonoid metabolic process (GO:0009812), and flavonoid biosynthetic process (GO:0009812); these results indicate that the purple formation of purple Pak-choi may be related to some genes of the specific GO pathway.

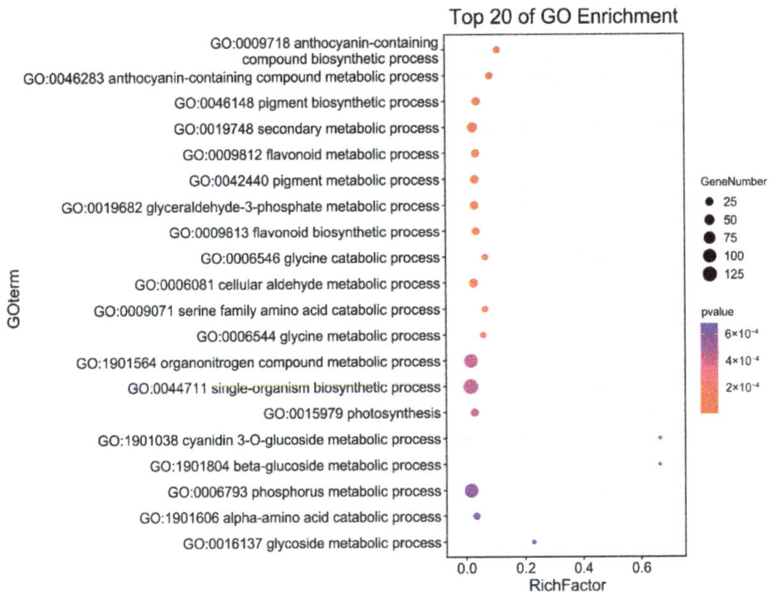

Figure 6. Analysis of GO enrichment for 738 DEGs.

KEGG pathways in G1, G2, and G3 were analyzed. There were 528, 694, and 386 DEGs annotated by KEGG pathways, while 87 pathway classes were enriched from 738 common DEGs. As shown in Figure 7, the flavonoid biosynthesis pathway (ko00941), anthocyanin biosynthesis (ko00942), phenylalanine metabolism (ko00940), and glutathione metabolism (ko00480) were the four pathways with the most significant differences. Song (2020) pointed out anthocyanin accumulation is the main reason for the formation of the purple leaf of Pak-choi, and the above four pathways are closely related to anthocyanin synthesis [15]. There were 6 DEGs related to the phenylalanine metabolic pathway, 10 DEGs related to flavonoid biosynthesis, 2 DEGs related to anthocyanin biosynthesis, and 8 DEGs related to glutathione metabolism.

The phenylalanine biosynthesis pathway is the precursor of the anthocyanin metabolism pathway. Phenylalanine synthetase gene *BraA04000661* (*BrPAL2.2)* and Cinnamate hydroxylase genes *BraA04002213* (*BrC4H3*) and *BraA03001710* (*BrC4H5*) were up-regulated in these three comparison groups. These differential genes may promote anthocyanin synthesis and accumulation by producing more anthocyanin biosynthesis precursors. Although the expression of the ligase gene *BraA05002651* (*Br4CL4*), which is used to synthesize 4-coumaryl CoA, is down-regulated in purple materials, *BraA05001886* (*Br4CL1*), *BraA05002646* (*Br4CL2*), and *BraA07003006* (*Br4CL3)* were expressed normally in the two materials. The expressions of early anthocyanin biosynthesis genes such as hydroxylase gene *BraA09004531* (*BrF3H*), chalcone isomase gene *BraA09004891* (*BrCHI1*), and chalcone synthase gene (*BraA10002265*, *BrCHS1*; *BraA03000633*, *BrCHS2*) were up-regulated in the three phases of purple material. Late anthocyanin biosynthesis genes such as anthocyanin dioxygenase genes (*BraA0100144*, *BrANS1*; *BraA03005399*, *BrANS2*), the dihydroflavonol-4-reductase gene *BraA09002044* (*BrDFR*), and the flavonoid 3-O-glucosidtransferase gene (*BraA10000963*, *BrUGT79B1.1*; *BraA06000554*, *BrUGT79B1.2*) were significantly up-regulated in the three comparison groups. The expression patterns of anthocyanin transport genes such as *BraA01003440* (*BrTT12*), *BraA10002029* (*BrTT19.1*), *BraA02000754* (*BrTT19.2*), *BraA08004000* (*BrGSTF6*), *BraA09000406* (*Br5MAT*), and *BraA08003959* (*Br3AT1*) are all up-regulated (Table 3).

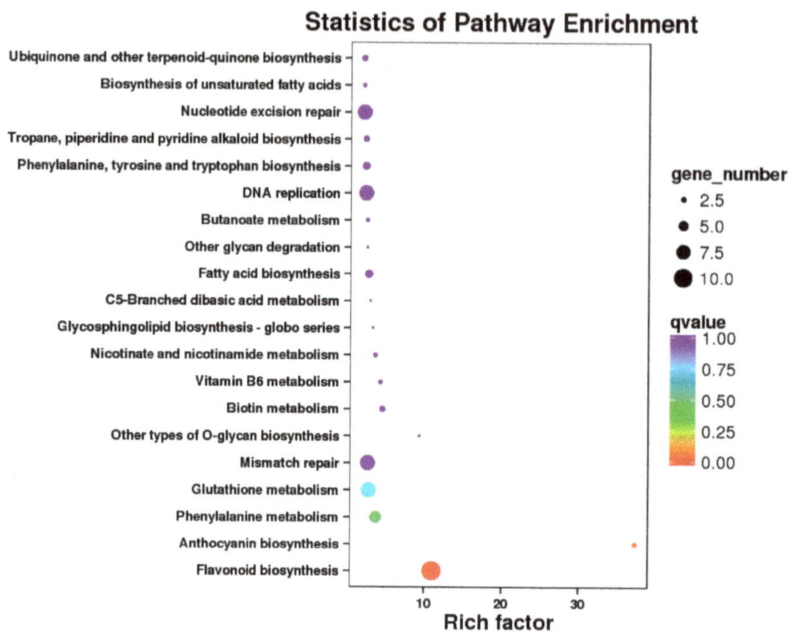

Figure 7. Analysis of KEGG pathway for 738 common DEGs.

3.7. Transcription Factor Identification

Several transcription factors (TFs) have been found to be closely related to anthocyanin synthesis in *Arabidopsis*. In the previous analysis, we found that some DEGs belong to genes that encode transcription factors. To better understand which transcription factors are involved in regulating anthocyanin biosynthesis, the differentially expressed transcription factors were analyzed in this study. By comparing 738 DEG sequences with PlantTFDB, a total of 28 transcription factors belonging to 18 transcription factor families were identified. Among them, MYB-related, bHLH, and NAC types are the main transcription factor families, each containing three or more DEGs (Table 4).

Table 3. Identification of structure genes and transportation genes of anthocyanin synthesis.

Gene ID	FPKM						HYYTC-T1 Vs PQC-T1	HYYTC-T2 Vs PQC-T2	HYYTC-T3 Vs PQC-T3	Gene Annotation
	PQC-T1	PQC-T2	PQC-T3	HYYTC-T1	HYYTC-T2	HYYTC-T3				
BraA04000661	30.43	42.55	137.89	3.74	11.90	23.96	Up	Up	Up	*BrPAL2.2*
BraA04002213	11.58	11.90	68.38	4.13	5.83	24.98	Up	Up	Up	*BrC4H3*
BraA03001710	33.88	36.42	154.03	2.87	4.41	11.99	Up	Up	Up	*BrC4H5*
BraA05002651	1.17	1.28	3.43	12.08	10.99	17.32	Down	Down	Down	*Br4CL4*
BraA08002920	25.22	12.86	25.15	8.83	5.78	5.52	Up	Up	Up	*BrCCoAOMT*
BraA09004531	108.59	85.32	306.09	18.84	23.84	96.13	Up	Up	Up	*BrF3H1*
BraA10002265	55.68	43.29	299.84	8.87	9.53	108.21	Up	Up	Up	*BrCHS1*
BraA03000633	14.07	14.17	139.38	0.70	1.72	14.58	Up	Up	Up	*BrCHS2*
BraA09004891	10.96	18.55	86.07	1.42	1.94	8.72	Up	Up	Up	*BrCHI1*
BraA09002044	160.86	115.98	443.50	5.45	7.75	7.46	Up	Up	Up	*BrDFR*
BraA01001444	109.91	85.16	393.62	3.57	5.76	36.51	Up	Up	Up	*BrANS1*
BraA03005399	12.68	13.28	60.08	0.99	0.76	0.81	Up	Up	Up	*BrANS2*
BraA10000963	21.80	16.19	50.35	0.60	0.66	1.34	Up	Up	Up	*BrUGT79B1.1*
BraA06000554	55.17	40.02	132.54	1.41	1.96	4.30	Up	Up	Up	*BrUGT79B1.2*
BraA01003440	17.90	21.31	39.02	1.80	3.30	0.94	Up	Up	Up	*BrTT12*
BraA02000754	83.23	70.92	205.42	3.90	3.33	23.59	Up	Up	Up	*BrTT19.2*
BraA10002029	22.01	19.89	80.82	0.44	0.88	0.56	Up	Up	Up	*BrTT19.1*
BraA08004000	4.66	8.18	34.57	0.48	0.51	0.30	Up	Up	Up	*BrGSTF6*
BraA08003959	59.03	38.90	149.81	1.72	1.26	4.69	Up	Up	Up	*Br3AT1*
BraA09000406	26.01	17.14	74.53	0.50	0.59	0.38	Up	Up	Up	*Br5MAT*

Table 4. Identification of transcription factors.

Gene ID	FPKM						HYYTC-T1 Vs PQC-T1	HYYTC-T2 Vs PQC-T2	HYYTC-T3 Vs PQC-T3	Gene Annotation
	PQC-T1	PQC-T2	PQC-T3	HYYTC-T1	HYYTC-T2	HYYTC-T3				
BraA08002374	1.43	2.48	2.02	2.68	6.98	6.28	Down	Down	Down	*MYB44*
BraA05004112	0.39	1.47	17.01	2.41	5.91	36.21	Down	Down	Down	*REVEILLE 8*
BraA05001641	0.33	1.02	0.29	13.47	10.87	4.40	Down	Down	Down	*REVEILLE 2*
BraA02002130	1.39	0.77	1.86	3.33	5.06	6.84	Down	Down	Down	*MYB1R1*
BraA05003486	4.19	7.22	4.97	1.55	1.81	0.73	Up	Up	Up	*ABCG29*

Table 4. *Cont.*

Gene ID	FPKM						HYYTC-T1 Vs PQC-T1	HYYTC-T2 Vs PQC-T2	HYYTC-T3 Vs PQC-T3	Gene Annotation
	PQC-T1	PQC-T2	PQC-T3	HYYTC-T1	HYYTC-T2	HYYTC-T3				
BraA10003119	3.05	1.97	2.87	0.04	0.15	0.29	Up	Up	Up	*FLA10*
BraA08002359	1.83	10.69	2.05	0.60	0.90	0.24	Up	Up	Up	*GARP-G2-like*
BraA01004300	2.70	3.61	3.35	13.46	12.6	9.96	Down	Down	Down	*TRIHELIX*
BraA09002835	2.27	2.91	11.05	0.55	0.75	1.45	Up	Up	Up	*TT8*
BraA01002210	10.20	15.05	11.72	2.97	4.44	2.44	Up	Up	Up	*bHLH3*
BraA03006442	0.20	0	0.11	21.55	12.22	11.01	Down	Down	Down	*BEE 2*
BraA09000011	0.33	0.19	0.78	1.13	1.77	4.05	Down	Down	Down	*UNE10*
BraA07001006	0	0	0	1.48	1.49	2.34	Down	Down	Down	*C3H*
BraA07003043	0.79	0.77	1.95	3.21	5.40	7.33	Down	Down	Down	*C2H2*
BraA02000673	0.19	1.09	7.61	1.70	3.55	22.18	Down	Down	Down	*C2C2-CO-like*
BraA04002035	1.51	3.18	4.07	0.44	0.13	1.04	Up	Up	Up	*C2H2*
BraA09002226	0	0	0	28.04	24.80	28.55	Down	Down	Down	*NAC69*
BraA03000407	0	0	0	5.17	5.02	5.83	Down	Down	Down	*NAC82*
BraA07000730	1.44	1.05	1.26	9.42	6.79	5.31	Down	Down	Down	*NAC101*
BraA02004460	12.68	11.73	19.29	4.94	4.55	7.41	Up	Up	Up	*MADS-MIKC*
BraA10001525	4.74	9.59	5.03	39.14	47.88	16.02	Down	Down	Down	*AP2/ERF-ERF*
BraA05004664	3.71	3.00	4.79	14.32	12.83	15.55	Down	Down	Down	*TCP*
BraA09002413	0.31	0.12	0.73	2.15	2.51	2.33	Down	Down	Down	*NF-YB*
BraA08001844	1.12	7.81	1.11	3.22	20.05	9.09	Down	Down	Down	*WRKY18*
BraA03002344	0.20	0.07	0.03	7.21	7.87	2.27	Down	Down	Down	*bZIP34*
BraA09005994	0.02	0	0.14	2.45	1.69	2.99	Down	Down	Down	*BES1*
BraA04002319	3.07	7.84	71.47	23.13	41.41	149.16	Down	Down	Down	*DBB*
BraA08002660	0.17	0.05	0.37	0.94	0.72	1.68	Down	Down	Down	*HB-HD-ZIP*

For the three types of transcription factors, MYB, bHLH, and WD40, a total of eight MYBs and three bHLHs were detected in this study. Of the eight DEGs that code MYB, *BraA08002374 (MYB44)*, *BraA05004112 (REVEILLE 8)*, *BraA05001641 (REVEILLE 2)*, *BraA02002130 (MYB1R1)*, and *BraA01004300 (TRIHELIX)* were down-regulated in the three comparison groups, while *BraA05003486 (FLA10)*, *BraA10003119 (ABCG29)*, and *BraA08002359 (GARP-G2-like)* were the opposite. Among the three DEGs encoding bHLH, *BraA09002835 (TT8)* and *BraA01002210 (bHLH3)* were up-regulated in the three comparison groups, while *BraA03006442 (BEE 2)* was the opposite. In addition, no differential genes annotated as the WD40 family were found.

In addition to MYB and bHLH transcription factors, 17 transcription factors were identified in this study. Only *BraA02004460* (*MADS*) and *BraA04002035* (*C2H2*) were up-regulated in three phases of purple material, while other genes were down-regulated. In particular, *BraA03002344* (*bZIP34*) and *BraA08001844* (*WRKY18*) are more down-regulated.

3.8. Identification of WGCNA Modules of Genes Related to Anthocyanin Metabolism

The weighted gene co-expression network analysis (WGCNA) is a method to analyze target genes at the network level, which mainly constructs weighted association networks with differentially expressed genes and further screens for target genes. In this study, eight modules were identified from the RNA-seq data (Figure 8A), and correlation analysis was conducted between the module feature genes of the eight modules and different samples. Correlation analysis between modules and traits showed that the absolute value of the correlation coefficient between module 'MEblack' and two true leaf stages of purple material (PQC-T3) was the highest ($r = 0.98$), and the correlation was significant $p = 7 \times 10^{-4}$, while the other modules were weakly correlated or not correlated with the treatment relationships (Figure 8B). Since purple material has the most purple leaf color and the highest anthocyanin content in the two true leaf stages, it is believed that the gene in the module "black" has a high correlation with the anthocyanin metabolism pathway. KEGG enrichment analysis was performed on 784 genes contained in module "Black". The enrichment was mainly in physiological metabolism, and a total of 50 pathway categories were enriched, among which phenylalanine metabolism and flavonoid biosynthesis were the most important pathways, and 16 and 13 genes were enriched, respectively (Figure 9). In order to further investigate whether the 784 genes discovered by WGCNA share differential genes with the previous 738 genes, it was found through Venn diagram analysis that there were 76 common DEGs and these genes were up-regulated in three comparison groups (Table S2). Interestingly, 76 genes contain all previously identified anthocyanin synthesis structure genes and transport genes, including two transcription factors *BraA09002835* (*TT8*) and *BraA02004460* (*MADS-MIKC*). In addition, 784 genes were newly identified that may be involved in anthocyanin biosynthesis. Genes such as *BraA01002675* (*BrPAL1*), *BraA09001715* (*BrPAL2.1*), *BraA02000559* (*BrCHS3*), *BraA05001212* (*Peroxidase* 19), *BraA10000387* (*Peroxidase* 3), *BraA10002815* (*Cytochrome P450*), and *BraA07003905* (*BrCOMT*) were up-regulated in the two comparison groups (G2 and G3). The anthocyanin accumulation-related gene enriched in module 'black' was identified by WGCNA, which confirms the accuracy of WGCNA data analysis.

In order to further evaluate the importance of 76 shared genes in the anthocyanin metabolism pathway, association nodes with connectivity (weight value) >0.60 were selected. Import data into Cytoscape software (version 3.10.2), co-expression network construction, and core gene mining were carried out, and the weight network analysis diagram of 14 genes was constructed. Cinnamate hydroxylase gene (*BraA04002213, BrC4H3*), flavanone-3-hydroxylase (*BraA09004531, BrF3H1*), and Chalcone synthetase (*BraA10002265, BrCHS1*) were selected as the core genes of anthocyanin biosynthesis pathway of purple Pak-choi (Figure 10).

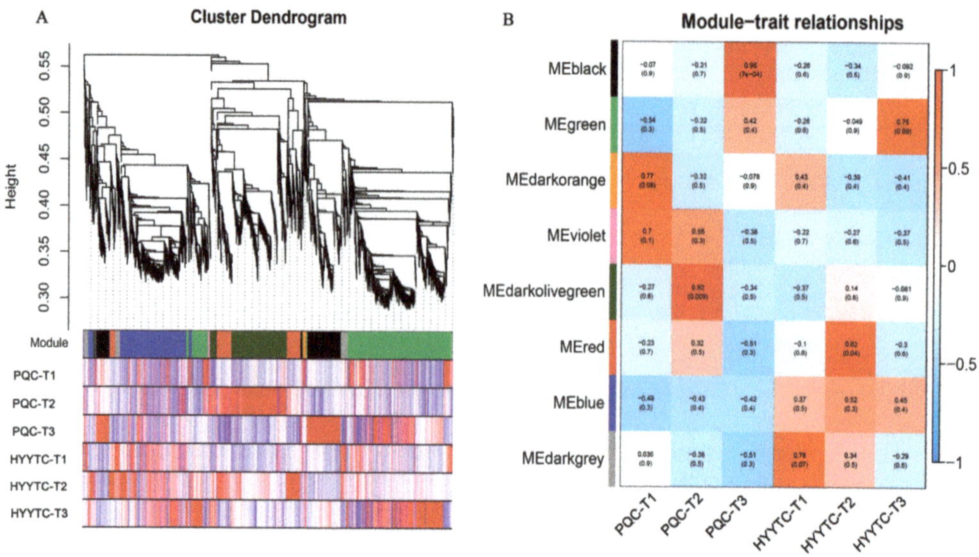

Figure 8. Weighted correlation network analysis of anthocyanin-related genes. (**A**) Hierarchical clustering tree showing co-expression modules. Each leaf in the tree represents one gene. The major tree branches constitute 8 modules labeled by different colors. (**B**) Module–trait relationship. The left lane indicates 8 module eigengenes. The right lane indicates the module–trait correlation from −1 to 1.

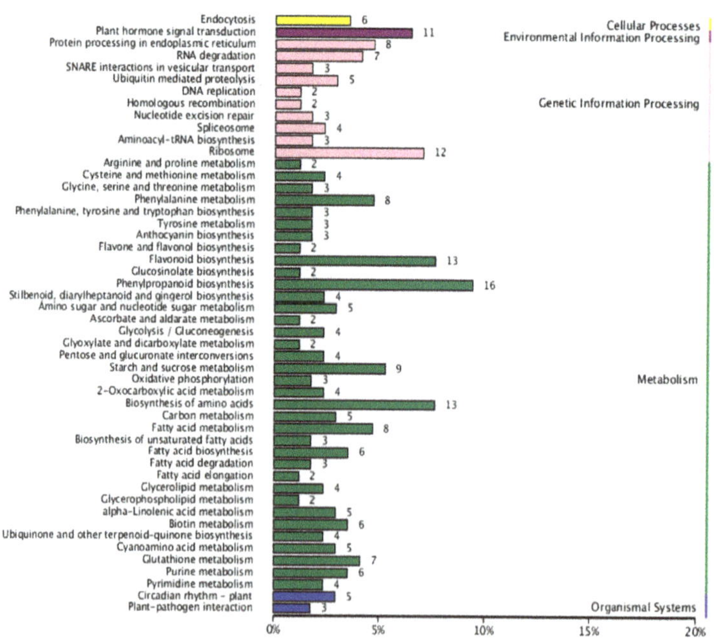

Figure 9. Analysis of KEGG pathway for 784 common DEGs.

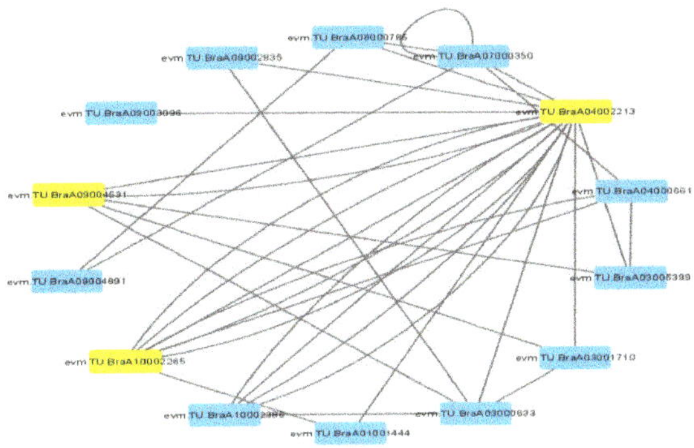

Figure 10. The gene network map of 14 genes from the MEblack module.

3.9. RT-qPCR Verification

To verify the results of RNA-seq sequencing, nine differentially expressed genes related to anthocyanin biosynthesis and regulation were selected for quantitative RT-qPCR detection in purple material 'PQC' and green material 'HYYTC'. The quantitative RT-qPCR results of the nine genes were similar to the expression patterns of transcriptome data. The results of transcriptome analysis were reliable (Figure 11).

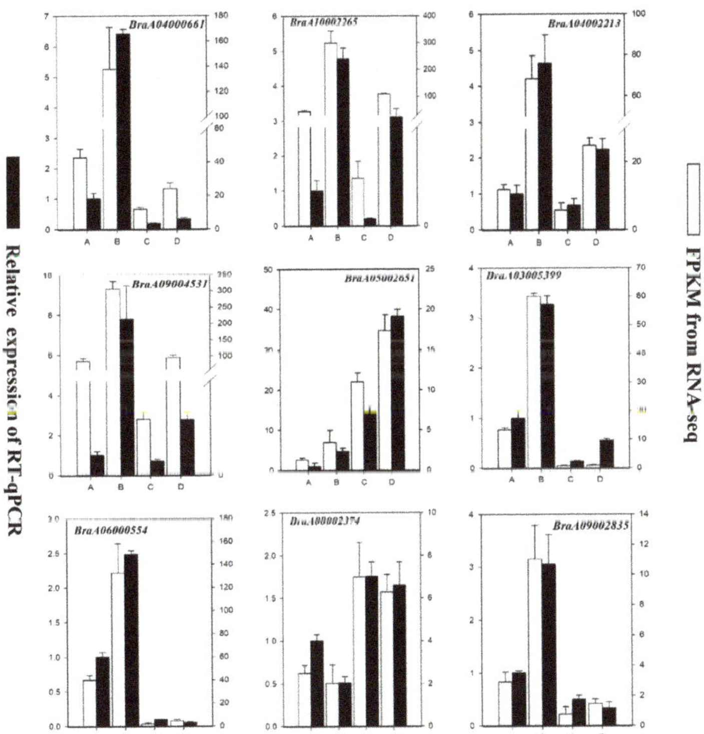

Figure 11. Relative expression of nine differentially expressed genes involved in the regulation of anthocyanin biosynthesis during different leaf growth stages in two lines. A: cotyledon flattening stage (PQC-T2); B: two true leaf stage (PQC-T3); C: cotyledon flattening stage (HYYTC-T2); D: two true leaf stage (HYYTC-T3).

4. Discussion

4.1. Structural Genes and Transport Genes Affecting Anthocyanin Synthesis in Purple Pak-Choi

A previous study showed the anthocyanin synthesis structural genes (*BrPAL1.2, BrPAL2.2, BrC4H5, BrFLS1, BrFLS3, BrDFR, BrANS1, BrANS2, BrCOMT1, BrCCoAOMT, BrUGT75C1, Brugt79B1.1*, and *BrUGT79B1.2*) and anthocyanin transport genes (*BrTT12, BrTT19.1, BrTT19.2, BrATPase 1*, and *BrATPase 10*) were differentially expressed in purple Pak-choi variety 'Zi Zuan' and green Pak-choi variety 'Jing Guan' in their five leaf stage [8]. However, in our study, *BrFLS3, BrCOMT1, BrUGT75C1*, and *BrATPase 1* were not differentially expressed, and *BrFLS1* was not even expressed in T1, T2, and T3. Meanwhile, eight differentially expressed genes, including early anthocyanin synthesis genes (*BrC4H3, BrF3H1, BrCHI1, BrCHS1*, and *BrCHS2*) and anthocyanin transport genes (*BrGSTF6, Br5MAT*, and *Br3AT1*), were newly identified. Although the expression of *BraA05002651* (*Br4CL4*) is down-regulated in purple materials, there was no differential expression of *Br4CL1, Br4CL2*, and *Br4CL3* in the two materials. *BraA05001886* (*Br4CL1*), *BraA05002646* (*Br4CL2*), and *BraA07003006* (*Br4CL3*) were expressed normally in the two materials. This indicates that the *Br4CL4* gene is not the rate-limiting enzyme gene for anthocyanin accumulation in purple materials. The stable expression of other genes can provide sufficient substrates for the synthesis of subsequent anthocyanin precursors. On this basis, the cinnamate hydroxylase gene (*BraA04002213, BrC4H3*), flavanone-3-hydroxylase gene (*BraA09004531* and *BrF3H1*), and chalcone synthetase gene (*BraA10002265* and *BrCHS1*) are the core genes in the module significantly related to anthocyanin biosynthesis by WGCNA and weight network analysis. *BraA10002265* was a candidate gene for the purple-red trait gene in purple-red Chinese cabbage by transcriptome sequencing and gene mapping [7]. In four purple Pak-choi materials with different anthocyanin content, the expression of the above two genes was significantly correlated with anthocyanin content. The correlation coefficients between gene expression of *BraA09004531, BraA10002265*, and anthocyanin content reached 0.9234 and 0.6904, which further indicated that *BrF3H1* plays a more critical role in anthocyanin metabolism [15].

4.2. Transcription Factors Affecting Anthocyanin Accumulation in Purple Pak-Choi

Five transcription factors related to anthocyanin synthesis and regulation were found in Pak-choi, including *MYB08, MYB12, CPC, TT8*, and *LBD39* in their five-leaf stage [8]. But only *BraA09002835* (*BrTT8*) was identified in this study, and the other four genes were not differentially expressed in three treatments. *BrTT8* was significantly up-regulated in purple Pak-choi material 'Zi He', and the allogenic expression promoted the transcription of some anthocyanin biosynthesis genes in tomato regenerated buds, further proving that *BrTT8* played an important role in anthocyanin biosynthesis [16]. bHLH often forms complexes with MYB and WD40 to coordinate the biosynthesis of anthocyanins, among the 8 MYB-related DEGs identified, especially the expression level of *BraA08002374* (*BrMYB44*) was significantly down-regulated in purple Pak-choi. The increase in the *StMYB44* expression level may down-regulate the expression of DFR promoter activity, suggesting that the transcription level of *BrMYB44* may play a similar role in the anthocyanin metabolism pathway of Pak-choi. Therefore, whether *BrTT8* and *BrMYB44* form a ternary complex to jointly regulate anthocyanin metabolism is worthy of further study.

In addition to being directly controlled by MBW, it has been reported that anthocyanin synthesis is also regulated by other transcription factors and regulatory genes, such as *PIF3, HY5, COP1, WRKY, WIP, MADS, JAZ, SPL*, and *bZIP* domains, which affect anthocyanin biosynthesis [17,18]. In this study, with the exception of *BraA02004460* (*MADS*) and *BraA04002035* (*C2H2*), the remaining transcription factors were identified as negative regulatory factors, and their expression was inhibited in purple materials. In particular, *BraA03002344* (*bZIP34*) and *BraA08001844* (*WRKY18*) were less expressed in purple materials. Studies have shown that the bZIP family acts primarily as positive regulators of anthocyanin biosynthesis. But in Tu's report, it was mentioned that that knocking out one allele of *VvbZIP36* promotes the accumulation of anthocyanins in grapevine [19]. Additionally, *BraA03002344* (*bZIP34*) may be involved in anthocyanin biosynthesis as a negative regulatory gene. WRKY is one of the most characteristic of plant TF, which regulates various plant processes related to development, physiology, metabolism, and plant defense. The overexpression of *MdWRKY11* in apple callus could significantly promote anthocyanin accumulation [20]. However, there are currently no studies on *WRKY18* in anthocyanins accumulation, which is worth investigating. This study shows that all of the above genes are potential transcription factors regulating anthocyanin biosynthesis.

5. Conclusions

In this study, via RNA-seq analysis of two Pak-choi materials with different anthocyanin contents, a total of 20 anthocyanin biosynthesis-related structural genes and 28 transcriptional regulatory genes were identified, and the weight network analysis of 14 genes was constructed. The results showed that the cinnamate hydroxylase gene (*BraA04002213* and *BrC4H3*), flavanone-3-hydroxylase (*BraA09004531* and *BrF3H1*), and chalcone synthetase (*BraA10002265* and *BrCHS1*) were the core genes involved in the anthocyanin synthesis pathway of purple Pak-choi. The above results identified the key genes and metabolic pathways controlling the purple character of leaves, laid a foundation for further in-depth analysis of the molecular mechanism of anthocyanin accumulation in purple Pak-choi, and provided theoretical support for its application in facility cultivation and breeding.

Supplementary Materials: The following supporting information can be downloaded at: https://www.mdpi.com/article/10.3390/horticulturae10101018/s1. Table S1: Overview of the Pak-choi transcriptome sequencing assembly. Table S2: Annotation for 76 common DEGs.

Author Contributions: H.X. conceived and supervised the work. B.S. analyzed the data and drafted the manuscript. Q.Y. (Qichang Yang) and Y.L. provided guidance for the test. L.C., Z.J., X.Y., L.Z., T.H., Q.Y. (Qinyu Yang) and W.Z. performed the experiments. All authors have read and agreed to the published version of the manuscript.

Funding: This study was supported by Sichuan Natural Science Foundation Project-Fine mapping and functional identification of purple genes in pak-choi (2023NSFSC0168); the Local Financial Project of the National Agricultural Science and Technology Center (NASC2023TD01 and NASC2023TD10); Agricultural Science and Technology Innovation Program of CAAS (ASTIP-IUA-2023002) and Basic Scientific Research Funds Project of IUA-CAAS (S2023007); JBGS [2021] 078, a leading project in the revitalization of seed industry in Jiangsu Province, Breeding of new pak-choi varieties with heat-resistant and high-quality; National Key Research and Development Program-Integration and Demonstration of Advantageous and Characteristic Planting Technology in Western Sichuan Plateau (2022YFD1601506).

Data Availability Statement: The original contributions presented in the study are included in the article/Supplementary Materials, further inquiries can be directed to the corresponding authors.

Acknowledgments: The authors gratefully acknowledge Ning Yu from the Institute of Vegetable Crops, Jiangsu Academy of Agricultural Sciences for his helpful advice and text Revision.

Conflicts of Interest: All authors declare no conflicts of interest.

References

1. Hu, K.; Han, K.T.; Dai, S.L. Mechanism of environmental factors regulating anthocyanin synthesis and coloration in plants. *Chin. Bull. Bot.* **2010**, *45*, 307–317.
2. Du, D.N.; Zhang, C.; Gao, S.L. Effects of low temperature on flower color and anthocyanin synthesis of cut peony. *J. Plant Genet. Resour.* **2016**, *17*, 295–302.
3. Zhang, B.B.; Cai, Z.X.; Shen, Z.J. Effects of light quality on leaf coloration of red-leaf peach. *J. Fruit Trees* **2013**, *30*, 602–607.
4. Wang, F.; Wang, X.J.; Zhao, S.N. Regulation of light on anthocyanin biosynthesis in horticultural plants. *China Agric. Sci.* **2020**, *53*, 4904–4917.
5. He, X.Z.; Li, Y.; Lawson, D.; Xie, D.Y. Metabolic engineer-ing of anthocyanins in dark tobacco varieties. *Physiol. Plant* **2017**, *159*, 2–12. [CrossRef]
6. Lim, W.; Li, J. Co-expression of onion chalcone isomerase in Del/Ros1-expressing tomato enhances anthocyanin and flavonol production. *Plant Cell Tissue Organ Cult.* **2016**, *128*, 113–124. [CrossRef]

7. Zhang, S.J. *Purple Red Gene Mapping and Candidate Gene Analysis in Chinese Cabbage*; Chinese Academy of Agricultural Sciences: Beijing, China, 2014.
8. Zhang, L.; Xu, B.; Wu, T.; Yang, Y.F.; Fan, L.X.; Wen, M.X.; Su, J.X. Transcriptomic profiling of two Pak Choi varieties with contrasting anthocyanin contents provides an insight into structural and regulatory genesin anthocyanin biosynthetic pathway. *BMC Genomics*. **2017**, *18*, 288.
9. Zhang, Z.S.; Lin, J.W.; Wang, Z.P.; Yu, H.J.; Liu, D.L. Determination of anthocyanin content in black soybean skin from different producing areas by colorimetry. *Food Res. Dev.* **2012**, *33*, 143–145.
10. Wang, X.K. *Plant Physiological and Biochemical Test Principle and Technology*; Higher Education Press: Beijing, China, 2015.
11. Mortazavi, A.; Williams, B.A.; McCue, K.; Schaeffer, L.; Wold, B. Mapping and quantifying mammalian transcriptomes by RNA-seq. *Nat. Methods* **2008**, *5*, 621–628. [CrossRef]
12. Leng, N.; Dawson, J.A.; Thomson, J.A.; Ruotti, V.; Rissman, A.I.; Smits, B.M.G. EBSeq: An empirical Bayes hierarchical model for inference in RNA-seq experiments. *Bioinformatics* **2013**, *29*, 1035–1043. [CrossRef]
13. Jiang, H.B.; Xia, L.F.; Tian, Y.P. Analysis of genes related to anthocyanin synthesis in purple bud tea plant based on transcriptome sequencing. *J. Plant Genet. Resour.* **2018**, *19*, 967–978.
14. Livak, K.J.; Schmittgen, T.D. Analysis of relative gene expression data using real-time quantitative PCR and the 2-DDCT method. *Methods* **2001**, *25*, 402–408. [CrossRef]
15. Song, B.; Xu, H.; Chen, L.Z.; Fan, X.X.; Jing, Z.G.; Chen, S.; Xu, Z.G. Study of the Relationship between Leaf Color Formation and Anthocyanin Metabolism among Different Purple Pakchoi Lines. *Molecules* **2020**, *25*, 4809. [CrossRef]
16. Zhang, Y.J.; Chen, G.P.; Dong, T.T.; Pan, Y.; Zhao, Z.P.; Tian, S.B.; Hu, Z.L. Anthocyanin accumulation and transcriptional regulation of anthocyanin biosynthesis in purple bok choy (*Brassica rapa* var. *chinensis*). *Food Chem.* **2014**, *62*, 12366–12376. [CrossRef]
17. Zhou, H.; Lin, W.K.; Wang, H.L.; Gu, C.; Dare, A.P.; Espley, R.V.; He, H.P.; Allen, A.C.; Han, Y.P. Molecular genetics of blood-leshed peach reveals activation of anthocyanin biosynthesis by NAC transcription factors. *Plant J.* **2015**, *82*, 105–121. [CrossRef] [PubMed]
18. Dong, T.T.; Han, R.H.; Yu, J.W.; Zhu, M.K.; Zhang, Y.; Gong, Y.; Li, Z.Y. Anthocyanins accumulation and molecular analysis of correlated genes by metabolome and transcriptome in green and purple asparaguses (*Asparagus officinalis*). *Food Chem.* **2019**, *271*, 18–28. [CrossRef] [PubMed]
19. Tu, M.X.; Fang, J.H.; Zhao, R.K.; Liu, X.Y.; Yin, W.C.; Wang, Y.; Wang, X.H.; Wang, X.P.; Fang, Y.L. CRISPR/Cas9-mediated mutagenesis of *VvbZIP36* promotes anthocyanin accumulation in grapevine (*Vitis vinifera*). *Hortic. Res.* **2022**, *9*, uhac022. [CrossRef]
20. Liu, W.J.; Wang, Y.C.; Yu, L.; Jiang, H.Y.; Guo, Z.W.; Xu, H.F.; Jiang, S.H.; Fang, H.C.; Zhang, J.; Su, M.Y.; et al. *MdWRKY11* Participates in Anthocyanin Accumulation in Red-Fleshed Apples by Affecting MYB Transcription Factors and the Photoresponse Factor *MdHY5*. *J. Agric. Food Chem.* **2019**, *67*, 8783–8793. [CrossRef]

horticulturae

MDPI

Article

The Transcription Factor BrNAC19 Acts as a Positive Regulator of the Heat Stress Response in Chinese Cabbage

Shuai Yuan [1], Xiaoping Yong [2,3], Yuxin Lu [1], Yuxin Lei [1], Weijian Li [1], Qiuli Shi [1] and Xiuhong Yao [1,*]

1 Solid-State Fermentation Resource Utilization Key Laboratory of Sichuan Province, Department of Agriculture Forestry and Food Engineering, Yibin University, Yibin 644000, China; yuanbio11@163.com (S.Y.); luyuxin50304@163.com (Y.L.); leiyuxin2024@163.com (Y.L.); liweijian2025@163.com (W.L.); 2023090001@yibinu.edu.cn (Q.S.)

2 Rice and Sorghum Research Institute, Sichuan Academy of Agricultural Sciences (Deyang Branch, Sichuan Academy of Agricultural Sciences), Deyang 618099, China; yongxp@scsaas.cn

3 Vegetable Germplasm Innovation and Variety Improvement Key Laboratory of Sichuan Province, Sichuan Province Engineering Technology Research Center of Vegetables, Sichuan Academy of Agricultural Sciences, Chengdu 610000, China

* Correspondence: yaoxiuhong@yibinu.edu.cn

Abstract: The frequent occurrence of excessive heat events driven by global warming poses a great threat to plant growth and food security. To survive in heat stress (HS) environments, plants have evolved sophisticated response mechanisms, and the transcriptional network that controls the expression levels of HS-inducible genes serves as an essential component of this process. NAC (NAM, ATAF1/2, and CUC2) transcription factors (TFs) play key regulatory roles in the abiotic stress responses of plants. However, the functional roles of NAC TFs in the heat stress response of Chinese cabbage remain elusive. In the present study, we identified the *Brassica rapa* NAC family transcription factor *BrNAC19* as a close homologue of Arabidopsis *NAC019* and found that it serves as a positive regulator of the HS response. BrNAC19 displayed inducible gene expression in response to HS, and its subcellular localization showed that it was localized in the nucleus. Heterologous expression of *BrNAC19* significantly enhanced the heat tolerance of plants and reduced the accumulation of reactive oxygen species (ROS) under HS conditions. Furthermore, our results demonstrated that BrNAC19 directly targeted and promoted the expression of superoxide dismutase 1 (*CSD1*) and catalase 2 (*CAT2*), two antioxidant-enzyme coding genes in Chinese cabbage. Altogether, our results suggest that BrNAC19 enhances heat stress tolerance by positively regulating the expression of genes involved in the HS response and ROS scavenging and exhibits potential as a target gene in Chinese cabbage breeding to increase heat stress tolerance.

Keywords: heat stress; BrNAC19; transcriptional regulation; ROS scavenging

1. Introduction

The frequent occurrence of excessive heat events caused by global warming has become a global problem, seriously affecting crop production and food security [1]. Therefore, studying the molecular mechanisms of the heat stress response (HS) will provide theoretical guidance for molecular breeding and cultivation of heat-resistant plants and is also an important issue that needs to be addressed in response to HS [2]. As sessile organisms, plants have evolved complex and diverse systems to cope with HS. These heat stress responses are divided into two types, namely, basal thermotolerance and acquired thermotolerance. Basal thermotolerance reflects the natural capacity of the plant to defend itself against extreme heat attacks, while acquired thermotolerance (also called thermopriming) involves the ability to remember and form conditioned reflexes to avoid danger, which is actually an adaptive response that occurs after a priori exposure to sublethal HS [3].

Heat stress disrupts protein folding and denatures correctly folded proteins, which activates molecular chaperones such as heat shock proteins (HSPs) to assist in protein folding. During this process, heat shock transcription factors (HSFs) are required to bind to heat shock element sequences of HSP promoters to enhance the expression of HSPs [4]. Constitutively expressed HSFA1s, including *HSFA1a*, *HSFA1b*, and *HSFA1d*, are activated under HS conditions, which enables HSFA1s to activate the expression of *HSFA2* [5]. On the other hand, HS causes the accumulation of reactive oxygen species (ROS), which causes damage to plant tissues [6]. When high temperatures trigger the heat stress response of plants, plant antioxidants including superoxide dismutase (SOD), ascorbate peroxidase (APX), and catalase (CAT) are activated and are responsible for the scavenging of ROS [7].

Recent studies in *Arabidopsis thaliana* (L.) *Heynh* elucidated complex transcriptional networks during HS comprising many transcriptional regulators, and portions of transcription factors from different families jointly construct a transcriptional cascade in response to heat stress [5]. These transcription factor families include No apical meristem, *Arabidopsis thaliana*-activating factor 1/2, Cup-shaped cotyledon 2 (NAC), WRKY, MYB, basic region/leucine zipper motif (bZIP), basic-Helix-Loop-Helix (bHLH), ethylene response factors (ERF), dehydration-responsive element binding (DREB), and more [8,9]. As plant-specific transcriptional factors, NACs widely modulate essential aspects of plant function, including responses to HS [10]. Several NACs have been implicated in stress responses, and these genes are associated with enhanced tolerance in crop plants, such as rice, peach, pepper, and maize [11–14]. In the model plant Arabidopsis, AtNAC019 interacts with regulators of C-repeat binding factors (RCFs) and undergoes dephosphorylation by RCF2. Dephosphorylated NAC019 binds to the promoters of *HSFA1b*, *HSFA6b*, *HSFA7*, and *HSFC1* to positively regulate the heat stress response [15]. The other three NAC members, JUNGBRUNNEN1 (JUB1), *Arabidopsis thaliana*-activating factor 1 (ATAF1), and NAC055, are involved in the heat stress memory response [16,17]. Therefore, precise modulation of NAC proteins is expected to enhance thermotolerance in plants. However, further investigation is required to elucidate the molecular mechanisms by which NAC regulates heat stress responses in crops and vegetables.

Chinese cabbage (*Brassica rapa* L. ssp. *pekinensis*), a member of the cruciferae family, is highly favored by consumers due to its nutritional value, high yield, and storage and transport resilience [18]. As a cool-season vegetable [19], it thrives at optimal temperatures ranging from 15 °C to 25 °C and exhibits sensitivity to elevated heat conditions. Particularly, temperatures exceeding 30 °C can adversely affect the quality of the vegetable [20,21]. Consequently, investigating the mechanisms of heat tolerance in Chinese cabbage is of significant importance to the agricultural industry.

In this study, we revealed a positive role for the Chinese cabbage NAC transcription factor *BrNAC19* in plant basal thermotolerance. Overexpression of *BrNAC19* in Arabidopsis enhanced plant thermotolerance and promoted the expression of HS-response-related genes. In addition, overexpression of *BrNAC19* decreased ROS accumulation under heat stress conditions. Correspondingly, BrNAC19 also directly bound to the promoter of Chinese cabbage antioxidant enzyme-coding genes for transcriptional activation, including *BrCSD1* and *BrCAT2*. Collectively, the findings from our study provide valuable information for genetic manipulation to cultivate heat-tolerant cabbage in future plant breeding programs.

2. Materials and Methods

2.1. Plant Materials and Growth Conditions

The *A. thaliana* plants used in this study were on a Col-0 ecotype background. *BrNAC19-OE* was obtained via *Agrobacterium tumefaciens*-mediated floral transformation. Homozygotes were screened for hygromycin B resistance in the T2 population. The Chinese cabbage species used in this study is *B. rapa* ssp. *Pekinensis*.

Arabidopsis and Chinese cabbage seeds were sterilized with 70% ethanol, sown on Murashige and Skoog medium (MS; Phyto Technology Laboratories, Lenexa, KS, USA) supplemented with 1.5% sucrose and 0.8% agar, and incubated at 4 °C in the dark for 3 days. The seedlings were then grown on LD (16/8) at 22 °C under white light (100 μmol/m^2/s).

2.2. RNA Extraction and RT-qPCR

Total RNA was extracted from Chinese cabbage or Arabidopsis seedlings using a TaKaRa MiniBEST Universal RNA Extraction Kit (TaKaRa, Kyoto, Japan) according to the manufacturer's protocol. The concentration of RNA was measured by a microspectrophotometer (Thermo Scientific NanoDrop, Waltham, MA, USA), and 2 μg of RNA was reverse-transcribed using M-MLV Reverse Transcriptase (Invitrogen, Carlsbad, CA, USA). RT-qPCR was performed using a Hieff® qPCR SYBR Green Master Mix (YEASEN, Shanghai, China) with a CFX96 Real-Time PCR Detection System (Bio-Rad, Hercules, CA, USA). A two-step PCR amplification program was adopted, which included a holding stage (95 °C 30 s), a cycling stage (step1: 95 °C 30 sec, step2: 95 °C 30 s, number of cycles: 40), and a melt curve stage. Gene expression was measured in three independent biological replicates, and each biological replicate contained three technical replicates. Transcript expression levels were normalized to those of *ACTIN2* (AT3G18780) for Arabidopsis and *BrEF-1α* (Bra031602) for Chinese cabbage. The specific primer sequences are listed in Supplemental Table S1.

2.3. Phylogenetic Analysis and Sequence Alignment

The BrNAC19 protein sequence was downloaded from the Brassicaceae Database (BRAD), and the phylogenetic tree was constructed using MEGA 11 software with the neighbor-joining method and 1000 bootstrap replicates. The protein alignment was performed using DNAMAN (version 5.2.2) software.

2.4. Subcellular Localization

The coding sequence (CDS) of *BrNAC19* was amplified and cloned into the pCambia1302-GFP vector, then transformed into *A. tumefaciens* GV3101 and infiltrated into the leaves of *Nicotiana benthamiana*. After incubation for 2–3 d, GFP fluorescence was detected under a fluorescence microscope (Leica, Wetzlar, Germany) with an argon laser at an excitation wavelength of 488 nm using a 500–520 nm band-pass filter.

2.5. Thermotolerance Assay

The basal thermotolerance assay for Arabidopsis was performed as previously described [22]. Briefly, seedlings grown in a chamber for 7 days were subjected to 44 °C for 90 min in a constant-temperature incubator for HS triggering. After triggering, seedlings were transferred to normal growth conditions for 5 days for recovery.

2.6. Physiological Phenotype Assay

For measurement of chlorophyll contents, Arabidopsis seedlings were soaked in 80% acetone and placed in darkness for 16 h at 4 °C, then the absorbance of the supernatant was measured at 647 and 663 nm. Chlorophyll quantification was performed as previously described [23].

For electrolyte leakage detection, seedlings were harvested and rinsed twice with deionized water. After incubation at room temperature for 24 h, the conductivity was measured using an electroconductivity meter (DDSJ-308A; Lei-ci, Shanghai, China). Samples were then boiled at 100 °C for 15 min and shaken at room temperature for another 1 h before the electrical conductivity was measured again. The relative ion leakage was calculated based on the ratio of electrical conductivity before and after boiling.

Histochemical staining with DAB and NBT was performed as previously described [24]. Briefly, Arabidopsis seedlings were vacuum-infiltrated with NBT (0.5 mg/mL) solutions for 2 h or DAB (2 mg/mL) solutions for 8 h. Samples were then decolorized in ethanol (95%) on the heater until the leaves' blading faded. H_2O_2 and O_2^- levels were analyzed by specific detection kits following the manufacturer's instructions (Beyotime Biotechnology Co., Ltd., Shanghai, China).

2.7. EMSA Assay

EMSA assays were performed as previously described [25]. The full-length CDS of *BrNAC19* was cloned into pGEX-4T-1 to express the glutathione S-transferase (GST)-BrNAC19 recombinant protein. GST-BrNAC19 was purified using glutathione agarose (Cat no. G4510; MERCK, Darmstadt, Germany). Purified GST-BrNAC19 was incubated with the DNA probe, and the mixture was reacted in binding buffer (25 mM HEPES-KOH, pH 8.0, 50 mM KCl, 1 mM dithiothreitol and 10% glycerol) for 30 min and then separated on a polyacrylamide gel. The probe and protein were detected using an EMSA Kit (Thermo Fisher, Waltham, MA, USA).

2.8. Y1H Assays

Y1H assays were performed as previously described [26]. The full-length CDS of *BrNAC19* was cloned into the PGAD-T7 vector, and the promoters of *BrCSD1* and *BrCAT2* were cloned into the pHIS2 vector. The constructs and empty vector controls were transformed in the yeast strain AH109, which were grown on SD/-Leu-Trp-His dropout plates for interaction analysis. 3-Amino-1,2,4-triazole (3-AT) was applied to inhibit self-activation.

2.9. Dual-Luciferase Assay

The promoters of *BrCSD1* and *BrCAT2* were amplified from Chinese cabbage DNA and cloned into the pGreenII 0800-LUC vector. The reporters and effectors were transformed into Agrobacterium and infiltrated into the leaves of *N. benthamiana*. Luciferase and Renilla activities were quantified using a Dual-Luciferase Reporter Assay System (Promega, Madison, WI, USA) and detected with a GloMax 20–20 luminometer (Promega).

Transformation of Cabbage protoplasts was performed as described in a recent report [27]. Briefly, true leaves of Chinese Cabbage were digested for 6 h in an enzyme solution containing 1.5% cellulose R-10, 0.4% macerozyme R-10, 20 mmol/L KCl, 20 mmol/L MES (pH 5.7), 0.6 mol/L mannitol, 10 mmol/L $CaCl_2$, and 0.1% bovine serum albumin (BSA). Then, W5 (154 mmol/L NaCl, 125 mmol/L $CaCl_2$, 5 mmol/L KCl, and 2 mmol/L MES; pH 5.7) solution was added. After filtration and precipitation, the protoplasts were resuspended in MMG solution (0.3 mol/L mannitol, 15 mmol/L $MgCl_2$, and 4 mmol/L MES, pH 5.7). Lastly, protoplasts were subjected to PEG-mediated transfection. After incubation for 24–36 h, Luciferase and Renilla activities were quantified.

2.10. Statistical Analysis

One-way analysis of variance (ANOVA), two-way ANOVA, Student's *t* test, and Tukey's post hoc test were performed using SPSS_26.0 software. The band intensity of Western blot bands was measured with ImageJ (version 1.51j8) software.

3. Results

3.1. BrNAC19 Plays a Positive Role in Plant Basal Thermotolerance

To identify the Chinese cabbage NAC transcription factor involved in HS, we analyzed the gene expression pattern of cabbage NACs under heat stress, especially homologous genes of NAC members which have been reported to regulate HS in other species (Figure S1). Among the NAC members we tested, *Bra018998* was identified as a candidate with highly inducible gene expression in response to high temperature shock, and especially was greatly induced in the leaves of Chinese cabbage, suggesting that *Bra018998* may play regulatory roles in mediating the HS response in cabbage leaves (Figure 1A). The phylogenetic tree of NAC proteins in rapeseed, Arabidopsis, Camelina sativa, rice, wheat, and maize revealed that the Bra018998 protein shares a close evolutionary relationship with rapeseed BanC06g05930D, Arabidopsis AtNAC019, and Camelina sativa CsNAC55. Despite the phylogenetic tree showing that Bra018998 is most closely related to BanC06g05930D, a protein homology search showed that Bra018998 shared the highest identity with AtNAC019, and therefore we named Bra018998 BrNAC19 (Figure 1B). Subcellular localization showed that the fluorescence signal of BrNAC19 appeared in the nucleus, whether at room temperature or high temperature, suggesting its function as a transcription factor (Figure 1D).

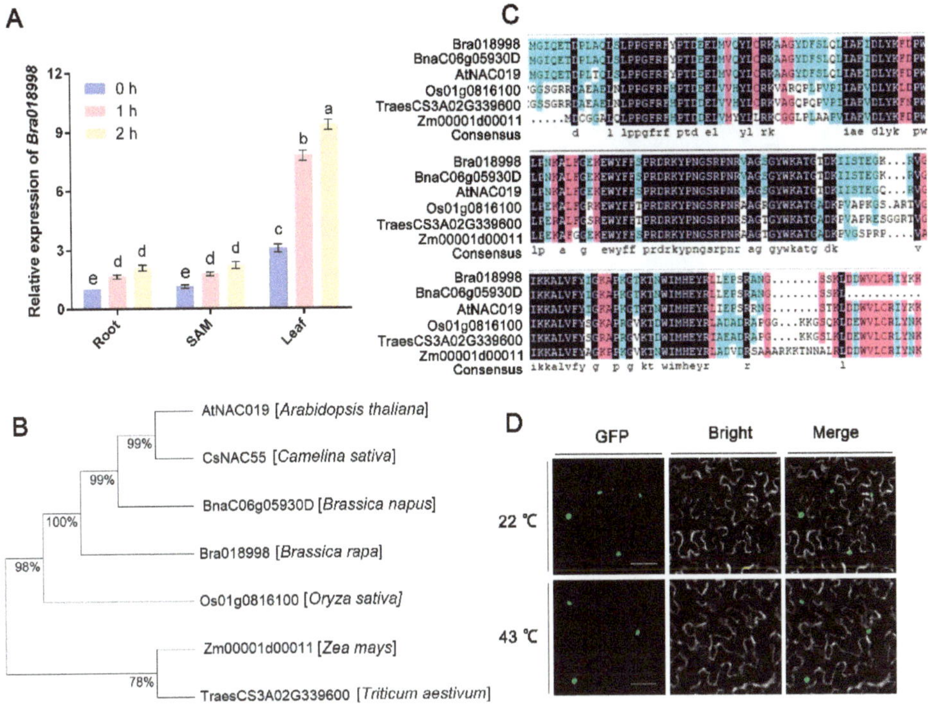

Figure 1. Heat-induced gene expression, homology analysis, and subcellular localization of *Bra018998*. (**A**) RT-qPCR detected the expression levels of *Bra018998* in the root, shoot apical meristems (SAM), and leaf of Chinese cabbage after high temperature exposure. Two-week-old Chinese cabbage seedlings grown under normal conditions were transferred to high temperature (43 °C) for the indicated time and then harvested for RNA extraction. The expression level of *Bra018998* in the roots at 0 h was set to one. Data are represented as the mean ± standard deviation (SD) of three biological replicates. (**B**) Phylogenetic analysis of Bra018998 with its orthologous genes based on their amino acid sequences. (**C**) Protein sequence multiple alignment of Bra018998 with its orthologous genes in other plant species. (**D**) Subcellular localization of Bra018998 in *N. benthamiana* leaf epidermis cells. Scale bars, 50 μm. The letters 'a' to 'e' above the bars indicate statistically significant differences between samples, and the presence of same letters between two groups indicates no significant differences (two-way ANOVA with Tukey's post hoc test; $p < 0.05$).

In Arabidopsis, NAC019 binds to the promoters of *HSFA1b*, *HSFA6b*, *HSFA7a*, and *HSFC1* for transcriptional activation, thus enhancing the thermotolerance of plants [15]. Since studies on NACs in crops have further confirmed the findings in Arabidopsis and supported the conserved function of stress response regulation [28,29], we speculated that BrNAC19 regulates heat tolerance similarly to AtNAC019. To explore its biological function, BrNAC19 was cloned and transformed in Arabidopsis under the control of the 35S promoter to generate the overexpression transgenic plants *BrNAC19-OE* (Figure S2). We employed a basal thermotolerance assay to elucidate the role of BrNAC19 in the heat stress response of Arabidopsis. First, 7-day-old seedlings were subjected to 44 °C for 90 min in a constant-temperature incubator and then allowed to recover in a growth chamber at 22 °C for 5 days before the survival rate of the plants was assessed. Seven-day-old *BrNAC19-OE* showed a similar phenotype to wild type Col-0 under normal conditions

(Figure 2A). However, after heat stress treatment and recovery, *BrNAC19-OE* exhibited enhanced thermotolerance, accompanied by an elevated survival rate, increased total chlorophyll content, and decreased electrolyte leakage (Figure 2B–D). Together, these results indicate that BrNAC19 positively confers basal thermotolerance in plants.

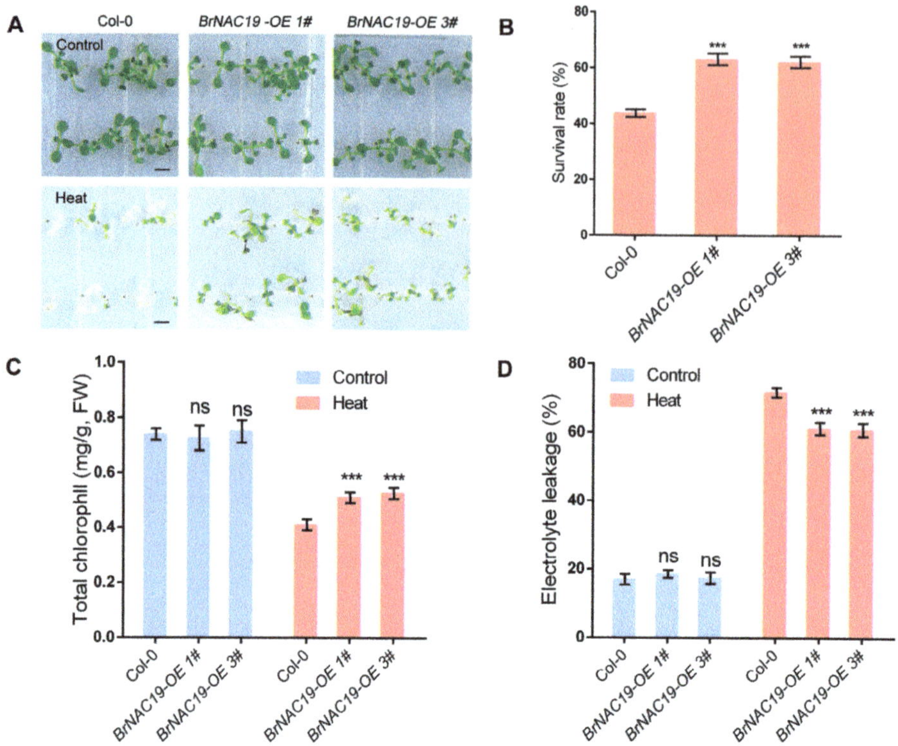

Figure 2. Overexpression of *BrNAC19* enhances plants thermotolerance in Arabidopsis. (**A**) Phenotypes of wild type (Col-0) and *BrNAC19-OE* seedlings in the basal thermotolerance assay (43 °C for 22 min and recovery at 22 °C for 5 days). The scale bar indicates 2 mm. 1# and 3# represent the numbering of different transgenic lines. (**B**) Survival rates of Col-0 and *BrNAC19-OE* seedlings in the basal thermotolerance assay. (**C**) Chlorophyll contents of the seedlings indicated in (**A**). (**D**) Electrolyte leakage assay of the seedlings indicated in (**A**). Data are presented as the mean ± standard deviation (SD) of three biological replicates. Significant differences compared with the wild type at same condition are noted (student's *t*-test, *** $p < 0.001$; ns indicates no significance).

3.2. *BrNAC19 Activates HS-Response Gene Expression*

Next, we investigated whether BrNAC19 modulates thermotolerance by transcriptionally regulating the key factors involved in the heat stress response. The HS response genes *CSD1/2*, *CAT1/2*, *HSF3*, and *HSFA1d* were selected for gene expression analysis in the wild type and overexpression lines of BrNAC19 [5,30,31]. Levels of these genes were upregulated in *BrNAC19-OE* plants after heat shock treatment (Figure 3A–F), indicating that BrNAC19 induces these genes to positively mediate the heat stress response.

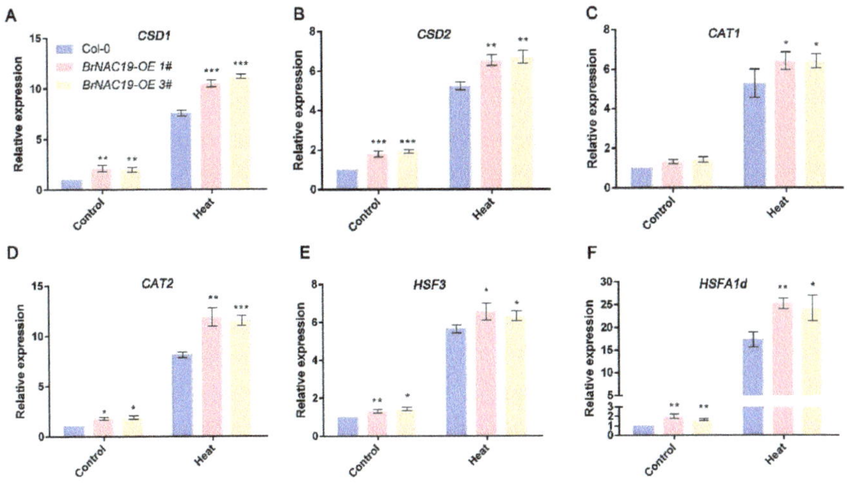

Figure 3. BrNAC19 promotes the expression of key regulators of HS response. (**A–F**) RT-qPCR detected the expression levels of *CSD1*, *CSD2*, *CAT1*, *CAT2*, *HSF3*, and *HSFA1d* in the indicated genotypes under different conditions. The expression level of each gene in Col-0 at 22 °C was set to one. All data are presented as means ± SD from three biological replicates. Significant differences compared with the wild type under the same conditions are noted (student's *t*-test, * $p < 0.05$, ** $p < 0.01$, *** $p < 0.001$).

3.3. Overexpression of BrNAC19 Decreases ROS Accumulation Under Heat Stress Conditions

Given that BrNAC19 activates the expression of transcription levels of genes encoding ROS-scavenging enzymes (Figure 3A–D), we supposed that BrNAC19-mediated heat stress tolerance was closely related to ROS scavenging. The production of ROS such as H_2O_2 and the superoxide anion (O_2^-) increases under heat stress [31], so we investigated the levels of H_2O_2 and O_2^- via diaminobenzidine (DAB) staining and nitroblue tetrazolium (NBT) staining. Under normal conditions, the accumulation of H_2O_2 and O_2^- in the wild type plants was only slightly higher than in the *BrNAC19-OE* plants. Conversely, significantly weakened staining was detected in *BrNAC19-OE* compared with WT under HS conditions (Figure 4A,B). To verify the histochemical staining, we quantified the accumulation of H_2O_2 and O_2^- in the wild type and *BrNAC19-OE* lines. Consistent with the histochemical staining results, H_2O_2 and O_2^- exhibited lower accumulation in the *BrNAC19-OE* plants (Figure 4C,D).

3.4. BrNAC19 Targets BrCSD1 and BrCAT2 for Activation

Since BrNAC19 promotes the expression of *CSD1/2* and *CAT1/2*, as well as decreasing the accumulation of ROS, in Arabidopsis under HS, we hypothesized that BrNAC19 may also play a role in H_2O_2 and O_2^- scavenging by transcriptionally regulated genes encoding ROS-scavenging enzymes in Chinese cabbage. Analysis of the promoter sequences of *BrCSD1/2* and *BrCAT1/2* revealed that the NAC-binding site (TNCGTG/A) to which NAC members bind was present 1000 bp upstream of the translation initiation site of *BrCSD1* and *BrCAT2* (Figure 5A). Therefore, we hypothesized that BrNAC19 activates *BrCSD1* and *BrCAT2* by interacting with its promoters, at least partially. Electrophoretic mobility shifts assay (EMSA) revealed that BrNAC19 bound to the fragments of the *BrCSD1* and *BrCAT2* promoters containing a binding site, whereas the binding was abolished when mutant probes were used (Figure 5B). Furthermore, we tested whether BrNAC19 activates *BrCSD1* and *BrCAT2* transcription via a yeast-one hybrid assay, and the results indicated that BrNAC19 directly interacted with the promoters of *BrCSD1* and *BrCAT2* (Figure 5C). Lastly, we performed transient expression in tobacco leaves to analyze direct transcriptional activation of *BrCSD1* and *BrCAT2* by BrNAC19 using a dual-luciferase experiment. When tobacco leaves were at 22 °C, BrNAC19 only mildly activated these two genes. However, when tobacco leaves were subjected to 43 °C heat shock, BrNAC19 strongly induced the expression of these two genes, indicating that BrNAC19 promotes the expression of *BrCSD1* and *BrCAT2* under thermal stimulation (Figure S3A,B). We further confirmed the transcriptional activation of *BrCSD1* and *BrCAT2* by BrNAC19 in Chinese cabbage protoplasts. When co-transfecting the reporter vector and BrNAC19-GFP, transcription of *BrCSD1* and *BrCAT2* was greatly induced (Figure 5D). In addition, we also detected the expression of *BrCSD1* and *BrCAT2* in Chinese cabbage, and the results indicated that the expression levels of *BrCSD1* and *BrCAT2* increased after heat treatment (Figure S4). All together, these results suggest that BrNAC19 directly targets *BrCSD1* and *BrCAT2* for transcriptional activation under heat stress.

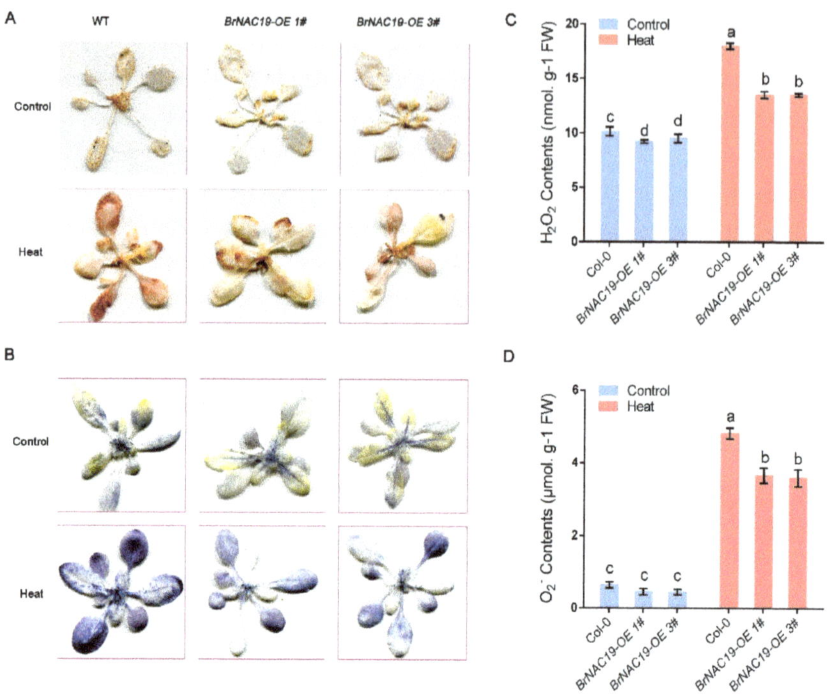

Figure 4. *BrNAC19-OE* rescues the ROS accumulation caused by HS. (**A**,**B**) Histochemical analysis of the generation of H_2O_2 and O_2^- by staining with DAB and NBT in WT and *BrNAC19-OE* plants. Brown precipitation and blue spots represent the presence of H_2O_2 (**A**) and O_2^- (**B**), respectively. (**C**,**D**) The levels of H_2O_2 (**C**) and O_2^- (**D**) in WT and *BrNAC19-OE* plants. Data are presented as the mean ± standard deviation (SD) of three biological replicates. The letters 'a' to 'd' above the bars indicate statistically significant differences between samples, and the presence of the same letters between two groups indicates no significant differences (two-way ANOVA with Tukey's post hoc test; $p < 0.05$).

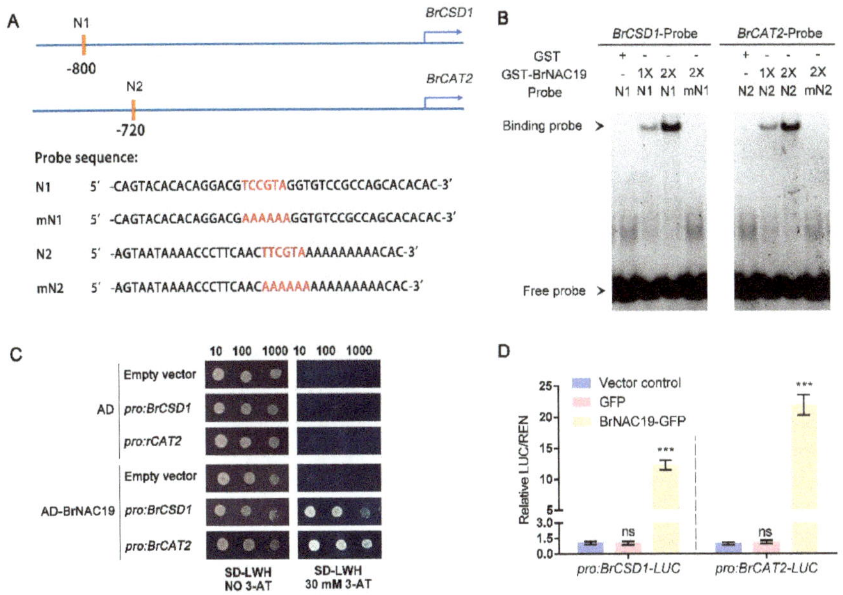

Figure 5. BrNAC19 directly induces the expression of *BrCSD1* and *BrCAT2*. (**A**) Schematic diagram of the binding sites in the *BrCSD1* and Br*CAT2* promoters, with the electrophoretic mobility shift assay (EMSA) probe sequences shown below the diagram. Red letters represent the NAC-binding site and mutation site. (**B**) EMSA revealed that BrNAC19 bound to the NAC-binding sites of the *BrCSD1* and Br*CAT2* promoters. The notation 2X indicates a twofold amount of glutathione S-transferase (GST)-BrNAC19 protein, and the probe sequence is shown in panel A. (**C**) Yeast one-hybrid (Y1H) assay revealed that BrNAC19 activates the *BrCSD1* and Br*CAT2* promoters. (**D**) Dual-luciferase assays indicated that BrNAC19 positively modulates transcription of *BrCSD1* and *BrCAT2* after heat shock. Data are presented as the mean ± standard deviation (SD) of three biological replicates (student's *t*-test, *** $p < 0.001$; ns indicates no significance).

4. Discussion

Heat shock poses a great threat to the growth, development, and production of crops and vegetables, thereby escalating as a serious challenge for food security as global warming progresses [32]. Plants enhance their tolerance through endogenous regulatory substances and heat-responsive transcriptional cascades [5,33]. In recent years, studies have elucidated the complex transcriptional regulatory networks involved in heat stress [5]. As plant-specific transcription factors, NAC family members play an important role in regulating plant growth, development, and stress response, including the heat stress response in multiple species [28]. In maize, a NAC transcription factor designated ZmNAC074 was identified that confers heat stress tolerance in transgenic Arabidopsis [14]. In cowpeas, overexpression of *VuNAC1/2* obviously enhance tolerance to drought, salinity, cold, and heat stress [34]. Upon analyzing the expression patterns of peach *NAC56* under abiotic stress, it was found that PpNAC56 responds to high-temperature stress, and overexpression of the *PpNAC56* gene in tomato plants conferred elevated heat tolerance [12]. *ONAC127* and *ONAC129* are responsive to heat stress and involved in the grain filling process of rice [35]. Using the CRISPR-Cas9 system, knockout of the *OsNAC006* transcription factor caused drought and heat sensitivity [36]. The membrane-associated NAC transcription factor OsNTL3 is involved in thermotolerance in rice, and inducible expression of the truncated form of *OsNTL3* without the transmembrane domain increases heat tolerance in rice seedlings [11]. In Arabidopsis, ANAC019 positively regulates plant heat stress response by activating HSFs [15]. However, it is still unclear how NAC transcription factors regulate the heat stress response in Chinese cabbage. In this study, we found that BrNAC19 responds to heat stimuli and positively modulates the heat stress response by promoting the expression of HSFs and genes encoding ROS scavengers (Figures 1–3). Additionally, we identified *BrCSD1* and *BrCAT2* as direct targets of BrNAC19 (Figure 5), suggesting that Chinese cabbage enhances thermotolerance through BrNAC19-enhanced antioxidant enzyme synthesis. Further genetic analysis in Chinese cabbage will more thoroughly reveal the role that NAC transcription factors played in regulating heat stress tolerance in Chinese cabbage [37].

ROS typically act as major signaling molecules in the stress response of plants; among them, H_2O_2 and O_2^- usually participate in the response to abiotic stresses such as heat stress to facilitate the transmission of stress signals [38]. However, the excessive accumulation of ROS causes lipid peroxidation, DNA damage, protein oxidation, and cell apoptosis, inflicting severe harm upon plants [7]. To eliminate the damage caused by ROS, plants have evolved an antioxidant defense system including diverse antioxidants like APX, SOD, and CAT [39]. To date, how ROS accumulation is regulated in Chinese cabbage during heat stress responses has remained poorly studied. In this study, we discovered that BrNAC19 promotes the expression of *CSD1* and *CAT2* both in Arabidopsis and Chinese cabbage, offering valuable insights into the mechanism by which Chinese cabbage utilizes NACs to regulate its adaptation to heat stress (Figures 3–5). Further investigation is needed to elucidate whether BrNAC19 contributes to the ROS scavenging system in Chinese cabbage under heat stress. In addition, whether BrNAC19 enhances antioxidant enzyme activity remains to be addressed.

5. Conclusions

Our results suggest that BrNAC19 plays a positive role in conferring thermotolerance in plants. When plants encounter high temperature stress, BrNAC19 induces the transcription of heat stress responsive genes and antioxidant enzyme-coding genes, thereby enhancing heat response intensity and preventing excessive accumulation of ROS (Figures 2–4). Importantly, our further molecular studies revealed the BrNAC19 targeted *BrCSD1* and *BrCAT2* for transcriptional activation (Figures 5 and S3). Overall, our study suggests that BrNAC19 induces the expression of heat stress response genes and ROS scavenging genes to enhance plant thermotolerance. Our data will help to develop detailed models of heat stress responses and clarify the roles that NAC transcription factors play in plants, which will facilitate the design of more heat stress-resilient crops and vegetables in response to global warming.

Supplementary Materials: The following supporting information can be downloaded at: https://www.mdpi.com/article/10.3390/horticulturae10121236/s1. Figure S1: RT-qPCR detected the expression of partial NACs tested under normal and high temperature conditions. Figure S2: Molecular characterization of *BrNAC19-OE* transgenic seedlings. Figure S3: BrNAC19 activated *BrCSD1* and *BrCAT2* transcription under heat. Figure S4: Heat treatment induced the expression of *BrCSD1* and *BrCAT2* in Chinese cabbage. Table S1. Primers used for RT-qPCR.

Author Contributions: Conceptualization, X.Y. (Xiuhong Yao); methodology, S.Y.; software, S.Y. and W.L.; validation, X.Y. (Xiaoping Yong); formal analysis, S.Y., X.Y. (Xiaoping Yong), and Y.L. (Yuxin Lei); resources, Y.L. (Yuxin Lu) and W.L.; data curation, Y.L. (Yuxin Lei); writing-original draft, X.Y. (Xiuhong Yao); writing-review and editing, Y.L. (Yuxin Lu); visualization, Y.L. (Yuxin Lu); supervision, S.Y.; project administration, X.Y. (Xiuhong Yao); funding acquisition, Q.S. and X.Y. (Xiuhong Yao). All authors have read and agreed to the published version of the manuscript.

Funding: This research was supported by the Joint Funds of the National Natural Science Foundation of China (Grant No. U22A20494), the Solid-State Fermentation Resource Utilization Key Laboratory of Sichuan Province (grant no. 2023GTZD03 to X.Y. and 2023GTZD05 to Q.S.), the Doctoral Research Launch Project of Sichuan Academy of Agricultural Sciences (NKYRCZX2024036), and the Project of Sichuan Province Engineering Technology Research Center of Vegetables (2023PZSC0303).

Data Availability Statement: The original contributions presented in the study are included in the article/supplementary material, and no other data were created.

Conflicts of Interest: The authors declare no conflict of interest.

References

1. Lesk, C.; Rowhani, P.; Ramankutty, N. Influence of extreme weather disasters on global crop production. *Nature* **2016**, *529*, 84–87. [CrossRef] [PubMed]
2. Huang, J.; Zhao, X.; Bürger, M.; Wang, Y.; Chory, J. Two interacting ethylene response factors regulate heat stress response. *Plant Cell* **2021**, *33*, 338–357. [CrossRef] [PubMed]
3. Zhao, J.; Lu, Z.; Wang, L.; Jin, B. Plant Responses to Heat Stress: Physiology, Transcription, Noncoding RNAs, and Epigenetics. *Int. J. Mol. Sci.* **2020**, *22*, 117. [CrossRef] [PubMed]
4. von Koskull-Döring, P.; Scharf, K.-D.; Nover, L. The diversity of plant heat stress transcription factors. *Trends Plant Sci.* **2007**, *12*, 452–457. [CrossRef] [PubMed]
5. Ohama, N.; Sato, H.; Shinozaki, K.; Yamaguchi-Shinozaki, K. Transcriptional Regulatory Network of Plant Heat Stress Re-sponse. *Trends Plant Sci.* **2017**, *22*, 53–65. [CrossRef] [PubMed]
6. Suzuki, N. Fine Tuning of ROS, Redox and Energy Regulatory Systems Associated with the Functions of Chloroplasts and Mitochondria in Plants under Heat Stress. *Int. J. Mol. Sci.* **2023**, *24*, 1356. [CrossRef]
7. Medina, E.; Kim, S.-H.; Yun, M.; Choi, W.-G. Recapitulation of the Function and Role of ROS Generated in Response to Heat Stress in Plants. *Plants* **2021**, *10*, 371. [CrossRef]
8. Baillo, E.H.; Kimotho, R.N.; Zhang, Z.; Xu, P. Transcription Factors Associated with Abiotic and Biotic Stress Tolerance and Their Potential for Crops Improvement. *Genes* **2019**, *10*, 771. [CrossRef]
9. Wang, C.; Shen, X.; Yang, T.; Yao, H.; Peng, X.; Xiong, C.; Cohen, H.; Hao, N.; Cao, J.; Wu, T. Genome-wide characterization and identification of root development and stress-related CsMYB36 genes. *Veg. Res.* **2023**, *3*, 19. [CrossRef]
10. Han, K.; Zhao, Y.; Sun, Y.; Li, Y. NACs, generalist in plant life. *Plant Biotechnol. J.* **2023**, *21*, 2433–2457. [CrossRef]
11. Liu, X.; Lyu, Y.; Yang, W.; Yang, Z.; Lu, S.; Liu, J. A membrane-associated NAC transcription factor OsNTL3 is involved in thermotolerance in rice. *Plant Biotechnol. J.* **2020**, *18*, 1317–1329. [CrossRef] [PubMed]
12. Meng, X.; Wang, N.; He, H.; Tan, Q.; Wen, B.; Zhang, R.; Fu, X.; Xiao, W.; Chen, X.; Li, D.; et al. Prunus persica transcription factor PpNAC56 enhances heat resistance in transgenic tomatoes. *Plant Physiol. Biochem.* **2022**, *182*, 194–201. [CrossRef] [PubMed]
13. Cai, W.; Yang, S.; Wu, R.; Cao, J.; Shen, L.; Guan, D.; Shuilin, H. Pepper NAC-type transcription factor NAC2c balances the trade-off between growth and defense responses. *Plant Physiol.* **2021**, *186*, 2169–2189. [CrossRef] [PubMed]
14. Xi, Y.; Ling, Q.; Zhou, Y.; Liu, X.; Qian, Y. ZmNAC074, a maize stress-responsive NAC transcription factor, confers heat stress tolerance in transgenic Arabidopsis. *Front. Plant Sci.* **2022**, *13*, 986628. [CrossRef] [PubMed]
15. Guan, Q.; Yue, X.; Zeng, H.; Zhu, J. The protein phosphatase RCF2 and its interacting partner NAC019 are critical for heat stress-responsive gene regulation and thermotolerance in Arabidopsis. *Plant Cell* **2014**, *26*, 438–453. [CrossRef]
16. Shahnejat-Bushehri, S.; Mueller-Roeber, B.; Balazadeh, S. Arabidopsis NAC transcription factor JUNGBRUNNEN1 affects thermomemory-associated genes and enhances heat stress tolerance in primed and unprimed conditions. *Plant Signal. Behav.* **2012**, *7*, 1518–1521. [CrossRef]
17. Alshareef, N.O.; Otterbach, S.L.; Allu, A.D.; Woo, Y.H.; de Werk, T.; Kamranfar, I.; Mueller-Roeber, B.; Tester, M.; Balazadeh, S.; Schmöckel, S.M. NAC transcription factors ATAF1 and ANAC055 affect the heat stress response in Arabidopsis. *Sci. Rep.* **2022**, *12*, 11264. [CrossRef]
18. Wang, W.; Zhao, J.F.; Sun, Y.D.; Luo, D.X.; Zhang, C.Y.; Wang, W.W. Research progress of storage of chinese cabbage. *J. Anhui Agric. Sci.* **2017**, *45*, 72–73, 77.
19. Li, L.; Guo, B.; Feng, C.; Liu, H.; Lin, D. Growth, physiological, and temperature characteristics in chinese cabbage pakchoi as affected by Cd- stressed conditions and identifying its main controlling factors using PLS model. *BMC Plant Biol.* **2022**, *22*, 571. [CrossRef]
20. Yue, L.; Li, G.; Dai, Y.; Sun, X.; Li, F.; Zhang, S.; Zhang, H.; Sun, R.; Zhang, S. Gene co-expression network analysis of the heat-responsive core transcriptome identifies hub genes in *Brassica rapa*. *Planta* **2021**, *253*, 111. [CrossRef]
21. Quan, J.; Zheng, W.W.; Wu, M.F.; Shen, Z.J.; Tan, J.R.; Li, Z.W.; Zhu, B.; Hong, S.B.; Zhao, Y.T.; Zhu, Z.J.; et al. Glycine Betaine and β-Aminobutyric Acid Mitigate the Detrimental Effects of Heat Stress on Chinese Cabbage (*Brassica rapa* L. ssp. *pekinensis*) Seedlings with Improved Photosynthetic Performance and Antioxidant System. *Plants* **2022**, *11*, 1213. [PubMed]
22. Yao, X.H.; Li, Y.L.; Chen, J.; Zhou, Z.X.; Wen, Y.; Fang, K.; Yang, F.B.; Li, T.T.; Zhang, D.W.; Lin, H.H. Brassinosteroids enhance BES1-required thermomemory in *Arabidopsis thaliana*. *Plant Cell Environ.* **2022**, *45*, 3492–3504. [CrossRef] [PubMed]
23. Zhang, Y.; Liu, Z.; Chen, Y.; He, J.-X.; Bi, Y. PHYTOCHROME-INTERACTING FACTOR 5 (PIF5) positively regulates dark-induced senescence and chlorophyll degradation in Arabidopsis. *Plant Sci.* **2015**, *237*, 57–68. [CrossRef] [PubMed]
24. Qiao, K.; Yao, X.; Zhou, Z.; Xiong, J.; Fang, K.; Lan, J.; Xu, F.; Deng, X.; Zhang, D.; Lin, H. Mitochondrial alternative oxidase enhanced ABA-mediated drought tolerance in *Solanum lycopersicum*. *J. Plant Physiol.* **2023**, *280*, 153892. [CrossRef] [PubMed]
25. Yao, X.; Fang, K.; Qiao, K.; Xiong, J.; Lan, J.; Chen, J.; Tian, Y.; Kang, X.; Lei, W.; Zhang, D.; et al. Cooperative transcriptional regulation by ATAF1 and HY5 promotes light-induced cotyledon opening in *Arabidopsis thaliana*. *Sci. Signal.* **2024**, *17*, eadf7318. [CrossRef]
26. Li, T.; Fang, K.; Tie, Y.; Lu, Y.; Lei, Y.; Li, W.; Zheng, T.; Yao, X. NAC transcription factor ATAF1 negatively modulates the PIF-regulated hypocotyl elongation under a short-day photoperiod. *Plant Cell Environ.* **2024**, *47*, 3253–3265. [CrossRef]

27. Yu, X.; Yu, J.; Lu, Y.; Li, W.; Huo, G.; Zhang, J.; Li, Y.; Zhao, J.; Li, J. An efficient and universal protoplast-based transient gene expression system for genome editing in Brassica crops. *Hortic. Plant J.* **2024**, *10*, 983–994. [CrossRef]
28. Singh, S.; Koyama, H.; Bhati, K.K.; Alok, A. The biotechnological importance of the plant-specific NAC transcription factor family in crop improvement. *J. Plant Res.* **2021**, *134*, 475–495. [CrossRef]
29. Chen, Y.; Li, X.; Xie, X.; Liu, L.; Fu, J.; Wang, Q. Maize transcription factor ZmNAC2 enhances osmotic stress tolerance in transgenic Arabidopsis. *J. Plant Physiol.* **2023**, *282*, 153948. [CrossRef]
30. Lämke, J.; Brzezinka, K.; Altmann, S.; Bäurle, I. A hit-and-run heat shock factor governs sustained histone methylation and transcriptional stress memory. *EMBO J.* **2016**, *35*, 162–175. [CrossRef]
31. Fonseca-González, I.; Velasquez-Agudelo, E.; Londoño-Mesa, M.H.; Álvarez, J.C. De novo transcriptome sequencing and annotation of the Antarctic polychaete *Microspio moorei* (Spionidae) with its characterization of the heat stress-related proteins (HSP, SOD & CAT). *Mar. Genom.* **2024**, *73*, 101085. [CrossRef]
32. Jagadish, S.K.; Way, D.A.; Sharkey, T.D. Plant heat stress: Concepts directing future research. *Plant Cell Environ.* **2021**, *44*, 1992–2005. [CrossRef]
33. Sharif, R.; Su, L.; Chen, X.; Qi, X. Involvement of auxin in growth and stress response of cucumber. *Veg. Res.* **2022**, *2*, 13. [CrossRef]
34. Srivastava, R.; Kobayashi, Y.; Koyama, H.; Sahoo, L. Cowpea NAC1/NAC2 transcription factors improve growth and tolerance to drought and heat in transgenic cowpea through combined activation of photosynthetic and antioxidant mechanisms. *J. Integr. Plant Biol.* **2023**, *65*, 25–44. [CrossRef] [PubMed]
35. Ren, Y.; Huang, Z.; Jiang, H.; Wang, Z.; Wu, F.; Xiong, Y.; Yao, J. A heat stress responsive NAC transcription factor heterodimer plays key roles in rice grain filling. *J. Exp. Bot.* **2021**, *72*, 2947–2964. [CrossRef]
36. Wang, B.; Zhong, Z.; Wang, X.; Han, X.; Yu, D.; Wang, C.; Song, W.; Zheng, X.; Chen, C.; Zhang, Y. Knockout of the *OsNAC006* Transcription Factor Causes Drought and Heat Sensitivity in Rice. *Int. J. Mol. Sci.* **2020**, *21*, 2288. [CrossRef]
37. Liu, T.K. An effective method for studying gene function of Chinese cabbage. *Veg. Res.* **2023**, *3*, 2. [CrossRef]
38. Liang, Y.; Kang, K.; Gan, L.; Ning, S.B.; Xiong, J.Y.; Song, S.Y.; Xi, L.Z.; Lai, S.Y.; Yin, Y.T.; Gu, J.W.; et al. Drought-responsive genes, late embryogenesis abundant group3 (LEA3) and vicinal oxygen chelate, function in lipid accumulation in *Brassica napus* and Arabidopsis mainly via enhancing photosynthetic efficiency and reducing ROS. *Plant Biotechnol. J.* **2019**, *17*, 2123–2142. [CrossRef]
39. Huchzermeyer, B.; Menghani, E.; Khardia, P.; Shilu, A. Metabolic Pathway of Natural Antioxidants, Antioxidant Enzymes and ROS Providence. *Antioxidants* **2022**, *11*, 761. [CrossRef]

horticulturae

MDPI

Article

Weighted Gene Co-Expression Network Analysis Identifies Hub Genes Governing Resistance to *Fusarium oxysporum* f. sp. *niveum* in Watermelon (*Citrullus lanatus*)

Chen Zhang [1,2,†], Xufeng Fang [1,2,†], Jing Zhang [1,2], Xinying Wang [1,2], Zhao Liu [3], Shusen Liu [3], Zhengfeng Song [3], Peng Gao [1,*] and Feishi Luan [2,*]

1 Key Laboratory of Biology and Genetic Improvement of Horticulture Crops (Northeast Region), Ministry of Agriculture and Rural Affairs, College of Horticulture and Landscape Architecture, Northeast Agricultural University, Harbin 150030, China; wkzc351504709@163.com (C.Z.); fangxufeng@neau.edu.cn (X.F.); jingzhang0411@163.com (J.Z.); wangxinying0731@163.com (X.W.)
2 College of Horticulture and Landscape Architecture, Northeast Agricultural University, Harbin 150030, China
3 Shandong Engineering Research Center for Watermelon and Melon Breeding Co., Ltd., Weifang 261000, China; sanmuliuzhao@163.com (Z.L.); sanmuseed@163.com (S.L.); wdlszf@163.com (Z.S.)
* Correspondence: gaopeng_neau@163.com (P.G.); luanfeishi@neau.edu.cn (F.L.)
† These authors contributed equally to this work.

Abstract: Watermelon (*Citrullus lanatus*), a vital economic crop, is severely threatened by Fusarium wilt (FW), which is caused by the soil-borne fungal pathogen *Fusarium oxysporum* f. sp. *niveum* (Fon). To elucidate the molecular mechanisms underlying FW resistance in watermelon, we tracked the infection process via microscopy, identifying three critical time points (1, 6, and 8 days post-inoculation) corresponding to spore germination, hyphal invasion of the xylem vascular system, and symptom onset, respectively. Transcriptional profiling at these stages revealed six disease-resistance-associated gene modules through differential expression analysis, expression pattern clustering, weighted gene co-expression network analysis, and functional enrichment. These modules exhibited strong correlations with distinct infection phases. Protein–protein interaction networks identified 35 hub genes, including receptor-like kinases; WRKY and ethylene-responsive factor transcription factors; and genes involved in cell wall reinforcement, hormone signaling, defense metabolism/detoxification, programmed cell death regulation, and antimicrobial compound biosynthesis. Differential expressions of these genes across infection stages likely underpin the observed phenotypic disparities. Five hub regulatory genes were identified by quantitative real-time PCR in the SRgreen and SRblack modules, namely, Cla97C01G014990 (WRKY transcription factor 42), *Cla97C02G042360* (calcium-transporting ATPase), *Cla97C08G155710* (AIG2), *Cla97C09G170380* (ethylene-responsive factor 1B-like), and *Cla97C06G121810* (receptor kinase, putative). These genes mediate early rapid defense responses via SRgreen and sustain long-term resistance through SRblack. By validating the expression patterns of hub genes, the study elucidated the watermelon resistance response and provided insights into transcriptional regulation during different stages of Fon–watermelon interactions. Additionally, it identified candidate genes that could enhance watermelon resistance to wilt disease.

Keywords: watermelon; Fusarium wilt; transcriptome analysis; WGCNA; hub gene

1. Introduction

Watermelon (*Citrullus lanatus*) is a globally important cucurbit crop, with China dominating its production and consumption, generating more than 60 million metric tons annually [1].

A critical constraint to watermelon cultivation is Fusarium wilt (FW), caused by the soil-borne pathogen *Fusarium oxysporum* f. sp. *niveum* (Fon), which reduces yields by 30–80% through vascular system colonization and water transport disruption [2]. Fon persists as chlamydospores for more than a decade in soil, germinating upon detecting host root exudates to develop infection hyphae that penetrate root tissues [3]. Comparative studies using confocal laser scanning microscopy revealed that both resistant and susceptible watermelon varieties permit initial root colonization by Fon, but defense mechanisms in resistant varieties restrict pathogen progression beyond cortical layers, preventing vascular invasion [4,5]. In susceptible plants, xylem occlusion by fungal hyphae, gums, and tyloses triggers hydraulic failure and wilting, underscoring the critical role of transcriptional reprogramming in vascular defense.

Recent genetic mapping efforts have identified key quantitative trait loci (QTLs) underlying Fon resistance in watermelon. *Fon-1*, a major QTL, was mapped to chromosome 1 in a recombinant inbred line (RIL) population derived from the cultivated variety 97103 and wild accession PI296341-FR, and this QTL explained 48.1% of the phenotypic variance (LOD = 13.2). This locus contains candidate genes, including a receptor kinase, glucan endo-1,3-β-glucosidase precursors, and three acidic chitinases, suggesting roles in pathogen recognition and cell wall remodeling [6]. A QTL mapping analysis was performed using F2:3 and RIL populations derived from the Fon race 1-resistant donor *Citrullus amarus* USVL246-FR2 and susceptible parent USVL114. This study identified a major QTL (*qFon1-9*) on chromosome 9 of USVL246-FR2 that confers resistance to Fon race 1 and represents an uncharacterized genetic resource for enhancing FW resistance in watermelon. Notably, seven receptor-like kinase (RLK) genes mapped within the 1.5-LOD confidence interval of *qFon1-9* emerged as strong candidates for broad-spectrum resistance. These RLKs might function as pattern recognition receptors to perceive conserved pathogen-associated molecular patterns (PAMPs), thereby activating basal defense responses independent of pathogen race specialization [7].

Transcription factors (TFs) are essential DNA-binding proteins that regulate gene expression by modulating transcriptional dynamics through interactions with cis-regulatory elements in the promoters of stress-responsive genes. These proteins act as central regulators of plant defenses by controlling the expression of stress-induced genes critical for activating host defense mechanisms [8]. Among plant TF families, the WRKY superfamily is a prominent regulator of immunity. Its members are characterized by a conserved WRKY domain that mediates DNA recognition, and they are widely distributed across plant species [9]. WRKY TFs govern diverse immune responses, functioning as both activators and repressors of defense pathways. They are integral to the regulation of PAMP-triggered immunity (PTI) and effector-triggered immunity (ETI). In cotton, group IIc WRKY TFs enhanced resistance to *F. oxysporum* f. sp. *vasinfectum* by activating the GhMKK2–GhNTF6 signaling cascade, which increases the GhMYC2-mediated expression of flavonoid biosynthesis genes. This pathway increased the levels of antimicrobial secondary metabolites, thereby restricting pathogen colonization [10]. PdpapWRKY28 overexpression significantly suppressed reactive oxygen species (ROS) accumulation in *Populus davidiana* and enhanced antioxidant enzyme activity, thereby conferring robust tolerance to environmental stresses [11]. In addition to WRKY TFs, APETALA2/ethylene-responsive factor (AP2/ERF) TFs also play central roles in regulating plant responses to diverse biotic and abiotic stresses, including salinity, drought, temperature extremes, and pathogenic infections [12]. AP2/ERF family members modulate defense responses through intricate molecular mechanisms, such as the transcriptional activation of stress-responsive genes and crosstalk with phytohormone signaling pathways. MPK-mediated phosphorylation of *Glycine max* ERF113 (GmERF113) enhances its transcriptional activity, thereby increasing the expression of defense-related genes and augmenting immunity against the oomycete pathogen *Phytophthora sojae* [13]. In *Gossypium hirsutum*, GbERF1 overexpression significantly increased lignin biosynthesis gene transcription, resulting in increased lignin deposition and enhanced resistance to the wilt pathogen *Verticillium dahliae* [14]. The *Arabidopsis* ERF protein ORA59 directly binds to the PDF1.2 promoter to activate jasmonic acid/ethylene (JA/ET)-dependent defense responses, thereby improving resistance to the necrotrophic fungus *Botrytis cinerea* [15].

Weighted gene co-expression network analysis (WGCNA), a method for identifying trait-related modules [16], has been widely applied to dissect hub genes associated with disease resistance [17–19]. In this study, we employed WGCNA to characterize genome-wide transcriptomic responses in two watermelon varieties, namely, the FW-resistant variety PI 296341 (R) and FW-susceptible variety W1-1 (S), following infection with Fon. This study performed a comprehensive analysis of compatible (susceptible interaction) and incompatible (resistant interaction) molecular mechanisms between Fon and watermelon, thereby generating genome-wide expression profiles of disease resistance-related genes. By integrating co-expression networks and pathogen challenge dynamics, our work provides a systems-level framework for unraveling genetic determinants of FW resistance.

Expression pattern analysis of selected hub genes was conducted to elucidate their potential roles in disease resistance mechanisms. These findings provide critical molecular insights into plant–pathogen interactions, establishing a robust framework for further exploration of *Fusarium oxysporum* defense mechanisms in host plants.

2. Materials and Methods

2.1. Plant Materials

FW-resistant variety PI 296341 (R) and FW-susceptible variety W1-1 (S) were obtained from the molecular genetic breeding research group for watermelons and melons of Northeast Agricultural University, China. Seeds were germinated and grown under controlled conditions (light/dark = 16 h/8 h, light intensity of 120 μmol m^{-2} s^{-1}, 28 °C, 70% relative humidity).

2.2. Fungal Transformation with GFP

The Fon (*Fusarium oxysporum* f. sp. *niveum* race 1) isolate was laboratory-preserved and maintained on potato dextrose agar (PDA) at 25 °C. Fon was transformed as previously described [20]. Briefly, *Agrobacterium tumefaciens* strain AGL1 (Coolaber Science & Technology Co., Ltd., Beijing, China) harboring pCAMBIA1303-GFP was resuscitated in lysogeny broth (LB) medium supplemented with 50 μg/mL kanamycin (Beijing Solarbio Science & Technology Co., Ltd., Beijing, China) and 50 μg/mL rifampicin (Beijing Solarbio Science & Technology Co., Ltd., Beijing, China) under standard incubation conditions (28 °C, 180 rpm). AGL1 (OD_{600} = 0.15) was mixed with Fon microconidia (1 × 10^6 conidia/mL) in equal volumes (100 μL each). The mixture (200 μL) was plated onto nitrocellulose membranes (Beijing Solarbio Science & Technology Co., Ltd., Beijing, China) and incubated in a co-cultivation medium supplemented with 5 mM glucose at 25°C for 48 h. Membranes were then transferred to PDA containing 120 μg/mL hygromycin B (Beijing Solarbio Science & Technology Co., Ltd., Beijing, China) and 300 μg/mL cefotaxime (Beijing Solarbio Science & Technology Co., Ltd., Beijing, China) for 7–10 days at 25 °C to permit transformant growth. Hygromycin-resistant transformants were isolated and maintained on PDA supplemented with hygromycin B (120 μg/mL).

2.3. Inoculation of Watermelon Plants

Fungal conidia were produced by culturing Fon in a 250 mL conical flask containing 150 mL of potato dextrose broth at 28 °C on a rotary shaker (150 rpm) for 5–7 days. The conidial suspension was adjusted to a final concentration of 1 × 10^6 conidia/mL using sterile distilled water for subsequent plant inoculation. Fifteen-day-old watermelon seedlings grown in a controlled growth chamber (16 h/8 h light/dark photoperiod, 28 °C, 70% relative humidity) were carefully removed from vermiculite substrate and gently rinsed with tap water to remove residual particles. Seedling roots were then immersed in freshly prepared spore suspensions (1×10^6 conidia/mL) for 20 min. Inoculated seedlings were replanted into fresh vermiculite, whereas control plants were subjected to identical procedures except they were immersed in sterile distilled water instead of the spore suspension.

2.4. Microscopic Analysis

Watermelon roots were sampled at 24 h intervals up to 8 days post-inoculation (dpi). At each time point, five plants were harvested, rinsed to remove the growth medium, and dissected to isolate the taproot and lower hypocotyl. Root segments adjacent to the root base were excised and sectioned into 5 mm pieces, whereas hypocotyl segments (0–1 cm above the main root) were hand-sectioned into 2–3 mm pieces. To determine the critical time points of Fon infection in watermelon roots, the colonization process of Fon-GFP was observed in watermelon roots at different time points, with five biological replicates per time point. Samples were mounted in water droplets on glass slides under coverslips for microscopic examination using a confocal laser scanning microscope (TCS SP8, Leica, Berlin, Germany) with a 10× objective. Roots treated with non-GFP-labeled Fon served as control samples to ensure that the observed fluorescence was attributable solely to the colonization of Fon-GFP. Fluorescence was excited at 488 nm, and emitted light was recorded between 520 and 540 nm.

2.5. RNA Sequencing (RNA-Seq) and Data Analysis

Root samples were collected from Fon-inoculated plants (S and R) and water-treated controls at 0, 1, 6, and 8 dpi. To explore nonspecific immune responses in the plants, they were exposed to killed pathogens (autoclaved at 121 °C for 5 min) at 1 dpi. In total, 48 samples (10–12 roots per time point) were rapidly frozen in liquid nitrogen for RNA extraction and subsequent RNA-seq. Raw sequencing data were processed using fastp v0.20.1 to trim low-quality bases and remove adapter sequences [21], yielding clean reads. Clean reads were then aligned to the reference genome (http://cucurbitgenomics.org/organism/21, accessed on 19 January 2024) using HISAT2 v2.1.1 [22]. StringTie was subsequently employed to assemble transcript isoforms from mapped reads for each sample, generating gene expression quantification matrices [23].

2.6. Data Integration and Network Construction

Differentially expressed genes (DEGs, $\log_2$ fold change > 1 and false discovery rate < 0.05) were clustered and visualized using the ClusterGVis package (https://github.com/junjunlab/ClusterGVis, accessed on 24 March 2024). Co-expression networks were constructed using the R package WGCNA (version 1.73). Subsequently, Cytoscape v3.9.1 was employed to visualize intramodule gene networks and highlight biological interactions among hub genes [24]. Networks were filtered to retain the top 100 edges with the highest weights. Hub genes were prioritized by degree, with the top 10 genes with the highest degrees designated as critical nodes and visualized as red vertices in the network.

2.7. Quantitative Real-Time PCR

qRT-PCR was used to verify the expression of select DEGs. Root samples were collected from Fon-inoculated plants (S and R) and water-treated controls at 0, 1, 6, and 8 dpi. Total RNA was extracted from watermelon root samples using the Spin Column Plant Total RNA Purification Kit (Sangon Biotech, Shanghai, China). Approximately 1 μg of the total RNA was used as a template to synthesize the cDNA with HiScript® II Q RT SuperMix for qPCR (+gDNA wiper) (Vazyme, Nanjing, China). qRT-PCR was performed using ChamQ™ Universal SYBR® qPCR Master Mix (Vazyme, Nanjing, China) and a LightCycler® 480 II Real-time PCR instrument (Roche, Basel, Switzerland). The relative mRNA level of each candidate gene was evaluated against watermelon *ClACTIN* (Gene Accession Number Cla97C02G038590) as a reference gene. The qPrimerDB website (https://biodb.swu.edu.cn/qprimerdb/, accessed on 1 April 2024) was used to design qRT-PCR-specific primers (Table S1). The gene expression level was calculated using the $2^{-\Delta\Delta CT}$ method with three biological replicates.

2.8. Statistical Analysis

Data came from three biological replications and were analyzed using Graphpad software (version 10.4.0) for statistical analysis. Statistically significant differences were assessed using one-way ANOVA, followed by a post hoc Tukey test at a significance level of $p < 0.05$.

3. Results

3.1. Analysis by Confocal Laser Scanning Microscopy

Following inoculation with Fon, S exhibited pronounced leaf drooping and chlorosis at 8 dpi (Figure 1A,B). In contrast, R displayed no discernible phenotypic alterations under identical infection conditions (Figure 1C,D). Within 1 dpi, most Fon-GFP spores germinated on the root surface, forming initial hyphae (green fluorescence; Figure 1E). By 2 dpi, the fungal hyphae developed densely across the root surface, indicating robust colonization (Figure 1F). Between 4 and 5 dpi, Fon-GFP hyphae penetrated the intercellular spaces along the edges of root cells (Figure 1G and Figure S1). At 6 dpi, Fon-GFP hyphae entered the root vascular system, as evidenced by their presence within xylem vessels (Figure 1H,I). By 8 dpi, extensive colonization of the root vascular system by Fon-GFP was observed, coinciding with the onset of wilt symptoms in watermelon seedlings (Figure 1B,J).

Combining observations of the infection process with phenotypic responses in S inoculated with Fon-GFP, we identified 1, 6, and 8 dpi as critical time points. Spore germination occurred at 1 dpi, initiating the formation of hyphae, which likely marks the early recognition phase of plant defense against fungal pathogens (Figure 1E). By 6 dpi, hyphae had invaded the root vascular system, as evidenced by their presence within xylem vessels (Figure 1H,I), suggesting a transition to necrotrophic growth and triggering a moderate defense response in the host. Extensive colonization of the vascular system at 8 dpi coincided with the onset of wilt symptoms in watermelon seedlings (Figure 1B,J), indicating an advanced stage of infection in which the pathogen disrupts cellular integrity. These key time points (1, 6, and 8 dpi) were pivotal for subsequent experiments aimed at elucidating the molecular mechanisms underlying plant–pathogen interactions during these critical phases of infection.

3.2. Identification of DEGs

To identify DEGs associated with Fon infection, mRNA libraries were constructed and divided into 16 groups, which comprised samples from two genotypes of infected plants and water-treated controls at 0, 1, 6, and 8 dpi, with three biological replicates per group ($n = 3$). In total, 2,397,660,805 raw reads were generated, and after quality control, 2,357,983,183 clean reads were obtained for downstream analysis (Table S2).

Comparative analysis of pathogen-treated and water-treated samples revealed distinct transcriptional patterns. In cultivar S, 1, 1, 1197, and 1250 upregulated DEGs were identified at 0, 1, 6, and 8 dpi, whereas 13, 18, 1561, and 1713 downregulated DEGs, respectively, were identified at these time points (Figure 2A). In cultivar R, pathogen-challenged plants exhibited 0, 219, 200, and 754 upregulated DEGs and 0, 365, 31, and 1171 downregulated DEGs at 0, 1, 6, and 8 dpi, respectively (Figure 2B).

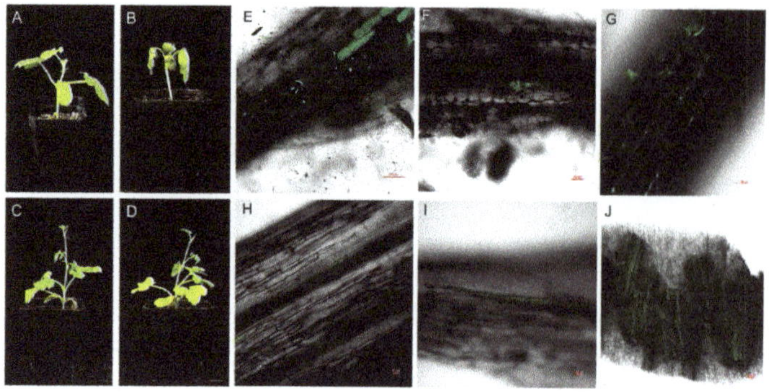

Figure 1. Phenotypes of watermelon seedlings (S: susceptible, FW-susceptible watermelon variety W1-1, and R: resistant, FW-resistant watermelon variety PI 296341) after 8 days of Fon (*Fusarium oxysporum* f. sp. *niveum* race 1) and water treatment and confocal laser scanning microscopic analysis of Fon-GFP colonization in S. (**A**) S seedlings after 8 days of water treatment. (**B**) S seedlings after 8 days of Fon inoculation. (**C**) R seedlings after 8 days of water treatment. (**D**) R seedlings after 8 days of Fon inoculation. Scale bar = 1.5 cm. (**E**) At 1 dpi, Fon-GFP attached to the surface of watermelon roots began to germinate. (**F**) At 2 dpi, Fon-GFP on the surface of watermelon roots began to exhibit dense hyphal proliferation. (**G**) Between 4 and 5 dpi, Fon-GFP hyphae penetrated the intercellular spaces along the edges of root cells. (**H**,**I**) Longitudinal section of the root. At 6 dpi, hyphae were observed to extend into the xylem vessels. (**J**) In transverse sections of the root, extensive colonization of hyphae in the xylem vessels was observed at 8 dpi. Fluorescence was excited at 488 nm and emitted light between 520 and 540 nm. Scale bar = 50 μm.

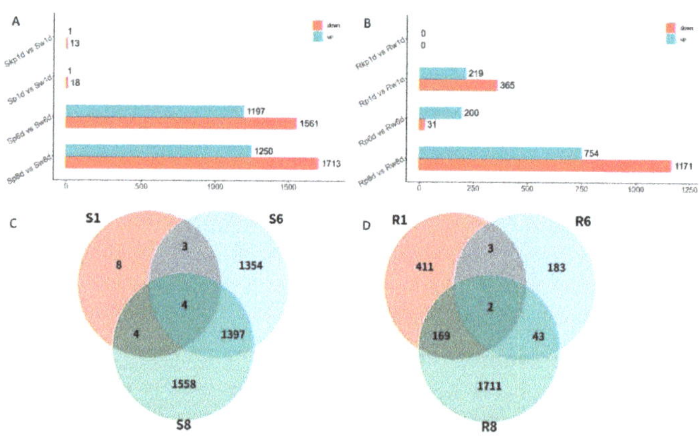

Figure 2. Graphical representation of gene expression changes following inoculation with *Fusarium oxysporum* f.sp. *niveum* (Fon). (**A**) DEGs between different treatments at each time point in susceptible watermelon (S). (**B**) DEGs between different treatments at each time point in resistant watermelon (R). (**C**) Venn diagram of overlapped genes in S. (**D**) Venn diagram of overlapped genes in R. R: FW-resistant watermelon variety PI 296341; S: FW-susceptible watermelon variety W1-1; w: water treatment; p: pathogen treatment (*Fusarium oxysporum* f. sp. *niveum* race 1); kp: killed pathogen treatment; S1: Sp1d vs. Sw1d; S6: Sp6d vs. Sw6d; S8: Sp8d vs. Sw8d; R1: Rp1d vs. Rw1d; R6: Rp6d vs. Rw6d; R8: Rp8d vs. Rw8d.

Given the minimal differential expression observed between water-treated and killed pathogen-treated samples at 1 dpi in both cultivars, data from killed pathogen-treated samples at this time point were excluded from subsequent analyses. A comparative overlap analysis was conducted across three critical infection stages (1, 6, and 8 dpi) in R and S. In R, 411, 183, and 1711 DEGs were uniquely expressed at 1, 6, and 8 dpi, respectively. Conversely, cultivar

S exhibited 8, 1354, and 1558 uniquely expressed DEGs, respectively, at these time points (Figure 2C). Notably, cultivar R displayed the temporal conservation of two DEGs across all stages, whereas cultivar S retained four persistently expressed DEGs (Figure 2D).

3.3. Determination of Key Clusters

To systematically characterize cultivar-specific defense responses, DEGs upregulated by pathogen exposure in the S and R cultivars were clustered using the ClusterGVis package (Version 0.1.2), yielding 12 distinct clusters in each cultivar. Subsequent functional annotation prioritized defense-associated clusters by integrating expression dynamics with functional enrichment analyses of Gene Ontology (GO) terms and Kyoto Encyclopedia of Genes and Genomes (KEGG) pathways. In R, 1058 upregulated DEGs were grouped into 12 clusters, five of which (R1, R5, R7, R9, R12) exhibited potential associations with disease resistance. DEGs in R1 (contained 59 genes) and R5 (contained 96 genes) were upregulated at 6 dpi, whereas those in R7 (contained 74 genes) displayed sustained upregulation at both 6 and 8 dpi. Genes in R9 (contained 132 genes) and R12 (contained 91 genes) were specifically activated at 8 and 1 dpi, respectively. Functional enrichment analysis revealed that these clusters were significantly associated with immune system processes, defense responses, phenylpropanoid biosynthesis, responses to biotic stimuli, glutathione metabolism, salicylic acid (SA) signaling, and NOD-like receptor pathways (Figure 3). In S, four clusters (S5, S6, S7, S10) displayed putative roles in disease resistance. Genes in S5 (contained 106 genes) were upregulated at 6 and 8 dpi, those in S6 (contained 136 genes) were upregulated at 8 dpi, and those in S7 (contained 123 genes) and S10 (contained 213 genes) were upregulated at 6 dpi. The enriched functions included immune system processes, responses to biotic stimuli, glutathione metabolism, defense responses, plant–pathogen interactions, hormone signaling, and phenylpropanoid biosynthesis (Figure 4). Merging these key clusters yielded 941 DEGs involved in various aspects of disease resistance.

3.4. Determination of Key Modules Associated with Fon Resistance Through Co-Expression Network Analysis

To systematically characterize the expression divergence of defense-related DEGs between S and R, WGCNA was performed on 941 defense-associated DEGs from key clusters (R1, R5, R7, R9, R12, S5, S6, S7, S10). A soft threshold power of $\beta = 17$ was selected to construct hierarchical clustering trees based on scale-free topology criteria (Figure S2). Following this, genes were partitioned into 10 distinct co-expression modules using dynamic tree cutting (mergeCutHeight = 0.25; Figure 5A). To delineate key modules linked to disease resistance, we analyzed correlations between co-expression modules and phenotypic traits associated with specific infection stages. In R, three modules demonstrated robust associations with pathogen responses. Specifically, SRgreen exhibited a significant positive correlation with Rp1d ($r^2 = 0.85$, $p = 3 \times 10^{-12}$), SRyellow exhibited a significant positive correlation with Rp6d ($r^2 = 0.64$, $p = 7 \times 10{-}^{6}$), and SRblue exhibited a significant positive correlation with Rp8d ($r^2 = 0.83$, $p = 2 \times 10^{-11}$). In S, two modules displayed strong pathogen-responsive patterns, with SRbrown being positively correlated with Sp8d ($r^2 = 0.66$, $p = 2 \times 10^{-6}$) and SRturquoise being positively correlated with Sp6d ($r^2 = 0.93$, $p = 2 \times 10^{-18}$). SRred and SRblack exhibited conserved expression trajectories in both S and R, suggesting shared regulatory mechanisms underlying basal pathogen responses (Figure 5B). Conversely, SRmagenta displayed divergent transcriptional dynamics between the two genotypes, with antagonistic expression patterns in S versus R, indicative of the cultivar-specific modulation of defense-related pathways. The SRpurple and SRpink modules exhibited strong correlations with the water treatment group, and thus, they were excluded from further analysis.

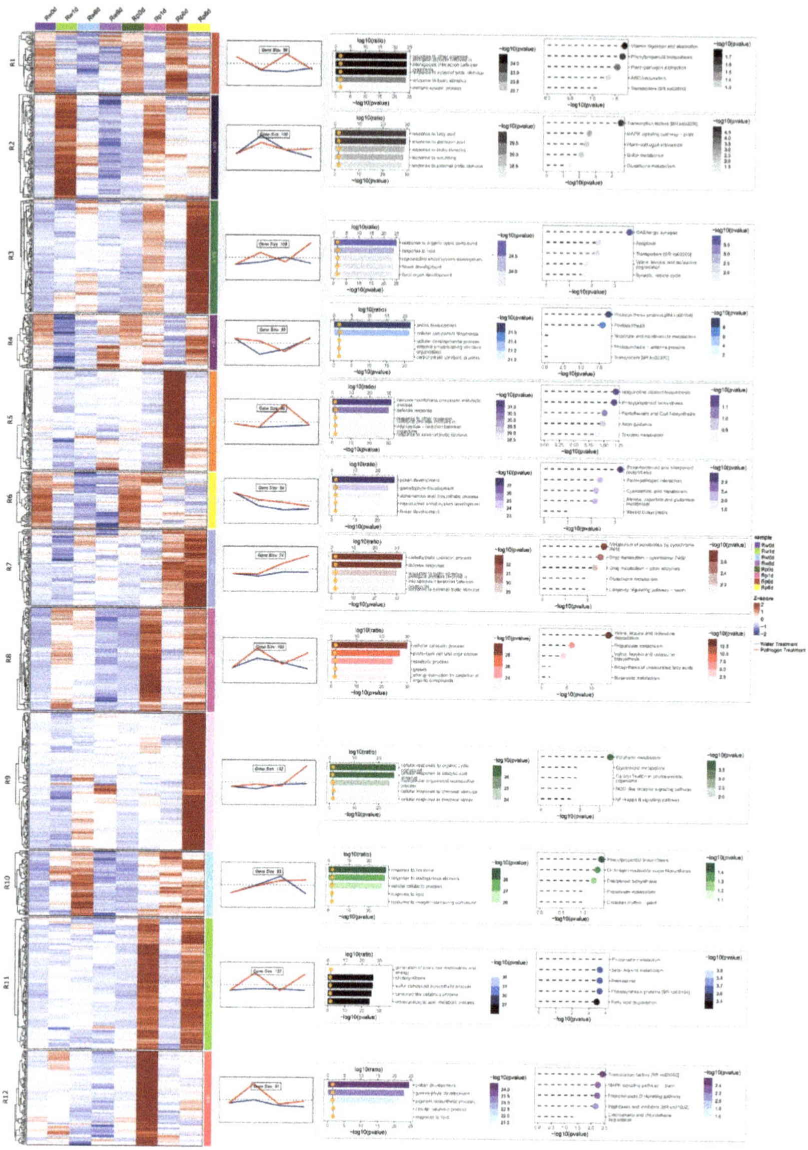

Figure 3. Co-expression cluster identification and functional annotation of DEGs in resistant (R) watermelon upregulated upon inoculation with *Fusarium oxysporum* f.sp. *niveum* (Fon). The Rw0d data are the same as the Rp0d data. From left to right are expression calorimetric maps, expression trend maps and the number of genes contained, GO enrichment results, and KEGG pathway enrichment results. R: FW-resistant watermelon variety PI 296341, w: water treatment, p: pathogen treatment (*Fusarium oxysporum* f. sp. *niveum* race 1). In expression calorimetric maps, red indicates upregulation and blue indicates downregulation, with white representing baseline expression (z-score normalized), and R1–R12 denote the 12 gene clusters obtained. In expression trend maps, the red line represents data from the pathogen treatment, the blue line represents data from the water treatment, and trend lines denote mean expression values per cluster. In the GO enrichment results, the top 5 enriched terms in Biological Process (BP) are shown. In the KEGG pathway enrichment results, the top 5 most significant terms from the KEGG enrichment analysis are shown.

Figure 4. Co-expression cluster identification and functional annotation of DEGs in susceptible (S) watermelon upregulated upon inoculation with *Fusarium oxysporum* f.sp. *niveum* (Fon). The Sw0d data are the same as the Sp0d data. From left to right are expression calorimetric maps, expression trend maps and the number of genes contained, GO enrichment results, and KEGG pathway enrichment results. S: FW-susceptible watermelon variety W1-1, w: water treatment, p: pathogen treatment (*Fusarium oxysporum* f. sp. *niveum* race 1). In expression calorimetric maps, red indicates upregulation and blue indicates downregulation, with white representing baseline expression (z-score normalized), and S1–S12 denote the 12 gene clusters obtained. In expression trend maps, the red line represents data from the pathogen treatment, the blue line represents data from the water treatment, and trend lines denote mean expression values per cluster. In the GO enrichment results, the top 5 enriched terms in Biological Process (BP) are shown. In the KEGG pathway enrichment results, the top 5 most significant terms from the KEGG enrichment analysis are shown.

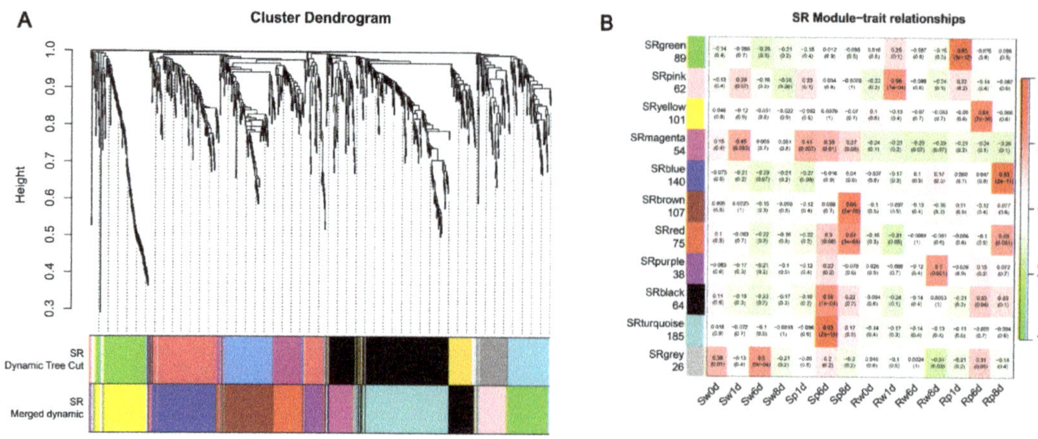

Figure 5. Co-expression network analysis of DEGs and the module–trait relationship in response to *Fusarium oxysporum* f.sp. *niveum* (Fon) infection of watermelon. (**A**) Cluster dendrogram of different genes in co-expression modules. Dynamic tree cut (minimum module size = 30 genes, merge cut height = 0.25). Color bars below the dendrogram represent assigned modules. SR denotes the data used in WGCNA from susceptible (S) and resistant (R) watermelon. (**B**) Heat map of the correlation between templates and groups. Each row represents a module eigengene, and each column represents a different characteristic. Each cell contains the corresponding correlation and *p*-value, and the table is color coded by correlation according to the color legend. The number below each module name denotes the number of genes within the module. Excluding the gray module, a total of 10 gene modules were identified.

3.5. GO and KEGG Enrichment Analyses of Key Modules

GO and KEGG pathway enrichment analyses were performed to delineate the biological significance of the identified modules. The SRgreen module exhibited significant enrichment for plant-specific signaling pathways, including the mitogen-activated protein kinase (MAPK) signaling pathway and plant and plant hormone signal transduction (Figure 6A). SRyellow was associated with metabolic reprogramming, particularly the amine metabolic process and tryptophan biosynthetic process (Figure S3). SRblue displayed enrichment in stress-responsive functions, such as cytokinin metabolic process and responses to toxic substances (Figure S4). In the S cultivar, SRbrown was associated with structural and interactive pathways, including cellulose biosynthetic processes and plant–pathogen interactions (Figure S5), whereas SRturquoise was linked to defense-related processes, including external encapsulating structure organization and phenylpropanoid biosynthesis (Figure S6). The conserved modules SRred and SRblack demonstrated significant enrichment for pathways associated with stress adaptation and secondary metabolism. SRred was specifically linked to cellular responses to abiotic stimuli and beta-alanine metabolism (Figure S7), whereas SRblack encoded functions critical to phytohormone dynamics and structural defense, including brassinosteroid biosynthesis and phenylpropanoid biosynthesis (Figure 6B). In stark contrast, the divergent SRmagenta module exhibited a unique enrichment of glutamine metabolic processes, reflecting genotype-specific metabolic reprogramming during pathogen challenge (Figure S8). Based on module correlation and GO and KEGG enrichment analyses, six key modules, namely, SRgreen, SRblue, SRbrown, SRturquoise, SRblack, and SRmagenta, were identified as being associated with disease resistance responses.

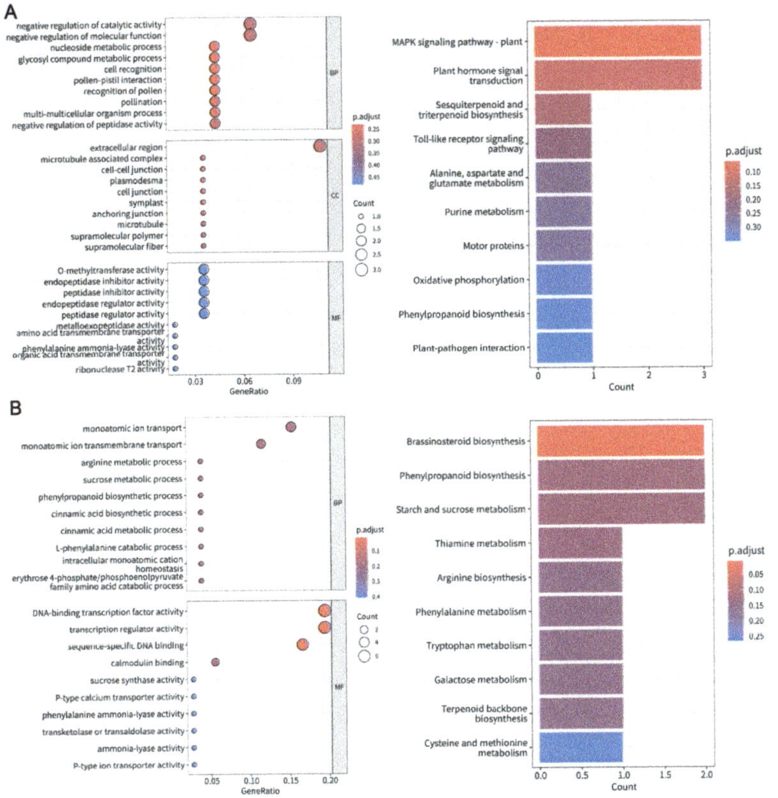

Figure 6. Functional enrichment analysis of key co-expression modules in pathogen-responsive genes in the *Fusarium oxysporum* f.sp. *niveum*–watermelon pathosystem. (**A**) GO and KEGG pathway enrichment analyses of SRgreen module genes. (**B**) GO and KEGG pathway enrichment analyses of SRblack module genes.

3.6. Identification of Hub Genes Involved in FW Resistance Through WGCNA

To identify hub genes within the defense-related modules, gene co-expression networks were constructed and analyzed using Cytoscape (Figure 7). Networks were filtered to retain the top 100 edges with the highest weights. Hub genes were prioritized according to the degree, with the top 10 highest degree genes designated as critical nodes and visualized as red vertices in the network (Table S3). Based on the annotated functions in the reference genome, 35 hub genes were selected (Table 1).

The SRblack module encompassed key regulatory and metabolic components, including WRKY TF 42, calcium-transporting ATPase, 1-deoxy-d-xylulose-5-phosphate synthase 2a, and benzyl alcohol O-benzoyltransferase-like. The SRgreen module featured signaling and hormone-related elements: basic-leucine zipper (bZIP) TF family, ethylene-responsive TF 1B-like, putative receptor kinase, expansin-like B1, patatin, and gibberellin receptor GID1B-like.

SRblue contained genes associated with cell wall dynamics and defense signaling, notably cellulose synthase-like protein H1, SA-binding protein 2-like, gibberellin 2-beta-dioxygenase 8, cellulose synthase-like protein B3, xylose isomerase, and detoxification-related proteins. SRbrown is enriched for structural remodeling and stress-responsive factors, including mannan endo-1,4-beta-mannosidase 1-like, receptor-like protein kinase

4, NINJA family transcriptional repressor, and bZIP TF 53. The SRturquoise module comprised defense execution machinery: metacaspase-9, FAD-binding berberine family protein, peroxidase, MADS-box TF, endo-1,4-beta-xylanase A-like, NAC domain-containing protein, thaumatin-like protein, and pectinesterase inhibitor. Conversely, SRmagenta encoded glutathione S-transferase and MADS-box TFs.

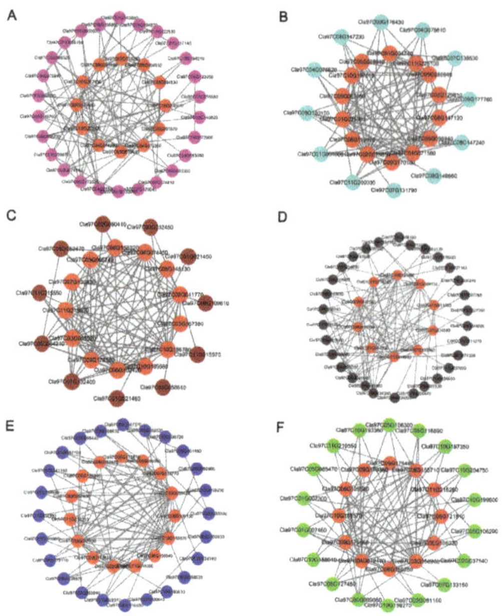

Figure 7. Identification of hub genes in the protein–protein interaction (PPI) networks following inoculation of watermelon with *Fusarium oxysporum* f.sp. *niveum*. (**A**) PPI network of SRmagenta. (**B**) PPI network of SRturquoise. (**C**) PPI network of SRbrown. (**D**) PPI network of SRblack. (**E**) PPI network of SRblue. (**F**) PPI network of SRgreen.

Table 1. Hub genes associated with watermelon resistance to *Fusarium oxysporum* f.sp. *niveum*.

ID	Degree	Description	Module
Cla97C07G140960	15	Cytochrome P450	
Cla97C01G014990	15	WRKY transcription factor 42	
Cla97C02G042360	11	Calcium-transporting ATPase	
Cla97C11G216240	9	Dynamin-related protein 4C	SRblack
Cla97C01G007750	8	Basic leucine zipper 43	
Cla97C02G034580	8	Benzyl alcohol O-benzoyltransferase-like	
Cla97C08G155710	20	Protein AIG2	
Cla97C05G105320	7	Basic-leucine zipper (BZIP) transcription factor family	
Cla97C09G170380	7	Ethylene-responsive transcription factor 1B-like	
Cla97C06G121810	7	Receptor kinase, putative	SRgreen
Cla97C09G175480	7	Expansin-like B1	
Cla97C08G159250	6	Patatin	
Cla97C11G215260	6	Gibberellin receptor GID1B-like	
Cla97C02G035430	28	Cellulose synthase-like protein H1	
Cla97C03G057720	19	Salicylic acid-binding protein 2-like	
Cla97C07G132770	15	Gibberellin 2-beta-dioxygenase 8	
Cla97C02G035420	13	Cellulose synthase-like protein B3	SRblue
Cla97C11G218300	5	Xylose isomerase	
Cla97C09G173070	4	Protein detoxification	

Table 1. *Cont.*

ID	Degree	Description	Module
Cla97C08G145130	8	Mannan endo-1,4-beta-mannosidase 1-like	SRbrown
Cla97C03G057380	8	Receptor-like protein kinase 4	
Cla97C07G130430	8	NINJA family protein AFP2-like	
Cla97C08G158320	8	bZIP transcription factor 53	
Cla97C04G071580	22	Metacaspase-9	SRturquoise
Cla97C01G025380	20	FAD-binding berberine family protein	
Cla97C05G089640	18	Peroxidase	
Cla97C05G099940	14	MADS-box transcription factor	
Cla97C05G098940	9	Endo-1,4-beta-xylanase 1	
Cla97C11G225100	8	NAC domain	
Cla97C10G197610	6	Thaumatin-like protein	
Cla97C06G125610	6	PMEI domain-containing protein	
Cla97C06G110100	6	Pectin methylesterase inhibitor superfamily protein	
Cla97C04G074260	16	Glutathione S-transferase U8-like	SRmagenta
Cla97C07G136250	9	MADS-box protein, putative	
Cla97C04G079260	7	Glutathione S-transferase	

3.7. *qRT-PCR Validation of RNA-Seq Data*

To validate the reliability of RNA-seq data and differential expression profiles, 15 hub genes were selected for analysis by qRT-PCR (Figure 8). These genes represented diverse expression patterns observed in the transcriptomic dataset. Among them, in the SRblack module, *Cla97C01G014990* (WRKY TF 42) and *Cla97C02G042360* (calcium-transporting ATPase) were upregulated at 6 and 8 dpi. By contrast, *Cla97C08G155710* (AIG2), *Cla97C09G170380* (ethylene-responsive TF 1B-like), and *Cla97C06G121810* (receptor kinase, putative) in the SRgreen module were upregulated at 1 dpi.

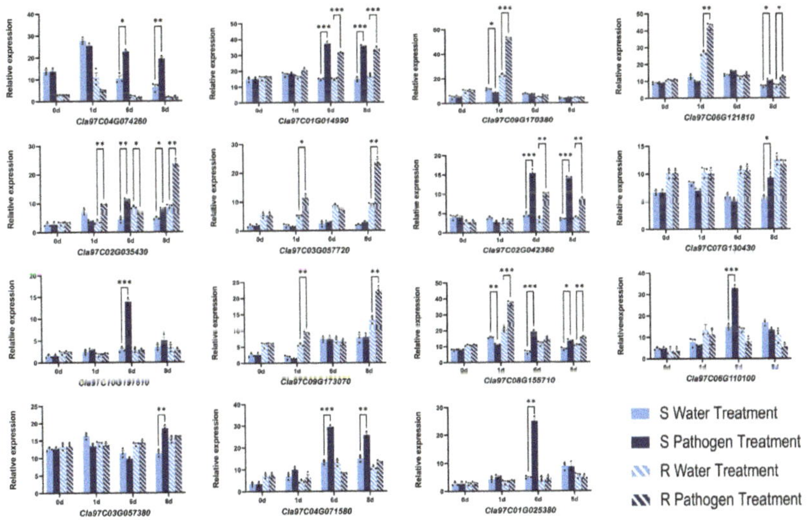

Figure 8. qRT-PCR analysis of 15 hub genes following inoculation of resistant (R) and susceptible (S) watermelon with *Fusarium oxysporum* f.sp. *niveum*. Error bars indicate the mean ± standard deviation, and the three data points in each column indicate three biological replicates. * $p < 0.05$, ** $p < 0.01$, *** $p < 0.001$.

4. Discussion

The interaction between Fon and watermelon involves distinct infection phases, including a biotrophic stage during which the pathogen obtains nutrients from living host cells and a necrotrophic stage characterized by cell death and tissue collapse. Our time-course microscopic observations revealed that Fon-GFP spores initiated germination and root colonization at 1 dpi, followed by hyphal penetration into xylem vessels at 6 dpi, culminating in wilt symptoms at 8 dpi (Figure 1). This progression aligns with the biotrophy-to-necrotrophy transition described in previous studies [4,5,25–27], establishing 1, 6, and 8 dpi as critical infection milestones for dissecting host defense mechanisms.

Transcriptomic profiling has emerged as a cornerstone for dissecting the molecular underpinnings of plant–*F. oxysporum* interactions, particularly in unraveling the dynamic interplay between host defense mechanisms and pathogen virulence strategies [11,28,29]. Differential expression analysis revealed distinct transcriptional strategies between R and S during Fon infection. In S, the majority of DEGs were observed at 6 and 8 dpi, coinciding with xylem colonization and symptom onset. Conversely, the R genotype exhibited high numbers of DEGs at all three critical infection stages (1, 6, and 8 dpi), suggesting an earlier and more dynamic defense response. This early transcriptional activation (1 dpi) in R likely reflects rapid pathogen recognition and signaling, which might suppress Fon development during the biotrophic phase [5]. Venn diagram analysis further highlighted stage-specific defense mechanisms (Figure 2). In R, the minimal overlap of DEGs across time points indicated distinct transcriptional programs tailored to each infection phase [early recognition (1 dpi), xylem defense (6 dpi), and necrosis-associated resistance (8 dpi)]. As the infection process progresses, *F. oxysporum* secretes different proteins to facilitate its progression [30]. This phased response likely enhances the ability to counter Fon progression at multiple stages [31]. By contrast, the S genotype displayed significant overlap in DEGs between 6 and 8 dpi, suggesting a delayed and generalized defensive response. The lack of early (1 dpi) DEGs in S might have permitted Fon to proliferate unchecked during the biotrophic phase, leading to uncontrolled xylem invasion and subsequent necrotrophic expansion [4].

The application of ClusterGVis for transcriptomic clustering and functional enrichment analysis provided a systematic framework to identify biologically meaningful gene sets underlying host–pathogen interactions [32]. Expression clustering analysis revealed distinct transcriptional strategies in the R and S cultivars during Fon infection. Using ClusterGVis, we identified nine clusters (R1, R5, R7, R9, R12, S5, S6, S7, S10) enriched for upregulated defense-related genes in Fon-treated samples. Notably, R8, R11, and R12 exhibited pronounced upregulation on 1 dpi in R, whereas S lacked significant upregulation at this stage, highlighting a critical temporal divergence in defense initiation.

The R8 cluster was enriched for pathways related to valine, leucine, and isoleucine degradation and cellular catabolic processes. Concurrently, upregulated genes in cluster R11 were involved in the generation of precursor metabolites and energy, sulfur compound biosynthetic processes, and carbohydrate catabolic processes. This metabolic rewiring likely provides precursors and energy for defense responses [33]. Cluster R12, activated at 1 dpi, was enriched for transcription factors (e.g., WRKY) and the MAPK signaling pathway. The MAPK cascade is a conserved regulator of PTI and ETI in plants [34]. The early upregulation of MAPK components in R aligned with rapid pathogen recognition and signal amplification, which are critical for deploying localized defense barriers (e.g., callose deposition, ROS bursts) to restrict Fon proliferation [35,36]. Contrasting these findings, S displayed minimal early transcriptional activation (1 dpi), with significant upregulation delayed until 6 and 8 dpi. This delay likely permitted Fon to establish biotrophic colonization unimpeded, leading to unchecked xylem invasion and eventual necrosis.

Integrating our findings, the SRblue module coordinated late-stage defense (8 dpi) against Fon through synergistic cell wall reinforcement and hormonal regulation. Hub genes, such as *Cla97C02G035430* (cellulose synthase-like H1) and *Cla97C02G035420* (cellulose synthase-like B3), likely enhance structural resistance by promoting cellulose deposition in papillae, analogous to barley, in which silencing a cellulose synthase-like gene (*HvCslD2*) increased fungal penetration because of reduced epidermal cellulose content [37]. Additionally, *Cla97C03G057720* (SA-binding protein 2-like) is implicated in SA transport, which is crucial for priming defenses, as demonstrated in *Arabidopsis*, where exogenous SA treatment increased PR1 expression and reduced *F. oxysporum*-induced necrosis [38]. The SRturquoise module established a robust immune barrier during late infection (6 dpi) by coordinating ROS signaling, programmed cell death (PCD), and antimicrobial metabolic pathways. For instance, *Cla97C04G071580* (metacaspase-9) regulates post-mortem cellular clearance, similar to *Arabidopsis AtMC9*, which facilitates the autolysis of xylem vessel contents post-vacuolar rupture, thereby restricting *F. oxysporum* spread by eliminating compromised host tissues [39]. Furthermore, *Cla97C10G197610* (thaumatin-like protein) exhibits direct antifungal activity, akin to *ClTLP27* in watermelon, which inhibits *Fo* mycelial growth through conserved cysteine-rich domains, thereby disrupting fungal cell wall integrity [40]. The SRbrown module, which was positively correlated with late-stage Fon infection (8 dpi), might orchestrate immune responses against Fon through the synergistic regulation of JA signaling. *Cla97C07G130430* (NINJA-family protein AFP2-like) potentially suppresses JA hyperproduction by recruiting TOPLESS/TPR corepressors to JAZ–MYC2 complexes, as demonstrated in *Arabidopsis*, in which NINJA-mediated repression fine-tunes defense–growth trade-offs [41]. The SRmagenta module, which was associated with the responses of the S cultivar to Fon infection, might coordinate oxidative stress management and transcriptional reprogramming to mitigate pathogenic impact. *Cla97C04G074260* (glutathione S-transferase U8-like) and *Cla97C04G079260* (glutathione S-transferase) potentially enhance ROS detoxification, as demonstrated in *Lilium regale*, in which *LrGSTU5* overexpression upregulated *PR1b*, *chitinase*, and antioxidant enzymes (GST, SOD, APX), thereby reducing superoxide anion accumulation and improving *Fo* resistance [42].

The SRgreen module coordinated early transcriptional defense responses in watermelon during *F, oxysporum* (Fo) infection, potentially through synergistic interactions of its constituent genes. *Cla97C08G155710* (protein AIG2) can fine-tune and balance SA and tryptophan-derived secondary metabolite chemical defense systems in response to nonpathogenic and pathogenic microbes [43]. *Cla97C09G170380* (ethylene-responsive TF 1B like) might enhance peroxidase activity and ROS scavenging, as demonstrated by *PdPapERF109* overexpression in *Populus*, which improved *F. oxysporum* resistance by reducing oxidative damage [44]. *Cla97C05G105320* (bZIP TF) likely upregulates antioxidant enzymes (e.g., SOD, CAT) and pathogenesis-related genes, as supported by studies demonstrating that *LrbZIP1* in *Lilium* amplified defense responses against *F. oxysporum* through stress signaling pathways [45]. *Cla97C06G121810* (receptor kinase, putative) could initiate PTI by suppressing *Fo*-induced ROS bursts, paralleling findings in which *VfLRR–RLK1* in *Vernicia* enhanced root architecture and *Fo* resistance via ROS detoxification [46]. *Cla97C09G175480* (expansin-like B1) might regulate structural defenses by modifying cell wall dynamics, as sugarcane *Expansin* genes were found to limit fungal entry by modulating stomatal apertures [47]. *Cla97C08G159250* (patatin) might balance localized cell death and defense through oxylipin metabolism, akin to *PLP2* in *Arabidopsis*, which controls lipid signaling to restrict pathogen spread [48]. *Cla97C11G215260* (gibberellin receptor GID1B-like) potentially prioritizes defense over growth by stabilizing DELLA proteins, consistent with SA-mediated GID1 degradation mechanisms that reallocate resources to promote *Fo* resistance [49].

The SRblack module, which was positively correlated with both the S and R genotypes during late-stage Fon infection, likely orchestrates multifaceted defense mechanisms through coordinated hormonal signaling, calcium dynamics, and metabolic reprogramming. *Cla97C07G140960* (cytochrome P450) might enhance SA/ET-dependent defenses, as demonstrated in cucumber, in which *CYP82D47* overexpression increased *PR1*, *PR2*, and *EIN3* expression, improving resistance to *F. oxysporum* and powdery mildew [50]. *Cla97C02G042360* (calcium-transporting ATPase) potentially regulates PCD by modulating intracellular calcium fluxes, akin to *NbCA1* in *Nicotiana*, which delays pathogen-induced cell death to balance immunity and tissue integrity [51]. *Cla97C11G216240* (dynamin-related protein 4C) might fine-tune callose deposition by regulating the vesicle trafficking of callose synthases, as observed in *Arabidopsis*, in which *DRP2B* restricts excessive callose accumulation to optimize pathogen containment [52]. *Cla97C02G034580* (benzyl alcohol O-benzoyltransferase-like) could contribute to SA biosynthesis via peroxisomal β-oxidation, similar to *HSR201* in tobacco, which cooperates with *NtCNL* and *NtCHD* to produce SA precursors critical for defense priming [53].

The SRgreen and SRblack modules represent distinct sources of early- and late-phase resistance, respectively. In the SRblack module, *Cla97C01G014990* (WRKY TF 42) and *Cla97C02G042360* (calcium-transporting ATPase) exhibited high node degree values, suggesting that sustained resistance is mediated through WRKY-dependent transcriptional regulation and calcium signaling pathways. By contrast, the SRgreen module harbored the high-degree genes *Cla97C08G155710* (AIG2), *Cla97C09G170380* (ethylene-responsive TF 1B-like), and *Cla97C06G121810* (receptor kinase, putative), which likely recognize Fon via receptor kinases and rapidly initiate defense responses through AIG2-mediated signaling and ethylene-responsive hormonal pathways, enabling early-stage pathogen resistance.

5. Conclusions

In this study, we identified 1, 6, and 8 dpi as key infection stages of Fon. Through the analysis of differential expression profiles, WGCNA, and functional enrichment, we revealed six gene modules closely associated with disease resistance, which contribute to resistance at different infection stages. Furthermore, PPI network analysis combined with RT-qPCR identified five hub genes in the SRgreen and SRblack modules, including *Cla97C01G014990*, *Cla97C02G042360*, *Cla97C08G155710*, *Cla97C09G170380*, and *Cla97C06G121810*. The identified candidate genes provide rapid resistance during early infection and long-term resistance during later infection stages. This study screens for hub genes related to disease resistance in the Fon–watermelon interaction at different stages, offering novel disease-resistant genes for breeding watermelon resistant to Fusarium wilt.

Supplementary Materials: The following supporting information can be downloaded at https://www.mdpi.com/article/10.3390/horticulturae11060625/s1, Figure S1. Confocal laser scanning microscopic analysis of Fon-GFP colonization in S (FW-susceptible watermelon variety W1-1). Between 4 and 5 dpi, Fon-GFP hyphae penetrated the intercellular spaces along the edges of root cells and single frames; Figure S2: Analysis of network topology for various soft-thresholding powers. The soft threshold was determined by the scale-free network index; R^2 was set to 0. 8, and the best soft threshold was 17; Figure S3: GO and KEGG pathway enrichment analysis of SRyellow module genes; Figure S4: GO and KEGG pathway enrichment analysis of SRblue module genes; Figure S5: GO and KEGG pathway enrichment analysis of SRbrown module genes; Figure S6: GO and KEGG pathway enrichment analysis of SRturquoise module genes; Figure S7: GO and KEGG pathway enrichment analysis of SRred module genes; Figure S8: GO and KEGG pathway enrichment analysis of SRmagenta module genes. Table S1: qRT-PCR-specific primers; Table S2: Quality control (QC) analysis of transcriptome data; Table S3: The top 10 hub genes (highest degree) in the modules.

Author Contributions: Conceptualization, P.G.; data curation, C.Z.; formal analysis, X.F.; funding acquisition, P.G.; investigation, C.Z.; methodology, X.F.; project administration, X.F.; resources, Z.L., S.L. and Z.S.; software, C.Z.; supervision, F.L.; validation, J.Z. and X.W.; visualization, X.F.; writing—original draft preparation, C.Z.; writing—review and editing, P.G. All authors have read and agreed to the published version of the manuscript.

Funding: This research was funded by the Science and Technology-based Small and Medium-sized Enterprises Innovation Capacity Enhancement Project of Shandong Province (grant number 2023TSGC0804), the Weifang City Science and Technology-based Small and Medium-sized Enterprises Innovation Capacity Enhancement Project (grant number 2023TS1072), the National Nature Science Foundation of China (grant number U21A20229), the "Young Leading Talents" support program of Northeast Agricultural University (grant number NEAU2023QNLJ-005), the major development program of Heilongjiang Province (grant number GA23B007), and the Natural Science Foundation of Heilongjiang Province (grant number ZL2024C011).

Data Availability Statement: The data presented in this study are available upon request from the corresponding author.

Acknowledgments: We appreciate all the people who have collaborated on this project.

Conflicts of Interest: Authors Zhao Liu, Shusen Liu and Zhengfeng Song were employed by the Shandong Engineering Research Center for Watermelon and Melon Breeding Co., Ltd.

References

1. Food and Agriculture Organization of the United Nations. Available online: https://www.fao.org (accessed on 9 May 2025).
2. Martyn, R.D. Resistance to Races 0, 1, and 2 of *Fusarium* Wilt of Watermelon in *Citrullus* sp. PI-296341-FR. *Hortscience* **1991**, *26*, 429. [CrossRef]
3. Yvonne Couteaudier, C.A. Survival and Inoculum Potential of Conidia and Chlamydospores of *Fusarium oxysporum* f. sp. Lini in Soil. *Dev. Agric. Manag. For. Ecol.* **1991**, *23*, 551–556. [CrossRef]
4. Lu, G.; Guo, S.; Zhang, H.; Geng, L.; Martyn, R.D.; Xu, Y. Colonization of *Fusarium* Wilt-Resistant and Susceptible Watermelon Roots by a Green-Fluorescent-Protein-Tagged Isolate of *Fusarium oxysporum* f. sp. Niveum. *J. Phytopathol.* **2014**, *162*, 228–237. [CrossRef]
5. Zhang, M.; Xu, J.H.; Liu, G.; Yao, X.F.; Li, P.F.; Yang, X.P. Characterization of the Watermelon Seedling Infection Process by *Fusarium oxysporum* f. sp. Niveum. *Plant Pathol.* **2015**, *64*, 1076–1084. [CrossRef]
6. Ren, Y.; Jiao, D.; Gong, G.; Zhang, H.; Guo, S.; Zhang, J.; Xu, Y. Genetic Analysis and Chromosome Mapping of Resistance to *Fusarium oxysporum* f. sp. Niveum (FON) Race 1 and Race 2 in Watermelon (*Citrullus lanatus* L.). *Mol. Breed.* **2015**, *35*, 100. [CrossRef]
7. Branham, S.E.; Levi, A.; Wechter, W.P. QTL Mapping Identifies Novel Source of Resistance to *Fusarium* Wilt Race 1 in *Citrullus amarus*. *Plant Dis.* **2019**, *103*, 984–989. [CrossRef]
8. Xu, X.P.; Chen, C.H.; Fan, B.F.; Chen, Z.X. Physical and Functional Interactions between Pathogen-Induced *Arabidopsis* WRKY18, WRKY40, and WRKY60 Transcription Factors. *Plant Cell* **2006**, *18*, 1310–1326. [CrossRef]
9. Agarwal, P.; Reddy, M.P.; Chikara, J. WRKY: Its Structure, Evolutionary Relationship, DNA-Binding Selectivity, Role in Stress Tolerance and Development of Plants. *Mol. Biol. Rep.* **2011**, *38*, 3883–3896. [CrossRef]
10. Wang, L.; Guo, D.; Zhao, G.; Wang, J.; Zhang, S.; Wang, C.; Guo, X. Group IIc WRKY Transcription Factors Regulate Cotton Resistance to *Fusarium oxysporum* by Promoting GhMKK2-Mediated Flavonoid Biosynthesis. *New Phytol.* **2022**, *236*, 249–265. [CrossRef]
11. Diao, J.; Wang, J.; Zhang, P.; Hao, X.; Wang, Y.; Liang, L.; Zhang, Y.; Ma, W.; Ma, L. Transcriptome Analysis Reveals the Important Role of WRKY28 in *Fusarium oxysporum* Resistance. *Front. Plant Sci.* **2021**, *12*, 720679. [CrossRef]
12. Zhang, G.; Chen, M.; Li, L.; Xu, Z.; Chen, X.; Guo, J.; Ma, Y. Overexpression of the Soybean *GmERF3* Gene, an AP2/ERF Type Transcription Factor for Increased Tolerances to Salt, Drought, and Diseases in Transgenic Tobacco. *J. Exp. Bot.* **2009**, *60*, 3781–3796. [CrossRef] [PubMed]

13. Gao, H.; Jiang, L.; Du, B.; Ning, B.; Ding, X.; Zhang, C.; Song, B.; Liu, S.; Zhao, M.; Zhao, Y.; et al. GmMKK4-Activated GmMPK6 Stimulates GmERF113 to Trigger Resistance to *Phytophthora sojae* in Soybean. *Plant J.* **2022**, *111*, 473–495. [CrossRef] [PubMed]
14. Guo, W.; Jin, L.; Miao, Y.; He, X.; Hu, Q.; Guo, K.; Zhu, L.; Zhang, X. An Ethylene Response-Related Factor, *GbERF1-like*, from *Gossypium barbadense* Improves Resistance to *Verticillium dahliae* via Activating Lignin Synthesis. *Plant Mol. Biol.* **2016**, *91*, 305–318. [CrossRef] [PubMed]
15. Huang, L.-J.; Zhang, J.; Lin, Z.; Yu, P.; Lu, M.; Li, N. The AP2/ERF Transcription Factor ORA59 Regulates Ethylene-Induced Phytoalexin Synthesis through Modulation of an Acyltransferase Gene Expression. *J. Cell. Physiol.* **2024**, *239*, e30935. [CrossRef]
16. Langfelder, P.; Horvath, S. WGCNA: An R Package for Weighted Correlation Network Analysis. *BMC Bioinform.* **2008**, *9*, 559. [CrossRef]
17. Xu, X.; Lu, X.; Tang, Z.; Zhang, X.; Lei, F.; Hou, L.; Li, M. Combined Analysis of Carotenoid Metabolites and the Transcriptome to Reveal the Molecular Mechanism Underlying Fruit Colouration in Zucchini (*Cucurbita pepo* L.). *Food Chem. Mol. Sci.* **2021**, *2*, 100021. [CrossRef]
18. Anees, M.; Gao, L.; Umer, M.J.; Yuan, P.; Zhu, H.; Lu, X.; He, N.; Gong, C.; Kaseb, M.O.; Zhao, S.; et al. Identification of Key Gene Networks Associated With Cell Wall Components Leading to Flesh Firmness in Watermelon. *Front. Plant Sci.* **2021**, *12*, 630243. [CrossRef]
19. Shen, Q.; Wu, X.; Tao, Y.; Yan, G.; Wang, X.; Cao, S.; Wang, C.; He, W. Mining Candidate Genes Related to Heavy Metals in Mature Melon (*Cucumis melo* L.) Peel and Pulp Using WGCNA. *Genes* **2022**, *13*, 1767. [CrossRef]
20. Mullins, E.D.; Chen, X.; Romaine, P.; Raina, R.; Geiser, D.M.; Kang, S. *Agrobacterium*-Mediated Transformation of *Fusarium oxysporum*: An Efficient Tool for Insertional Mutagenesis and Gene Transfer. *Phytopathology* **2001**, *91*, 173–180. [CrossRef]
21. Chen, S. Ultrafast One-Pass FASTQ Data Preprocessing, Quality Control, and Deduplication Using Fastp. *iMeta* **2023**, *2*, e107. [CrossRef]
22. Kim, D.; Paggi, J.M.; Park, C.; Bennett, C.; Salzberg, S.L. Graph-Based Genome Alignment and Genotyping with HISAT2 and HISAT-Genotype. *Nat. Biotechnol.* **2019**, *37*, 907. [CrossRef] [PubMed]
23. Pertea, M.; Pertea, G.M.; Antonescu, C.M.; Chang, T.-C.; Mendell, J.T.; Salzberg, S.L. StringTie Enables Improved Reconstruction of a Transcriptome from RNA-Seq Reads. *Nat. Biotechnol.* **2015**, *33*, 290. [CrossRef] [PubMed]
24. Lopes, C.T.; Franz, M.; Kazi, F.; Donaldson, S.L.; Morris, Q.; Bader, G.D. Cytoscape Web: An Interactive Web-Based Network Browser. *Bioinformatics* **2010**, *26*, 2347–2348. [CrossRef] [PubMed]
25. Li, C.; Shao, J.; Wang, Y.; Li, W.; Guo, D.; Yan, B.; Xia, Y.; Peng, M. Analysis of Banana Transcriptome and Global Gene Expression Profiles in Banana Roots in Response to Infection by Race 1 and Tropical Race 4 of *Fusarium oxysporum* f. sp. Cubense. *BMC Genom.* **2013**, *14*, 851. [CrossRef]
26. Wang, X.; Zhu, Z.; Zhang, C.; Song, J.; Wang, Q.; Luan, F.; Gao, P. Evaluation of Differential miRNA Expression between *Fusarium* Wilt-Resistant and -Susceptible Watermelon Varieties. *Sci. Hortic.* **2024**, *332*, 113189. [CrossRef]
27. Sun, Y.; Yang, H.; Li, J. Transcriptome Analysis Reveals the Response Mechanism of *Frl*-Mediated Resistance to *Fusarium oxysporum* f. sp. Radicis-Lycopersici (FORL) Infection in Tomato. *Int. J. Mol. Sci.* **2022**, *23*, 7078. [CrossRef]
28. Kaushal, M.; Mahuku, G.; Swennen, R. Comparative Transcriptome and Expression Profiling of Resistant and Susceptible Banana Cultivars during Infection by *Fusarium oxysporum*. *Int. J. Mol. Sci.* **2021**, *22*, 3002. [CrossRef]
29. Lu, G.; Guo, S.; Zhang, H.; Geng, L.; Song, F.; Fei, Z.; Xu, Y. Transcriptional Profiling of Watermelon during Its Incompatible Interaction with *Fusarium oxysporum* f. sp. Niveum. *Eur. J. Plant Pathol.* **2011**, *131*, 585–601. [CrossRef]
30. Thatcher, L.F.; Williams, A.H.; Garg, G.; Buck, S.-A.G.; Singh, K.B. Transcriptome Analysis of the Fungal Pathogen *Fusarium oxysporum* f. sp. *Medicaginis* during Colonisation of Resistant and Susceptible *Medicago truncatula* Hosts Identifies Differential Pathogenicity Profiles and Novel Candidate Effectors. *BMC Genom.* **2016**, *17*, 860. [CrossRef]
31. Luo, J.; Zhang, A.; Tan, K.; Yang, S.; Ma, X.; Bai, X.; Hou, Y.; Bai, J. Study on the Interaction Mechanism between *Crocus Sativus* and *Fusarium oxysporum* Based on Dual RNA-Seq. *Plant Cell Rep.* **2023**, *42*, 91–106. [CrossRef]
32. Kuang, W.; Huang, J.; Yang, Y.; Liao, Y.; Zhou, Z.; Liu, Q.; Wu, H. Identification of Markers Correlating with Mitochondrial Function in Myocardial Infarction by Bioinformatics. *PLoS ONE* **2024**, *19*, e0316463. [CrossRef] [PubMed]
33. Manzo, D.; Ferriello, F.; Puopolo, G.; Zoina, A.; D'Esposito, D.; Tardella, L.; Ferrarini, A.; Ercolano, M.R. *Fusarium oxysporum* f. sp. *Radicis-Lycopersici* Induces Distinct Transcriptome Reprogramming in Resistant and Susceptible Isogenic Tomato Lines. *BMC Plant Biol.* **2016**, *16*, 53. [CrossRef] [PubMed]
34. Wang, D.; Wei, L.; Liu, T.; Ma, J.; Huang, K.; Guo, H.; Huang, Y.; Zhang, L.; Zhao, J.; Tsuda, K.; et al. Suppression of ETI by PTI Priming to Balance Plant Growth and Defense through an MPK3/MPK6-WRKYs-PP2Cs Module. *Mol. Plant* **2023**, *16*, 903–918. [CrossRef] [PubMed]
35. Pitzschke, A.; Djamei, A.; Bitton, F.; Hirt, H. A Major Role of the MEKK1-MKK1/2-MPK4 Pathway in ROS Signalling. *Mol. Plant* **2009**, *2*, 120–137. [CrossRef]

36. Shubchynskyy, V.; Boniecka, J.; Schweighofer, A.; Simulis, J.; Kvederaviciute, K.; Stumpe, M.; Mauch, F.; Balazadeh, S.; Mueller-Roeber, B.; Boutrot, F.; et al. Protein Phosphatase AP2C1 Negatively Regulates Basal Resistance and Defense Responses to *Pseudomonas syringae*. *J. Exp. Bot.* **2017**, *68*, 1169–1183. [CrossRef]
37. Douchkov, D.; Lueck, S.; Hensel, G.; Kumlehn, J.; Rajaraman, J.; Johrde, A.; Doblin, M.S.; Beahan, C.T.; Kopischke, M.; Fuchs, R.; et al. The Barley (*Hordeum vulgare*) Cellulose Synthase-like D2 Gene (*HvCslD2*) Mediates Penetration Resistance to Host-Adapted and Nonhost Isolates of the Powdery Mildew Fungus. *New Phytol.* **2016**, *212*, 421–433. [CrossRef]
38. Edgar, C.I.; McGrath, K.C.; Dombrecht, B.; Manners, J.M.; Maclean, D.C.; Schenk, P.M.; Kazan, K. Salicylic Acid Mediates Resistance to the Vascular Wilt Pathogen *Fusarium oxysporum* in the Model Host *Arabidopsis thaliana*. *Austral. Plant Pathol.* **2006**, *35*, 581–591. [CrossRef]
39. Bollhoner, B.; Zhang, B.; Stael, S.; Denance, N.; Overmyer, K.; Goffner, D.; Van Breusegem, F.; Tuominen, H. *Post mortem* Function of AtMC9 in Xylem Vessel Elements. *New Phytol.* **2013**, *200*, 498–510. [CrossRef]
40. Zhang, M.; Xu, J.; Liu, G.; Yang, X. Antifungal Properties of a Thaumatin-like Protein from Watermelon. *Acta Physiol. Plant.* **2018**, *40*, 186. [CrossRef]
41. Pauwels, L.; Barbero, G.F.; Geerinck, J.; Tilleman, S.; Grunewald, W.; Perez, A.C.; Chico, J.M.; Vanden Bossche, R.; Sewell, J.; Gil, E.; et al. NINJA Connects the Co-Repressor TOPLESS to Jasmonate Signalling. *Nature* **2010**, *464*, 788–791. [CrossRef]
42. Han, Q.; Chen, R.; Yang, Y.; Cui, X.; Ge, F.; Chen, C.; Liu, D. A Glutathione *S*-Transferase Gene from *Lilium regale* Wilson Confers Transgenic Tobacco Resistance to *Fusarium oxysporum*. *Sci. Hortic.* **2016**, *198*, 370–378. [CrossRef]
43. Wang, Z.; Yang, L.; Jander, G.; Bhawal, R.; Zhang, S.; Liu, Z.; Oakley, A.; Hua, J. AIG2A and AIG2B Limit the Activation of Salicylic Acid-Regulated Defenses by Tryptophan-Derived Secondary Metabolism in Arabidopsis. *Plant Cell* **2022**, *34*, 4641–4660. [CrossRef] [PubMed]
44. Diao, J.; Li, M.; Zhang, P.; Zong, C.; Ma, W.; Ma, L. Overexpression of the *PdpapERF109* Gene Enhances Resistance of *Populus davidiana* x *P. alba* Var. *Pyramidalis* to *Fusarium oxysporum* Infection. *J. For. Res.* **2022**, *33*, 1925–1937. [CrossRef]
45. Zhang, N.; Liu, D.; Zheng, W.; He, H.; Ji, B.; Han, Q.; Ge, F.; Chen, C. A bZIP Transcription Factor, *LrbZIP1*, Is Involved in *Lilium regale* Wilson Defense Responses against *Fusarium oxysporum* f. sp. Lilii. *Genes Genom.* **2014**, *36*, 789–798. [CrossRef]
46. Yue, Z.-L.; Tian, Z.-J.; Zhang, J.-W.; Zhang, S.-W.; Li, Y.-D.; Wu, Z.-M. Overexpression of Lectin Receptor-Like Kinase 1 in Tomato Confers Resistance to *Fusarium oxysporum* f. sp. Radicis-Lycopersici. *Front. Plant Sci.* **2022**, *13*, 836269. [CrossRef]
47. Liu, Z.; Yu, Z.; Li, X.; Cheng, Q.; Li, R. Two Sugarcane *Expansin* Protein-Coding Genes Contribute to Stomatal Aperture Associated with Structural Resistance to Sugarcane Smut. *J. Fungi* **2024**, *10*, 631. [CrossRef]
48. La Camera, S.; Balague, C.; Goebel, C.; Geoffroy, P.; Legrand, M.; Feussner, I.; Roby, D.; Heitz, T. The *Arabidopsis* Patatin-Like Protein 2 (PLP2) Plays an Essential Role in Cell Death Execution and Differentially Affects Biosynthesis of Oxylipins and Resistance to Pathogens. *Mol. Plant-Microbe Interact.* **2009**, *22*, 469–481. [CrossRef]
49. Yu, X.; Cui, X.; Wu, C.; Shi, S.; Yan, S. Salicylic Acid Inhibits Gibberellin Signaling through Receptor Interactions. *Mol. Plant* **2022**, *15*, 1759–1771. [CrossRef]
50. Wang, H.; Li, P.; Wang, Y.; Chi, C.; Jin, X.; Ding, G. Overexpression of Cucumber *CYP82D47* Enhances Resistance to Powdery Mildew and *Fusarium oxysporum* f. sp. Cucumerinum. *Funct. Integr. Genom.* **2024**, *24*, 14. [CrossRef]
51. Zhu, X.; Caplan, J.; Mamillapalli, P.; Czymmek, K.; Dinesh-Kumar, S.P. Function of Endoplasmic Reticulum Calcium ATPase in Innate Immunity-Mediated Programmed Cell Death. *Embo J.* **2010**, *29*, 1007–1018. [CrossRef]
52. Leslie, M.E.; Rogers, S.W.; Heese, A. Increased Callose Deposition in Plants Lacking *DYNAMIN-RELATED PROTEIN 2B* Is Dependent upon *POWDERY MILDEW RESISTANT 4*. *Plant Signal. Behav.* **2016**, *11*, e1244594. [CrossRef]
53. Kotera, Y.; Komori, H.; Tasaki, K.; Takagi, K.; Imano, S.; Katou, S. The Peroxisomal β-Oxidative Pathway and Benzyl Alcohol O-Benzoyltransferase HSR201 Cooperatively Contribute to the Biosynthesis of Salicylic Acid. *Plant Cell Physiol.* **2023**, *64*, 758–770. [CrossRef]

horticulturae

MDPI

Article

Transcriptomic Analysis of Resistant and Susceptible Eggplant Genotypes (*Solanum melongena* L.) Provides Insights into *Phytophthora capsici* Infection Defense Mechanisms

Hesbon Ochieng Obel [1,2], Xiaohui Zhou [1,2], Songyu Liu [1,2], Yan Yang [1,2], Jun Liu [1,2] and Yong Zhuang [1,2,*]

[1] Institute of Vegetable Crops, Jiangsu Academy of Agricultural Sciences, Nanjing 210014, China; 20230024@jaas.ac.cn (H.O.O.); 20100029@jaas.ac.cn (X.Z.); 20180052@jaas.ac.cn (S.L.); 20170007@jaas.ac.cn (Y.Y.); 20110024@jaas.ac.cn (J.L.)

[2] Jiangsu Key Laboratory for Horticultural Crop Genetic Improvement, Nanjing 210014, China

* Correspondence: 19960033@jaas.ac.cn; Tel.: +86-2584391752

Abstract: Phytophthora fruit rot caused by *Phytophthora capsici* is a devastating disease in many solanaceous vegetables, resulting in tremendous yield and economic losses. However, the underlying resistance or susceptibility to *P. capsici* in eggplant remains obscure. In this study, the transcriptomic analysis was performed between the resistant (G42) and susceptible (EP28) eggplant genotypes at 0, 1, 3 and 5 days post-inoculation (dpi). Taking 0 dpi as the control, a total of 4111, 7496 and 7325 DEGs were expressed at 1, 3 and 5 dpi, respectively, in G42 and 5316, 12675 and 12048 DEGs were identified at 1, 3 and 5 dpi, respectively, in EP28. *P. capsici* infection induced substantial transcriptional changes in the inoculated fruits. The analysis of the Kyoto Encyclopedia of Genes and Genomes (KEGG) identified defense-related pathways including 'plant-pathogen interactions', 'mitogen-activated protein kinase (MAPK)' and 'hormone biosynthesis and signal transduction'. The hormone-related genes encompassing ethylene, abscisic acid, auxins and gibberellins showed differential expression between G42 and EP28 eggplant genotypes, signifying their important roles in plant disease resistance. *P. capsici* infection induced the expression of major transcription factors such as MYB, NAC/NAM, bHLH, WRK, HSF, HD-ZIPAP2/ERF and Mad-box. qRT-PCR validation of the selected genes corroborates with RNA-seq, depicting the precision and consistency of the transcriptomic data. According to qRT-PCR and RNA-seq analyses, the expression of the pathogenesis-related gene transcriptional activator, *SmPTI6* (*Smechr0603020*), is upregulated in G42 and downregulated in EP28. This differential expression suggests a potential role in the resistance to *P. capsici*. Functional analysis via a virus-induced gene silencing (VIGS) system found that silencing *SmPTI6* in G42 enhanced infection by *P. capsici*, indicating that *SmPTI6* performs a critical role in response to pathogen attack. The comprehensive results obtained in this study provide a valuable resource for understanding the molecular mechanisms underlying eggplant resistance to *P. capsici* and for establishing breeding resistant eggplant genotypes to *P. capsici*.

Keywords: eggplant; disease-resistant; *Phytophthora capsici*; transcriptomics; virus-induced gene silencing

1. Introduction

Eggplant (*Solanum melongena* L.) is an important solanaceous vegetable crop cultivated worldwide [1]. China alone produces over 60 percent of the total eggplant produced globally annually [2]. There is a continuous increase each year in planting hectarage and yield, and various commercial gains are experiencing steady growth. Eggplant fruit contains multiple nutrients, such as vitamins, phenolics and antioxidants, which benefit human health [3]. However, its potential yield is affected by environmental stresses and biotic factors, including Phytophthora fruit rot [4,5]. Phytophthora fruit rot is a disease caused by the pathogenic oomycete *Phytophthora capsici*, which is found worldwide [6,7]. *P. capsici* can occur and infect eggplant at any growth and developmental period and affect different parts of the plant, causing damping off at the seedling stage, root rot, stem rot, collar rot, fruit rot and blight on the foliage [8,9]. *P. capsici* fruit rot is the most common and devastating disease in eggplant under suitable environmental conditions, comprising temperatures of 25–28 °C and 90% relative humidity [10,11]. By studying plant responses to pathogen infection, it is possible to gain insights into the mechanisms by which plants interact with pathogens, thereby elucidating the functional dynamics of plant immune systems, which are crucial for advancing disease-resistance breeding efforts and variety improvement. The immune response in plants is typically divided into two fundamental stages [12]. The primary layer of defense in plants is characterized by the immune response elicited by pathogen-associated molecular patterns (PAMPs), known as PAMP-triggered immunity (PTI). This phase consists of a range of immune responses initiated by pattern recognition receptors (PRRs) located on the surface of plant cells, which can detect PAMPs. In turn, pathogens utilize various tactics to undermine PTI, including releasing harmful effectors. To counteract these strategies, plants have developed nucleotide-binding leucine-rich repeat receptor (NLR) proteins that detect these effectors and inhibit their function, thereby bolstering their resistance. Consequently, this aspect of immunity is referred to as effector-triggered immunity (ETI) [12,13]. In plants, proteins associated with disease resistance operate as key immune receptors, tasked with detecting pathogens and activating strong defense responses [13]. For example, pathogenesis-related (PR) proteins (such as chitinases and glucanases) degrade microbial cell walls [14]. Nucleotide-binding leucine-rich repeat (NLR) proteins act as intracellular immune receptors [15]. Antimicrobial peptides (AMPs) like defensins disrupt pathogen membranes [16].

Significant yield losses occur in field production where conditions are suitable for *P. capsici* to thrive [17]. Some cultivated germplasms of eggplant are susceptible to Phytophthora fruit rot, highlighting the necessity for future breeding programs to develop eggplant varieties that can resist this affliction. While there are cultivars susceptible to *P. capsici*, there are also resistant cultivars, which need to be studied to understand their defense mechanisms. Analyzing the complete gene expression can yield important information regarding the molecular basis of the interactions between host plants and Phytophthora pathogens [18,19]. This approach provides valuable insights into the complex mechanisms that control resistance and basal defense responses, elucidating the complex interactions between these two entities. Previous studies on transcriptomics have yielded significant genes that have allowed the elucidation of plant–pathogen interactions. For example, transcriptomic responses of two wild tomato germplasms to *Phytophthora parasitica* infection revealed that certain genes associated with protease inhibitors, chitinases, defensins and PR-1 exhibited significant upregulation in the resistant germplasms [20]. In cucumber, the genetic factors that contribute to age-related resistance against *P. capsici* were investigated. The findings indicated that a significant number of genes associated with flavonoid and terpenoid biosynthesis were upregulated in the peels of resistant fruit at 16 days post-pollination [21]. Tobacco leaves infected with *P. parasitica* identified genes related to jasmonic acid and ethylene signaling pathways, as well as receptor-like kinases, pathogenesis-related (PR) genes, and various transcription factors [22]. In pepper, transcriptomic analysis revealed a greater number of genes linked to the phenylpropanoid biosynthesis pathway in the resistant *Piper flaviflorum* when compared to the susceptible *Piper nigrum* cv. Reyin-1 during *P. capsici* infections [23]. The comparative transcriptome and metabolome analysis of *P. flaviflorum* and *P. nigrum* in response to *P. capsici* infection demonstrated that the altered genes showed significant enrichment in pathways related to plant–pathogen interactions, phytohormone signal transduction, and secondary metabolic pathways, notably phenylpropanoid biosynthesis [24]. Therefore, it is crucial to analyze the responses of both resistant and susceptible plant lines to *P. capsici* infection to understand the mechanisms of defense.

In this study, two eggplant genotypes, G42 resistant and EP28 susceptible to *P. capsici*, were used. Phenotypic and transcriptomic analyses were conducted on the fruits of G42 and EP28 in response to *P. capsici* at 0, 1, 3 and 5 days post-inoculation (dpi). Our results unveiled the key Kyoto Encyclopedia of Genes and Genomes (KEGG) pathway that could contribute to the resistance mechanism to *P. capsici* infections. Identifying pathways of differentially expressed genes, expression patterns and functional analysis via VIGs of the *SmPTI6* gene involved in defense response will contribute to a deeper understanding of the molecular mechanism underlying resistance to *P. capsici* in eggplant and contribute to eggplant genotype improvement.

2. Materials and Methods

2.1. Plant Materials

The experimental plant materials comprised two contrasting eggplant genotypes involving G42, which is a resistant, and EP28, a type susceptible to *P. capsici*. The resistant genotype (G42) maintains green fruit coloration at all developmental stages, while the susceptible genotype (EP28) develops dark purple pigmentation. These eggplant germplasm materials were obtained from the Vegetable Institute, Jiangsu Academy of Agricultural Sciences, Nanjing, China. The plants were cultivated in a greenhouse during the spring–summer period, with air temperatures of 28–33 °C during the day and 14–20 °C at night. The fruits were collected at the commercial stage (30 days after anthesis). Eggplant fruits without skin blemishes were collected for in vitro inoculation with *P. capsici*.

The G42 and EP28 fruits were inoculated following the previous protocols [25–27]. Briefly, the collected fruits were subjected to surface disinfection using a 10% bleach solution for 5 min, followed by rinsing with distilled water. The fruits were arranged in plastic boxes and then inoculated with a mycelium V8 agar plug (6 mm in diameter) containing actively growing colonies of *P. capsici* on the fruit surface derived. The V8 Agar plug was placed was upside down on the eggplant fruit surface to ensure the part with active *P. capsici* mycelium was in direct contact with the fruit. The boxes were covered with polythene film and additionally with the box cover. The eggplant fruits were maintained in the humidity chambers under continuous light at room temperature (25 °C) (Figure S1). Evaluations of the fruits were conducted at 0, 1, 3 and 5 days post-inoculation (dpi). The evaluation time, 0–5 days, was based on the preliminary disease progression time of *P. capsici* and a previous study protocol [25]. The samples were obtained from the inoculated fruit parts for sequencing at each time point. Samples were referred herein as G42_0, G42_1, G42_3 and G42_5 for the resistant line, and EP28_0, EP28_1, EP28_3 and EP28_5 for the susceptible line. The sampled fruits were wrapped in tinfoil, immediately snap-frozen in liquid nitrogen, and stored at −80 °C. Three biological replications, three fruits for each replicate, were sampled and pooled for RNA extraction.

2.2. Total RNA Isolation and Illumina Sequencing

RNA was isolated from a total of 24 samples, with three replicates for each treatment, using Trizol reagent (Invitrogen, Waltham, MA, USA). RNA quantification was carried out using the NanoDrop 2000 (Thermo Fisher Scientific, Wilmington, DE, USA) and evaluated with the RNA Assay Kit (Nano 6000) on the Agilent Bioanalyzer 2100 System (Agilent Technologies, Santa Clara, CA, USA). A sequencing library was then constructed using the NEBNext Ultra-TM-RNA Library Preparation Kit (NEB, Ipswich, MA, USA). After completing quality checks and inspections, sequencing was performed on an Illumina sequencing platform (Beijing Baimaike Biotechnology Co., Ltd., Beijing, China).

2.3. Mapping of RNA Sequencing Reads and Differentially Expressed Genes

To obtain high-quality reads, raw fastQ data files were subjected to a cross-check for quality, with a threshold set at $\leq$20. Inadequate raw reads were trimmed using Trimmomatic-0.30 [28]. The resultant filtered reads were aligned to the eggplant genome (http://www.eggplant-hq.cn/, accessed on 7 July 2024) utilizing HISAT2 software (V2.2.1), and gene expression levels were quantified using FPKM (fragments per kb per million fragments) [29]. DEGs were identified through the $DESeq_2$ method [30]. The *p*-value threshold was determined via the false discovery rate (FDR) control approach, with significance criteria for DEGs established at an FDR < 0.01 and an absolute $\log_2$ ratio $\geq$ 1. Gene ontology (GO) and KEGG pathway enrichment analyses were conducted using the topGO method and KOBAS 2.0 (http://www.biostars.org/p/200126, accessed on 7 July 2024), respectively.

2.4. qRT-PCR Validation of DEGs Associated with P. capsici Infection

To validate the findings from the RNA-seq analysis, quantitative real-time PCR (qRT-PCR) was performed on 12 DEGs potentially associated with resistance to *P. capsici*. First-strand cDNA was synthesized from total RNA extracted from fruit samples of G42 and EP28 at four distinct time points (0, 1, 3 and 5 days post-inoculation), utilizing the ReverTran Ace qPCR RT kit (Toyobo, Shanghai Biotech Co., Ltd., Shanghai, China). Specific primers were designed with the aid of Primer 3 (https://www.primer3plus.com/, accessed on 23 August 2024. The actin gene of eggplant served as an internal control. The qRT-PCR was executed on the Roche LightCycler 960 II system employing a SYBR Green-based PCR assay. Each sample underwent three independent biological replicates, with three technical replicates for each biological sample included in the qRT-PCR analysis. The reaction mixture comprised 10 µL of diluted cDNAs (20 ng/reaction of cDNA) and 0.4 µL of each primer, adjusted to a final volume of 20 µL. The PCR protocol involved an initial denaturation at 95 °C for 30 s, followed by 40 cycles of 95 °C for 5 s and 60 °C for 30 s and PCR elongation at 72 °C for 10 min. Relative gene expression levels were determined using the $2^{\Delta\Delta Ct}$ method [31]. The primer information of the genes used for qRT-PCR is listed in Table S1.

2.5. Virus-Induced Gene Silencing (VIGS)

The VIGS procedure was executed following the previously outlined protocols [32,33], albeit with some alterations. In essence, a 300 bp cDNA fragment of *SmPTI6* was inserted into the pTRV2 vector and transformed into the *Agrobacterium* strain GV3101. Following this, the *Agrobacterium* containing the pTRV2–*SmPTI6* construct was mixed in equal volumes with *Agrobacterium* that contained pTRV1. pTRV1 produces viral RNA1, which contains genes for viral replication and movement, while pTRV2 encodes viral RNA2, which carries the gene of interest (or a fragment of it) that will be silenced. The resulting bacterial mixture was resuspended in an infiltration buffer consisting of 10 mM $MgCl_2$, 200 mM acetosyringone and 10 mM MES (2-(N-morpholino) ethanesulfonic acid) at a pH of 5.6, and it was subsequently employed for VIGS. Eggplant fruits of G42 were collected for VIGS analysis 25 days post-anthesis. After the fruits were surface sterilized with 75% ethanol, they were injected with the previously described agrobacterial mixture. The fruits were then placed in a dark chamber at 21 °C with 90% humidity for 24 h. This was followed by 5 days of photoperiod (16/8 h), during which the temperature and humidity were kept stable. A control experiment was executed using the pTRV2 empty vector instead of pTRV2–*SmPTI6*. *P. capsici* inoculation was performed by applying a mycelium plug to the fruit's surface 7 days post-VIGS silencing, and the inoculated fruits were incubated at 25 °C under dark conditions with 90% humidity. Using qRT-PCR, the expression of *SmPTI6* was measured at 0, 3 and 5 dpi.

2.6. Data Analysis

The quantitative experimental data were conducted in triplicates to ensure both accuracy and accountability. One-way ANOVA was used for data comparison, and the least significant difference (LSD) test at $p < 0.05$ and 0.01 was used for multiple comparisons.

Figures were drawn using Excel, Origin 8.0 and SRplot (http://www.bioinformatics.com.cn/en, accessed on 15 September 2024), and heatmaps were generated using TBtools-II (Ver 2.326) [34].

3. Results

3.1. Phenotypic Observations of P. capsici-Infected Eggplant Plants

We initially conducted disease assays to confirm the response of the two eggplant breeding lines (G42 and EP28) to *P. capsici* infection. The progression of the infection was observed over a period of five days in both genotypes. Symptoms indicative of *P. capsici* infection first manifested on the fruit surface at three days post-inoculation (dpi), presenting as light brown lesions around the inoculation site, which expanded as the disease advanced (Figure 1). By the fifth dpi, significant tissue colonization was evident on the fruit surface of the EP28 and showed extensive girdling due to *P. capsici* infection. Conversely, the resistant line (G42) exhibited only small, localized lesions with a reddish hue at the inoculation point (Figure 1). By5 dpi, it was clear that the G42 line had effectively limited fungal proliferation on the fruit surface (EP28) (Figure 1).

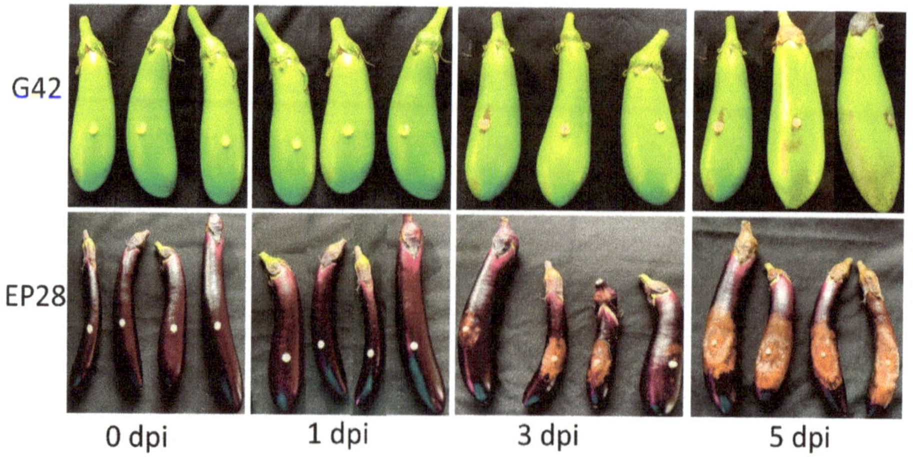

Figure 1. Disease symptoms in the genotypes G42 and EP28 at different times post-inoculation by *Phytophthora capsici*. In G42, no symptoms were observed, while in EP28, brown necrosis enlarged forming water-soaked lesions.

3.2. *Transcriptome Analysis of Eggplant Fruits Inoculated with P. capsici*

Twenty-four cDNA libraries were constructed in our study, generating more than 976 million raw reads (approximately 146 GB of data), with each RNA-seq library producing an average of 40 million raw reads (Table 1). The clean reads exhibited a proportion that spanned from 98.05% to 98.72% and a total of 144 GB clean bases. The average data size for each sample was up to 6.00 GB. The quality scores (Q30 were above 97%, and GC contents ranged from 42 to 47% (Table 1). The total reads that were mapped to the reference genome of the eggplant exhibited a percentage between 60.6% and 96.43%; uniquely mapped reads to the reference genome ranged from 58.33% to 93.28% and 2.34% to 5.17% and were aligned to multiple locations (Table S2). The proportion of reads aligned to exons ranged from 89.37% to 92.03% (Table S3). Utilizing the FPKM values, the computed correlation coefficients for each sample, derived from the three biological replicates of samples were determined. The correlation coefficients for all comparisons exceeded 0.94, demonstrating a high level of reproducibility (Figure S2). These findings suggest that the RNA sequencing data is highly dependable and appropriate for subsequent analyses.

Table 1. Summary of RNA sequencing from resistant (G42) and susceptible (EP28) eggplant genotypes inoculated with *P. capsici*.

Sample	Rep	Raw Data		Valid Data		Valid Ratio	Q20%	Q30%	GC%
		Read	Base (GB)	Read	Base (GB)				
G42_0	1	41,051,052	6.16	40,436,514	6.07	98.5	99.98	97.85	42
	2	40,999,126	6.15	40,340,254	6.05	98.39	99.98	97.85	42
	3	39,280,282	5.89	38,690,924	5.80	98.5	99.99	97.97	42
G42_1	1	39,052,176	5.86	38,398,142	5.76	98.33	99.98	97.42	42.5
	2	38,639,304	5.80	37,957,974	5.69	98.24	99.98	97.57	42.5
	3	36,404,544	5.46	35,741,318	5.36	98.18	99.98	97.42	42.5
G42_3	1	37,814,104	5.67	37,132,650	5.57	98.2	99.99	97.91	42.5
	2	39,094,046	5.86	38,449,424	5.77	98.35	99.98	97.78	42.5
	3	39,001,266	5.85	38,336,968	5.75	98.3	99.98	97.8	42.5
G42_5	1	40,394,562	6.06	39,606,844	5.94	98.05	99.98	97.53	43
	2	43,639,846	6.55	42,898,002	6.43	98.3	99.98	97.66	43
	3	40,953,524	6.14	40,237,462	6.04	98.25	99.98	97.64	43
EP28_0	1	41,697,682	6.25	41,002,602	6.15	98.33	99.98	97.85	42.5
	2	41,118,736	6.17	40,498,054	6.07	98.49	99.98	97.4	42.5
	3	37,345,316	5.60	36,784,428	5.52	98.5	99.98	97.29	42.5
EP28_1	1	43,245,042	6.49	42,493,762	6.37	98.26	99.98	97.74	43
	2	42,040,238	6.31	41,303,792	6.20	98.25	99.98	97.74	43
	3	41,146,058	6.17	40,476,920	6.07	98.37	99.98	97.76	43
EP28_3	1	39,956,998	5.99	39,330,384	5.90	98.43	99.98	97.63	44
	2	40,799,306	6.12	40,093,008	6.01	98.27	99.98	97.59	45
	3	41,843,078	6.28	41,097,570	6.16	98.22	99.98	97.65	44
EP28_5	1	43,525,606	6.53	42,887,580	6.43	98.53	99.97	97.84	47
	2	41,692,496	6.25	41,159,808	6.17	98.72	99.99	98.18	47
	3	45,737,468	6.86	45,111,530	6.77	98.63	99.98	98.14	46

3.3. *Identification of DEGs in Response to Phytophthora capsici Infection*

The gene expression profiles of healthy eggplant fruits of G42 and EP28 were utilized as the control group. Any gene that exhibited a two-fold or higher change in expression levels in infected fruits compared to the control group ($p < 0.001$) was identified as a differentially expressed gene (DEG). For G42 inoculated with *P. capsici*, there were a total of 4111 DEGs (1761 upregulated and 2350 downregulated) at G42_1, and the number of DEGs increased to 7496 (2701 upregulated and 4795 downregulated) at G42_3, then slightly decreased to 7325 (2924 upregulated and 4401 downregulated) at G42_5. In comparison to EP28, there were a total of 5316 DEGs (2241 upregulated and 3075 downregulated) at EP28_1, and the number of DEGs increased to 12,675 (3866 upregulated and 8809 downregulated) at EP28_3, then slightly decreased to 12,048 (3659 upregulated and 8389 downregulated) at EP28_5 (Figure 2A). Throughout all time points, a fewer number of genes displayed altered expression levels in resistant genotypes compared to susceptible genotypes. The results imply the stress caused by the infection is more severe in the susceptible genotype compared to the resistant genotype, and that this level of stress is accompanied by a greater change in gene expression. The results indicated that 3 dpi corresponded to the peak number of DEGs observed in the two genotypes. This suggests that the infection by *P. capsici* at this specific time point is pivotal for examining the defense response mechanisms.

The Venn diagram representing the DEGs following the inoculation of *P. capsici* at distinct point times effectively conveyed the unique and co-expressed DEGs among the samples (Figure 2B). A cumulative count of 665 upregulated genes and 1,116 downregulated genes was identified in the (G42) samples at the three time intervals relative to the control. A total of 1420 genes exhibited upregulation, while 2272 genes demonstrated downregulation, with co-expression observed at all time points in EP28. Intersection analysis of DEGs across various time points revealed a greater number of DEGs at 3 days post-inoculation (dpi) compared to 5 dpi in both genotypes. This finding aids in comprehending the persistent DEGs resulting from pathogen infection over time, thereby facilitating a deeper analysis of the dynamic responses associated with plant–pathogen interactions in eggplant. To illustrate the evolving trends among the sample groups, all DEGs were categorized into nine clusters utilizing the K-means method in MeV software (Figure 3). Genes classified within clusters 1, 5 and 7 exhibited the highest expression levels in the G42 group across various time points. Conversely, genes from clusters 2, 3, 4, 6 and 9 demonstrated peak expression levels in the EP28 group at designated time intervals. The expression profiles of genes in cluster 4, comprising 2304 DEGs, remained relatively stable before and after the inoculation with *P. capsici* for both genotypes, with the exception of EP28_5. In cluster 5, which includes 2452 genes, most exhibited stability throughout all time points in both genotypes, with the notable exception of G42-1. Cluster 7 maintained stability across all time points in the EP28 genotype, while variations were observed in G42. Cluster 8, consisting of 2388 DEGs, showed stability in both cultivars, except at G42-0. Lastly, cluster 9, containing 1689 DEGs, remained stable across all time points, although fluctuations were noted among different time points, including both upregulated and downregulated genes in EP28.

Because exposure to pathogens is known to involve the induction of defense-response genes [25], emphasis was placed on identifying DEGs that are expressed to a higher degree in the G42 and determining their expression in the EP28. A list of the top 25 of the highest-scoring genes with increased expression in G42 and the relative expression, in EP28, at the three time points in comparison to 0 dpi are shown in Table S4. Among the top DEGS are transcription factor MYB92, protein PELKI, Cysteine-rich receptor-like protein kinase, ethylene-responsive transcription factor_ERF098, probable glutathione S-transferase parA and homeobox-leucine zipper protein, among others. Moreover, in this study, two E3 ubiquitin ligases were also found to be continuously expressed differentially in the two cultivars after pathogen infection.

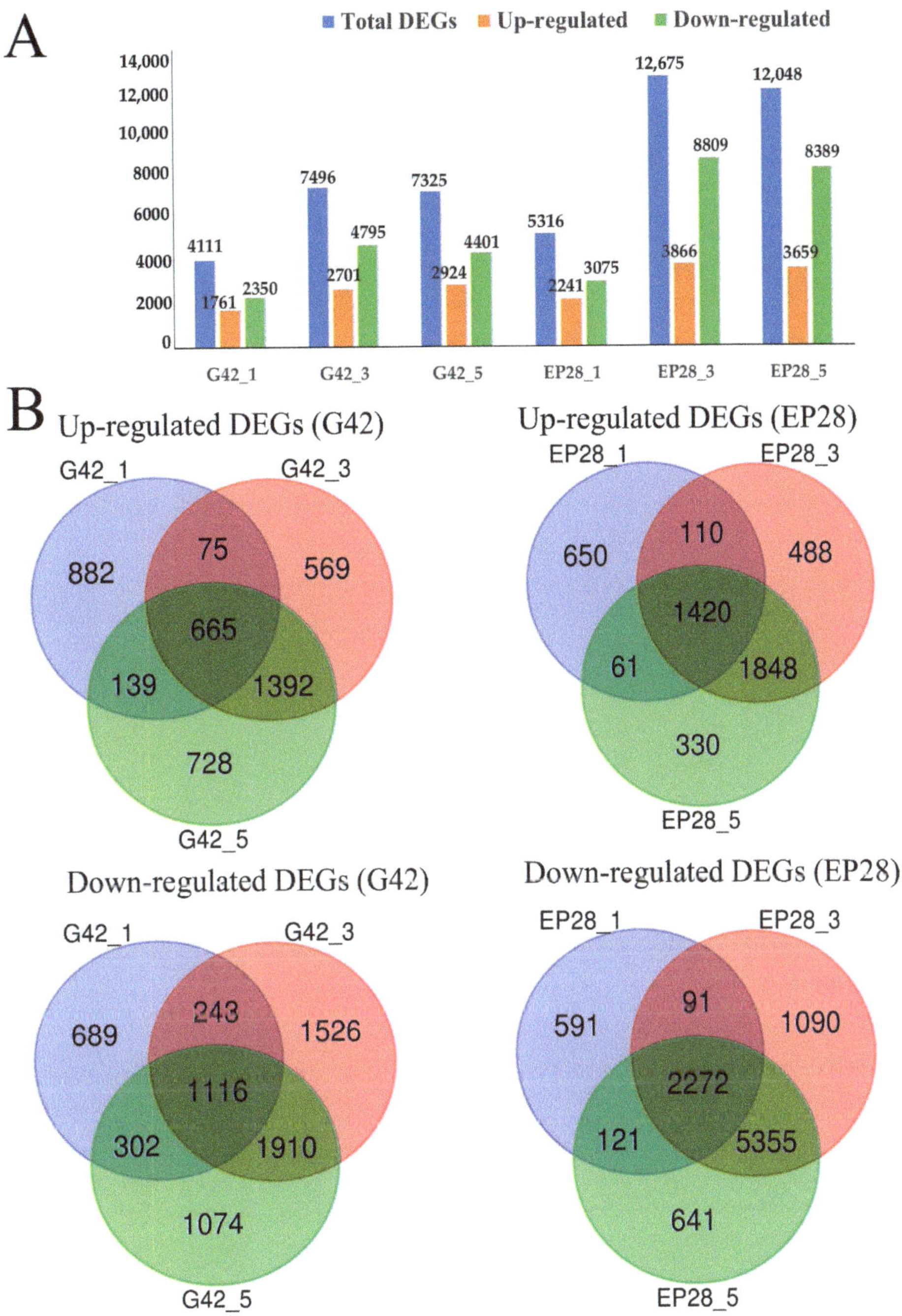

Figure 2. DEGs involved in *P. capsici* inoculated eggplant materials G42 and EP28 at different post-inoculation times. (**A**) Distribution of identified DEGS in all time points. (**B**). Venn diagrams of DEGs in pairwise comparison with 0 dpi as the control.

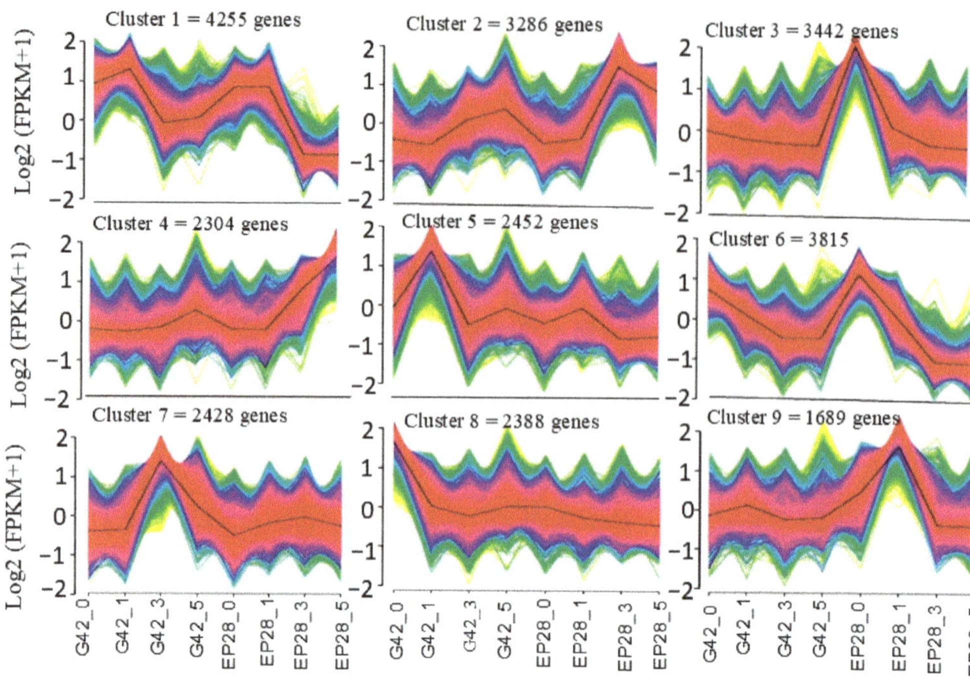

Figure 3. Cluster analysis of DEGs between the G42 and EP28 genotypes at time points based on the K-means method or hierarchical clustering. G42_0 and EP28_0 are control samples. In all panels, grey lines indicate the expression levels of individual genes, while blue and green lines indicate a consensus of all the DEGs within a specific cluster.

3.4. Gene Ontology Enrichment Analysis of Differentially Expressed Genes

The Venn diagram established that 1190 DEGs were specifically expressed in G42, while 4371 DEGs were expressed in EP28. A total of 13,649 DEGs were constitutively co-expressed between G42 and EP28 at all time points (Figure 4A). The results demonstrate that the inoculation of *P. capsici* induces significant changes in gene expression in eggplant fruits, indicating considerable differences in the responses of the resistant and susceptible eggplant lines utilized in this investigation. The DEGs at various time points for G42 and EP28 were analyzed using GOseq, with a corrected *p*-value cut-off of 10 ($-\log_{10}$ *p*-value). This analysis was conducted to discern the differences in gene ontology (GO) term enrichment between the two genotypes post-inoculation. In total, 29 GO terms were identified for G42, while EP28 revealed 34 GO terms (Figure 4B), which were categorized into biological processes (BP), cellular components (CC) and molecular functions (MF).

In BP terms, expression of genes related to 'response to chitin', 'glycolytic process' and 'response to cold' were remarkably more significant in G42, while the expression of 'intracellular signal transductions', 'microtubule-based movement', 'RNA splicing' and 'maturation of LSU-Rrna' were prominent in EP28. 'Photosynthesis' and 'defense response to bacterium processes' were present in both genotypes. In CC terms, 'chloroplast', 'chloroplast thylakoid membrane' and 'chloroplast envelope', were present in both genotypes. EP28 expressed GO terms related to 'kinesin complex', 'Golgi body', 'mitochondria' and 'peroxisomes'. In MF terms, 'protein kinase activity' and 'ATP binding' were remarkably more significant in G42 than in EP28. In contrast, 'molecular function', 'nucleotide binding', 'peptidyl-protyl-cis trans isomerase', and 'guanosine triphosphate (GTP)' were more significant in G42 than in EP28.

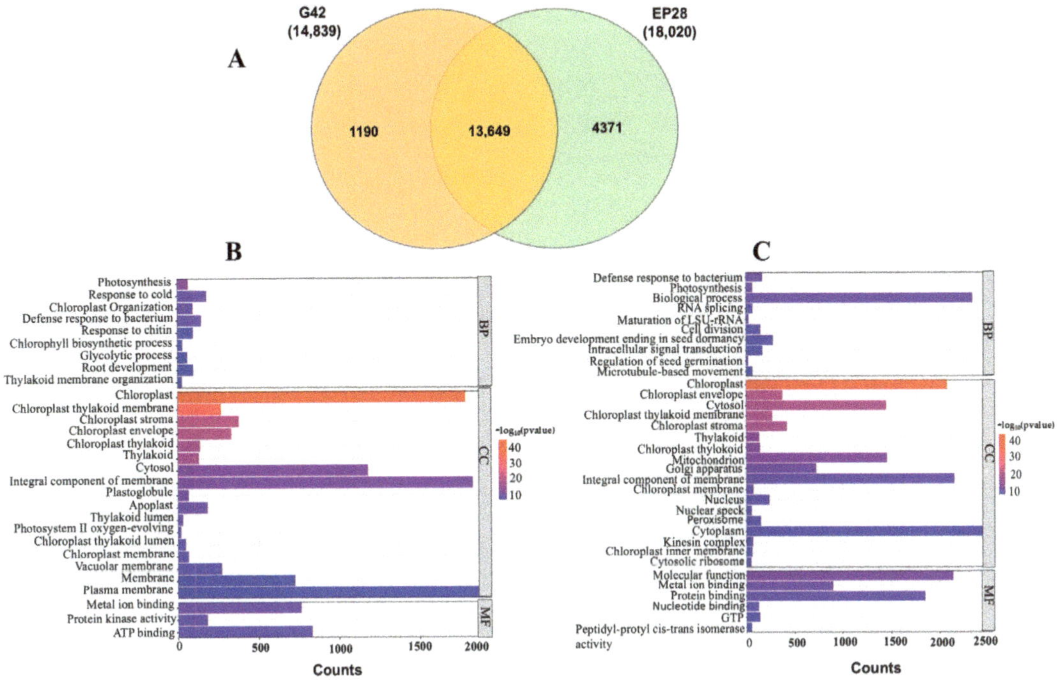

Figure 4. Gene distribution and functional analysis of specifically regulated genes after inoculation with *P. capsici*. (**A**) Venn diagram of DEGS in G42 and EP28. The figures in parentheses are the total DEGs in each genotype, while shared DEGs are shown within the Venn diagram. (**B**) Enriched GO terms associated with specifically regulated genes in G42 after *P. capsici* inoculation. (**C**) Enriched GO terms associated with specifically regulated genes in EP28 after *P. capsici* inoculation. The GO terms are categorized into biological processes (BP), cellular components (CC) and molecular functions (MF). The abscissa represents the number of genes associated with each process.

3.5. *Kyoto Encyclopedia of Genes and Genomes Enrichment (KEGG) Analysis of DEGS*

We performed a KEGG pathway enrichment analysis of the DEGs to investigate the response and resistance mechanism of eggplant in response to *P. capsici* inoculation. The top 20 enriched KEGG pathways were used to predict the biochemical pathways of eggplant in reactions induced by *P. capsici* infections. In G42, we found that the DEGs that appeared at G42_0 vs. _1 were significantly enriched in the photosynthetic antenna proteins, MAPK signaling pathway–plant, porphyrin metabolism, alpha-linolenic acid metabolism, plant–pathogen interactions and glycerophospholipid metabolism (Figure 5A). At 3 dpi (Figure 5B), the eggplant exhibited DEGs significantly enriched in plant hormone signal transduction, starch and sucrose metabolism, MAPK signaling pathway–plant and phenylpropanoid metabolism, among others. At the late stage, 5 dpi after *P. capsici* inoculation, most DEGs were significantly enriched in phenylpropanoid biosynthesis, phenylalanine biosynthesis, starch and sucrose metabolism, plant hormone signal transduction, ABC transporters, isoquinoline and glutathione metabolism, among others (Figure 5C). Interestingly, most of these pathways occur in all time points with varying intensity, thus signifying their importance in disease infection responses.

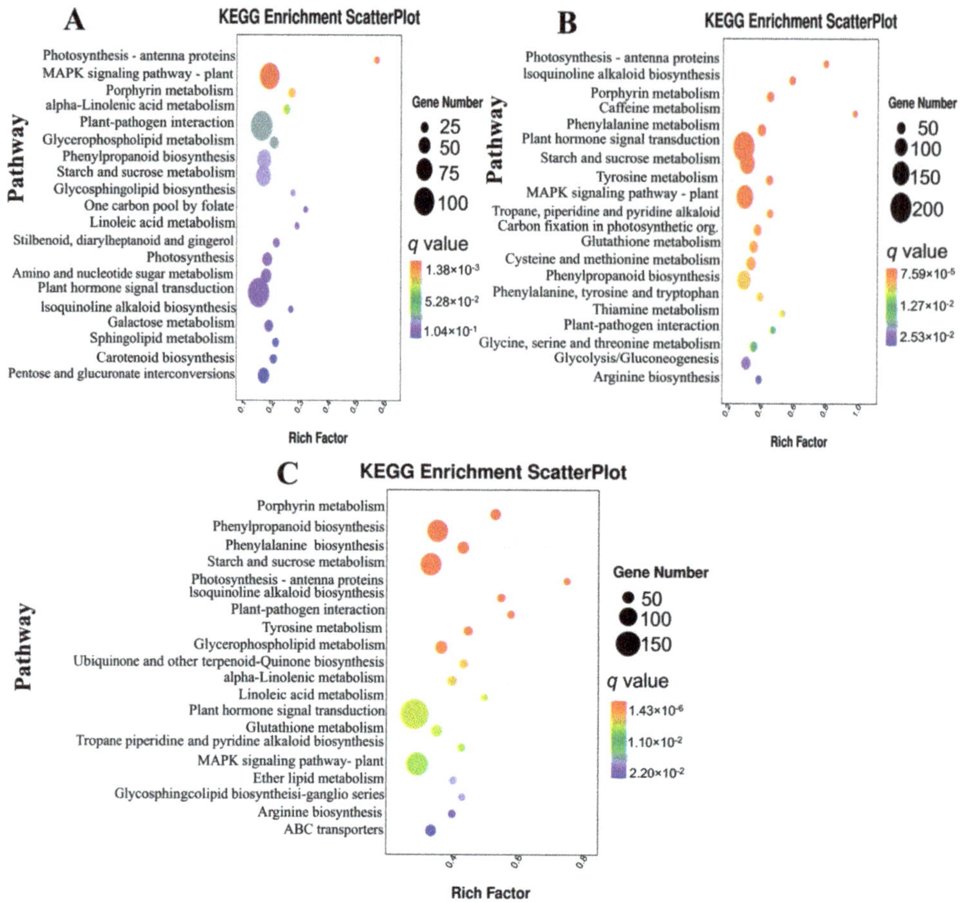

Figure 5. Scatterplot of KEGG pathways enrichment analysis for DEGs in genotype G42. (**A**) G42_0 vs. G42-1. (**B**) G42_0 vs. G42-3. (**C**) G42_0 vs. _5. The rich factor represents the proportion of differentially expressed genes (DEGs) that are associated with a specific pathway term relative to the total number of genes annotated within that pathway. An elevated rich factor signifies a higher level of enrichment. The *q* value, which is the adjusted *p* value, varies between 0 and 1, where a smaller *q* value reflects a higher degree of significance. The diameter of the circles corresponds to the quantity of genes involved. The twenty most-enriched pathway terms from the KEGG database are presented.

In the susceptible genotype (EP28), the results showed that the DEGs during the early stage of *P. capsici* inoculation (1 dpi) were significantly enriched in plant hormone signaling transductions, phenylpropanoid biosynthesis, starch and sucrose metabolism, MAPK signaling pathway and plant–pathogen interaction, among others (Figure 6A). At 3 dpi, DEGs were significantly enriched in glutathione metabolism, Arachidonic acid metabolism, glycosphingolipid metabolism, starch, sucrose metabolism and plant hormone signal transductions, among others (Figure 6B). At 5 dpi in EP28, the DEGs were enriched in glycosphingolipid biosynthesis, photosynthesis antenna proteins, porphyrin metabolism, plant hormone signal transduction, sphingolipid metabolism, tyrosine metabolism, starch and sucrose metabolism, among others (Figure 6C). In both genotypes, the rich factor of the metabolic pathways varied considerably and was highest at late stage 5 dpi.

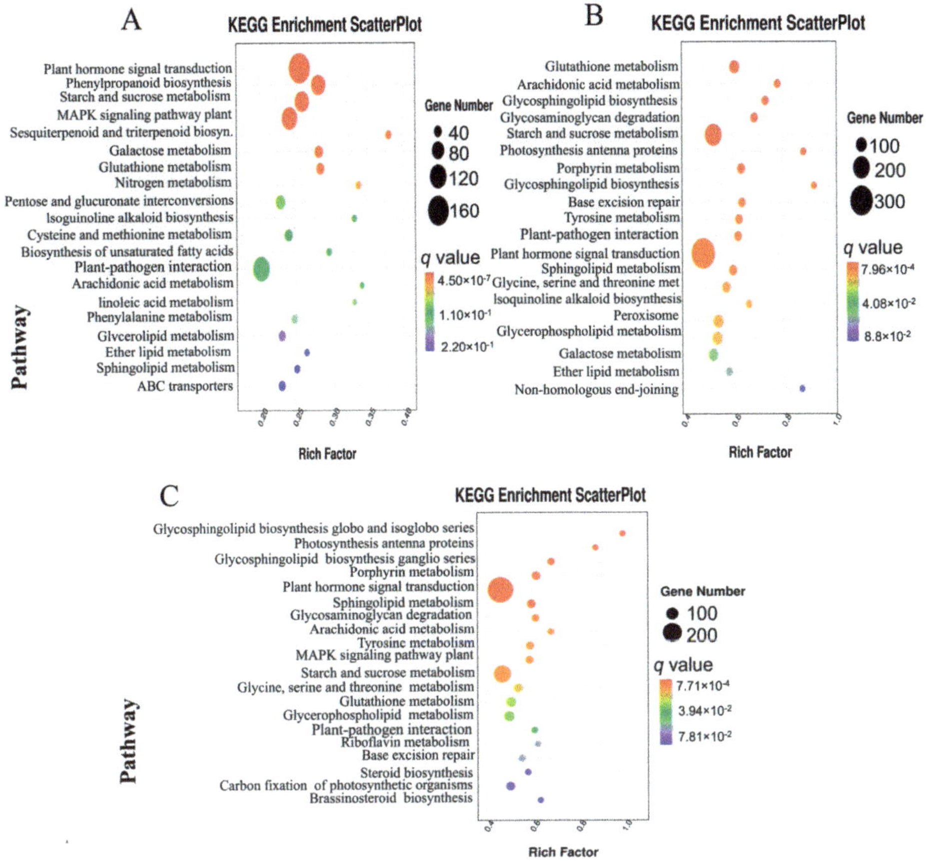

Figure 6. Scatterplot of KEGG pathways enrichment analysis for DEGs in genotype EP28. (**A**) EP28_0 vs. _1 G42-1. (**B**) EP28_0 vs. _3. (**C**) EP28_0 vs. _5. The rich factor represents the proportion of DEGs that are associated with a specific pathway term relative to the total number of genes annotated within that pathway. An elevated rich factor signifies a higher level of enrichment. The q value, which is the adjusted p value, varies between 0 and 1, where a smaller q value reflects a higher degree of significance. The diameter of the circles corresponds to the quantity of genes involved. The twenty most-enriched pathway terms from the KEGG database are presented.

3.6. Identification of DEGs Involved in Defense Response to P. capsici Infections

The integration of KEGG pathway and GO enrichment analyses revealed a varied array of pathways associated with defense mechanisms following eggplant *P. capsici* infections in G42 and susceptible EP28 eggplant genotypes. The plant–pathogen interaction pathway plays a significant role in plant disease resistance. In this study, the plant–pathogen interaction showed the expression of DEGs related to autophagy-related proteins, 18 calcium-binding proteins, 16 calcium-dependent protein kinase (CDPK) 3 calmodulin-like proteins, 3 CBS-domain-containing proteins, 10 cyclic nucleotide-gated ion channels (CNGC), 5 disease-resistance proteins, 8 ethylene-responsive transcription factor, 9 late blight-resistance protein homolog, 2 LRR-receptor like serine, 10 WRKY transcription factor, 4 respiratory burst oxidase homolog (Rboh) and pathogenesis-related genes transcriptional activator (PTI6) (Figure 7; Table S5). In the plant–pathogen interaction pathway, one disease-resistance protein (*Smechr0402283)* was upregulated in EP28 compared to G42 (Table S5). In this pathway, some calcium-binding proteins such as CML15 (*Smechr0103759*), CML49 (*Smechr1002778*), *CML19* (*Smechr1201626*) and *CML21* (*Smechr0400107*) were upregulated in EP28 at 3 and 5 dpi compared to G42. In G42, the calcium-binding proteins *CML16* (*Smechr1102012*), *CML48* (*Smechr1201717*), *CML50* (*Smechr1102453*) and *CML41* (*Smechr300575*) were upregulated, while *CML18* (*Smechr0200838*) and *CML19* (*Smechr1201626*) were downregulated at 3 and 5 dpi (Figure 7).

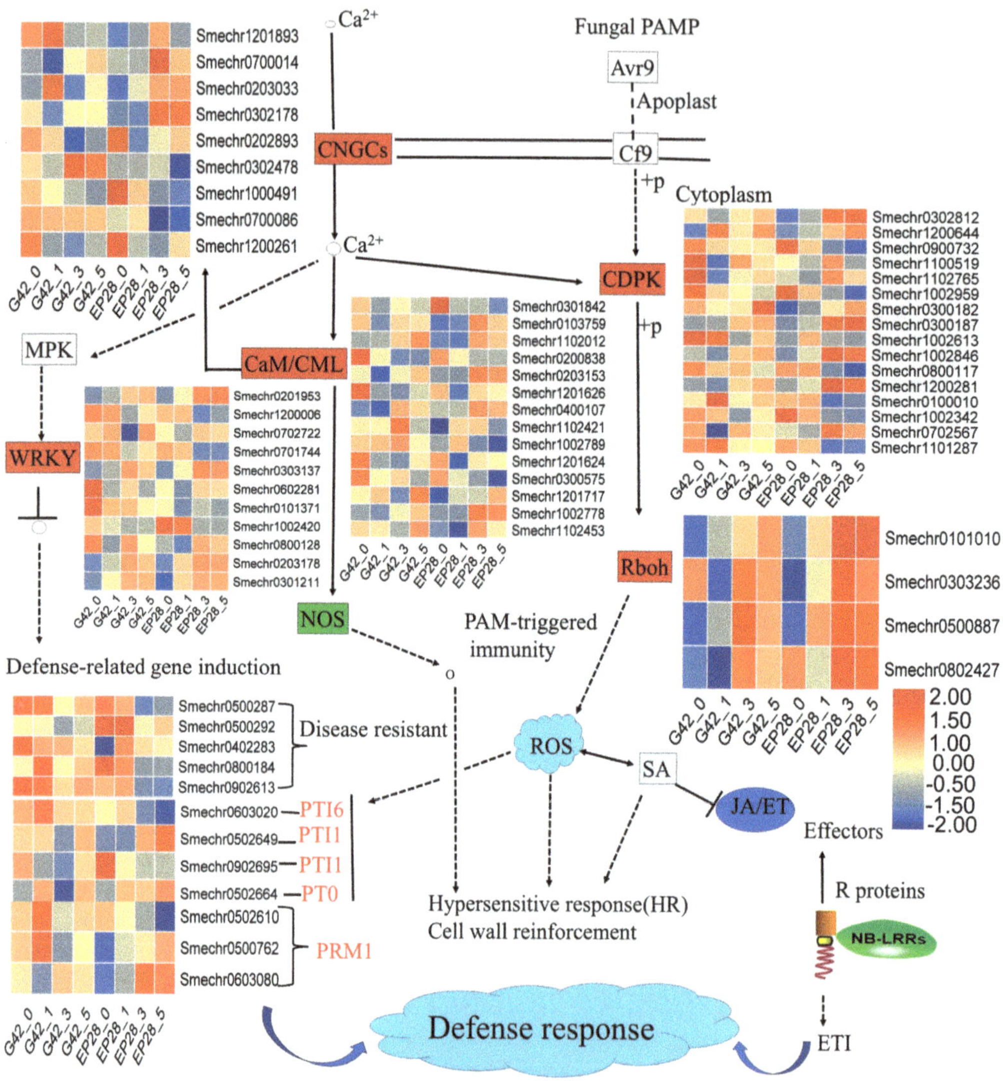

Figure 7. Differentially expressed genes (DEGs) linked to the plant–pathogen interaction pathway were analyzed. A threshold for $\log_2$ FPKM values was established at ±2.0, with red indicating upregulation and blue indicating downregulation. The gene IDs are displayed in horizontal rows, while the vertical columns correspond to various time points.

The MAPK signaling pathway plays a crucial role in plant signal transduction networks, through a tertiary kinase manner involving MAPKKK, MAPKK and MAPK signaling cascades. Through transcription factors such as WRKY or directly, MAPK signaling cascade induces pathogenesis-related protein 1 (PR1) and protein kinase-C-related kinases (PRK). The analysis of DEGs identified 38 DEGs associated with the MAPK signaling pathway (Table S5). The expression profiles of differential genes involved in the MAPK signaling pathway post-pathogen infection were assessed (Figure 8A; Table S5), highlighting significant upregulation of *Smechr0101057* in G42 and downregulation in EP28. Eight MAPK genes (*Smechr0303698*, *Smechr0100663*, *Smechr0800498*, *Smechr1000008*, *Smechr0101444*, *Smechr0500071*, *Smechr0302543* and *Smechr0200932*) were distinctively upregulated in EP28 at 3 and 5 dpi, hence depicting their significant role in eggplant susceptibility to *P. capsici* infection.

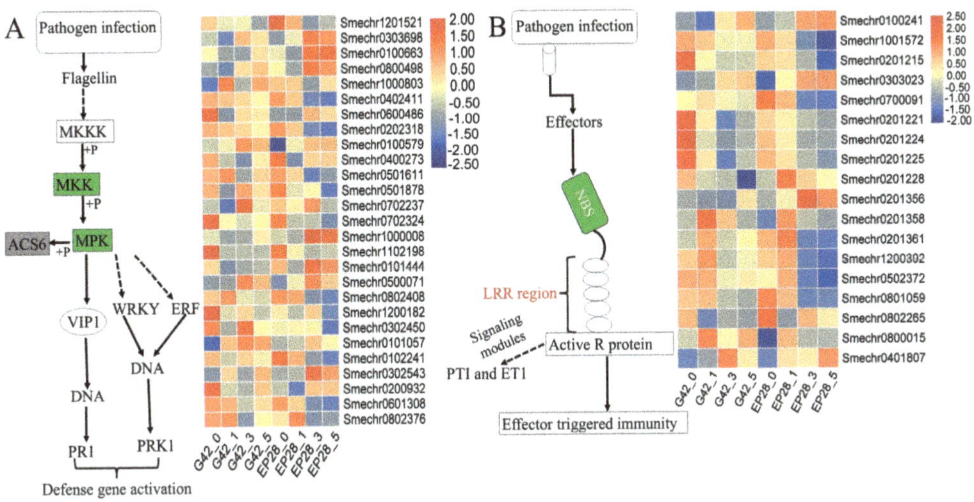

Figure 8. Expression of MAPK signaling pathway and NBS-LRR genes in response to *P.capsici* infection using a heatmap. Colors represent the $\log_2$ fold change values. (**A**) MAPK signaling pathway genes. (**B**) NBS-LRR related genes.

Calcium-dependent protein kinase (CDPK) plays a significant role in plant defense. CDPK DEGs showed significant differential expressions between G42 and EP28 at different time points. In G42, the expression of CDPK genes such as *Smechr1200644* and *Smechr0300182* were upregulated, while *Smechr1102765* showed downregulation upon *P. capsici* infection. In (EP28), the expressions of *Smechr0302812*, *Smechr1200644*, *Smechr0300187*, *Smechr1002846* and *Smechr1200281* were upregulated, while *Smechr0900732*, *Smechr0800117*, *Smechr0100010 and Smechr1002342* were downregulated at 3 and 5 dpi. The cyclic nucleotide-gated ion channels CNGCs such as *Smechr0700086* and *Smechr03024478* were upregulated in (G42) but showed downregulation in EP28 at 3 and 5 dpi. In EP28, *Smechr0700014*, *Smechr0203033 and Smechr0302178* and *Smechr0202893* were upregulated at 3 and 5 dpi, thus suggesting the role in *P. capsici* defense. Rboh (respiratory burst oxidase homologs, *Smechr0101010*, *Smechr0303236* and *Smechr0500887*) were significantly upregulated in the reactive oxygen species (ROS)-related pathway during the *P. capsici* infection, especially at 3 and 5 dpi stages in EP28 more than G42 (Figure 7).

The plant–pathogen interaction pathway also presents WRKY and ERF transcription factors. WRKY40 (*Smechr0602281*) was upregulated in EP28, while it was downregulated in

G42. WRKY6 (*Smechr0201953*) and WRKY75 (*Smechr0301211*) were upregulated in both G42 and EP28 after *P. capsici* inoculation. WRKY20/1/41 and 44 (*Smechr1200006, Smechr0701744, Smechr0101371* and *Smechr1002420*) were specifically downregulated in EP28 at 3 and 5 dpi (Figure 7). The ERF1 (*Smechr0500241*) and ERF119 (*Smechr1000134*) are upregulated in both genotypes, but ERF1 is more upregulated in EP28. The ERF118 (*Smechr1000134*) is downregulated in G42, while it shows upregulation at 1 dpi and downregulation at 3 and 5 dpi in EP28 (Table S5). The expression of the WRKY transcription factors induced the downstream defense-related R genes, PTI (pattern-triggered immunity) and RPM (plant resistance modulators)-interacting proteins. The pathogen-related gene PTI6 (*Smechr0603020*) was highly expressed in the G42-resistant genotype but downregulated in EP28 at 3 and 5 dpi (Figure 7). Two of the RPM1-interacting protein 4 (*Smechr0500762*, and *Smechr063080*) were significantly upregulated in EP28 at 3 and 5 dpi compared to G42, signifying their importance in response to *P. capsici* infections in eggplant fruit.

The nucleotide-binding site (NBS) and leucine-rich repeat (LRR) domains form the most important R genes in response to pathogen infection. The active R protein signaling induces pattern-triggered immunity (PTI) and effector-triggered immunity (ETI). The analysis of NBS-LRR yielded a total of 202 related genes in our study (Table S5). The expression of the selected 18 NBS-LRR genes (Figure 8B) shows that 13 genes were downregulated at 3 and 5 dpi in EP28, while 5 were upregulated at later stages. In G42, NBS-LRR-related genes (*Smechr0800015*) showed upregulation at 3 and 5 dpi in response to *P. capsici* infections but had a downregulation in EP28, hence offering resistance by recognizing receptors within pathogens.

Plant hormone signal transduction plays an important role in plant growth, development and defense systems. In this study, plant hormones including ethylene, abscisic acid, auxin (Aux), cytokinin (CK) and gibberellins (GA) were the majority and showed differential expression between G42 and EP28 at different time points upon *P. capsici* infection (Figure 9, Table S5). Auxin, in the tryptophan metabolic pathway, plays a significant role in pathogen defense systems. Thirty-seven auxin-related genes comprising AUX, SAUR and IAA showed differential expression in the two genotypes upon *P. capsici* infection. Most of the auxin-related genes were highly downregulated in EP28 than in G42 at 3 and 5 dpi (Figure 9A), while only six genes (*Smechr0901825, Smechr091897, Smechr0302266, Smechr0302267, Smechr032402* and *Smechr061419*) showed upregulation in EP28. One auxin gene (*Smechr062826*) showed significant upregulation at 5 dpi in G42. Gibberellin, synthesized via the diterpenoid pathway, plays a crucial role in regulating plant disease resistance by promoting the degradation of DELLAs, which are a group of nuclear proteins that inhibit growth and serve as key suppressors of gibberellin signaling. In response to pathogen attacks, plants develop a diverse array of defensive mechanisms. Here, 31 GA-related genes were identified (Figure 9B), with 11 and 17 upregulated while 10 and 14 downregulated in G42 and EP28, respectively, at 3 and 5 dpi. The ABA signaling pathway, derived downstream of the carotenoid biosynthetic pathway, influences the synthesis of various defense-related secondary metabolites to enhance plant disease resistance during pathogen invasion. A total of 15 genes associated with the ABA signal transduction pathway were identified, including PYL protein-related genes, abscisic acid-insensitive 5-like proteins and abscisic acid 8′-hydroxylase 4-like (Table S5). The selected nine DEGs (Figure 9C) showed two ABA-related genes (*Smechr0103889* and *Smechr0103890*) were specifically upregulated in EP28 at 3 and 5 dpi, signifying their response to *P. capsici* infections. The ethylene-related genes, synthesized through the cysteine and methionine, were found to be present in the plant–pathogen interaction pathway and the MAPK signaling cascade pathway, thus signifying their critical role in response to pathogen defense. Most ethylene-related genes were upregulated in G42, whereas many were downregulated in EP28 at the 3 and 5 dpi.

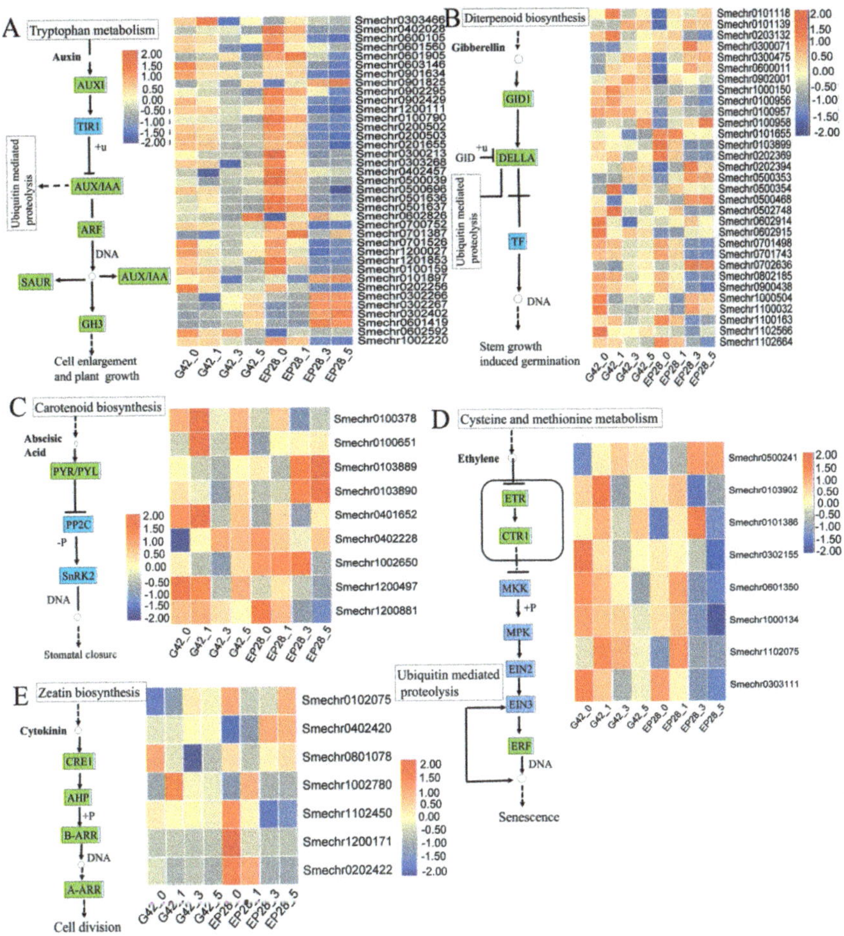

Figure 9. Differentially expressed genes (DEGs) that participate in phytohormone signaling pathways include those related to (**A**) Auxin, (**B**) Gibberellin, (**C**) Abscisic acid (ABA), (**D**) Ethylene (ETH) and (**E**) Cytokinin (CK). A heat map illustrates the expression levels of these DEGs, with red indicating upregulation and blue indicating downregulation of hormone-related genes. The blue boxes represent divergent regulation, while the green boxes represent regulation of interest in this study.

3.7. Transcription Factors Involved in Eggplant Fruit P. capsici Infection

Transcription factors play a crucial role in regulating the plant immune defense system in response to biotic and abiotic stresses, facilitating the activation or deactivation of these protective mechanisms. A comprehensive analysis has identified 703 transcription factors across 37 different families (Figure 10A, Table S6). The top 15 abundant transcription factor families included ERF/ETH (Ethylene; 121), MYB (myeloblastosis-related proteins; 93), NAM/NAC 64, bHLH (Basic helix–loop–helix; 58), HSF (Heat shock TFs; 49, WRKY; 48), NTFY (nuclear transcription factor Y subunit; 34), HD-ZIP (homeobox-leucine zipper protein; 32), GATA 30, bZIP (basic leucine zipper; 29), Mad-box (16), ASIL (Arabidopsis 6b-interacting protein 1-like; 15), Dof (Dof domain, zinc finger family protein; 13) and TCP (Teosinte branched1/Cincinnata/proliferating cell factor; 9) (Figure 10A). After ethylene, MYB genes form the second largest number of genes with 93 DEGs. In EP28, most of the MYB were downregulated at 3 and 5 dpi compared to their expression in G42 (Figure 10B). This result indicated that MYB TFs significantly mediate plant defense responses against *P. capsici* infection. A similar trend was observed in bHLH (Figure 10D) and HDZIP (Figure 10F).

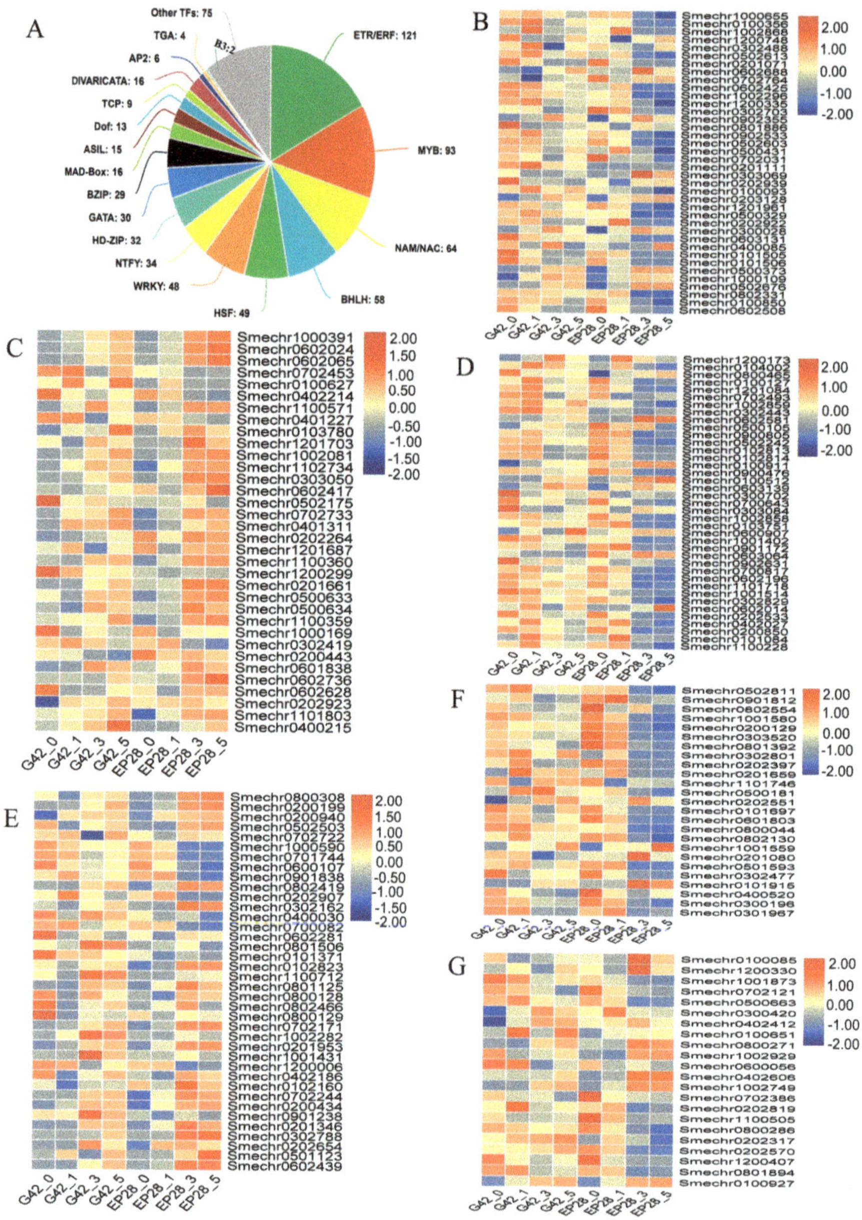

Figure 10. Comparison of transcription factor-related genes expressed in the G42 and EP28 eggplant genotypes in response to *P. capsici* infection. (**A**) Number of DEGs belonging to different transcription factor (TF) families. (**B**) MYB. (**C**) NAM/NAC. (**D**) Basic helix-loop-helix (bHLH). (**E**) WRKY. (**F**) HD-ZIP. (**G**) bZIP. Colors represent the $\log_2$ fold change values.

NAM/NAC plays a significant role in plant immunity in response to pathogen attack. Here, a significant number of NAC TFs were induced in EP28 compared to the G42 at 3 and 5 dpi (Figure 10C). Of the selected 34 NAM/NAC TFs, 26 genes were upregulated, while 8 were downregulated in EP28 at 3 and 5 dpi. The result depicts the importance of NAM/NAC transcription factors in response to *P. capsici* infection in eggplant fruits. WRKY transcription factors are involved in abiotic and biotic stress responses. The WRKY TFs were found to be present in the "plant-pathogen interaction pathway" and "MAPK signaling cascade pathway". Thirty-eight selected WRKY genes showed varied expression patterns between G42 and EP28 eggplant genotypes (Figure 10E), with 21 and 23 upregulated while 15 and 17 WRKY genes were downregulated in G42 and EP28, respectively. Specifically, 4 WRKY genes (*Smechr0801506*, *Smechr1100712*, *Smechr1002282* and *Smechr0901238*) showed distinctive upregulation in G42 and downregulation in EP28 at 3 and 5 dpi. These suggest their potential role in resistance to *P. capsici* infection.

3.8. qRT-PCR Validation of Differentially Expressed Genes

Twelve genes across the four stages of *P. capsici* infection were selected from the RNA-Seq data for qRT-PCR analysis to verify the expression patterns. Among these, expression levels of DEGs involved in LRR (*Smechr0202100*, *Smechr0502372*, *Smechr0201272*, *Smechr01358*, *Smechr0801732*, *Smechro1102377* and *Smechr0100358*), WRKY7/29 (*Smechr0402186* and *Smechr0802419*), pathogenesis-related genes transcriptional activator (PTI6, *Smechr0603020*), membrane kinase regulator (MAKR1, *Smechr0902144*) and ethylene-responsive TF (*Smechr0802118*) were analyzed. The results obtained from the qRT-PCR analysis aligned with the RNA-Seq data, revealing similar patterns of gene expression, whether upregulated or downregulated (Figure 11). This alignment suggests that both analytical approaches, RNA-Seq and qRT-PCR, can provide consistent and trustworthy insights into the molecular mechanisms underlying resistance to Phytophthora fruit rot.

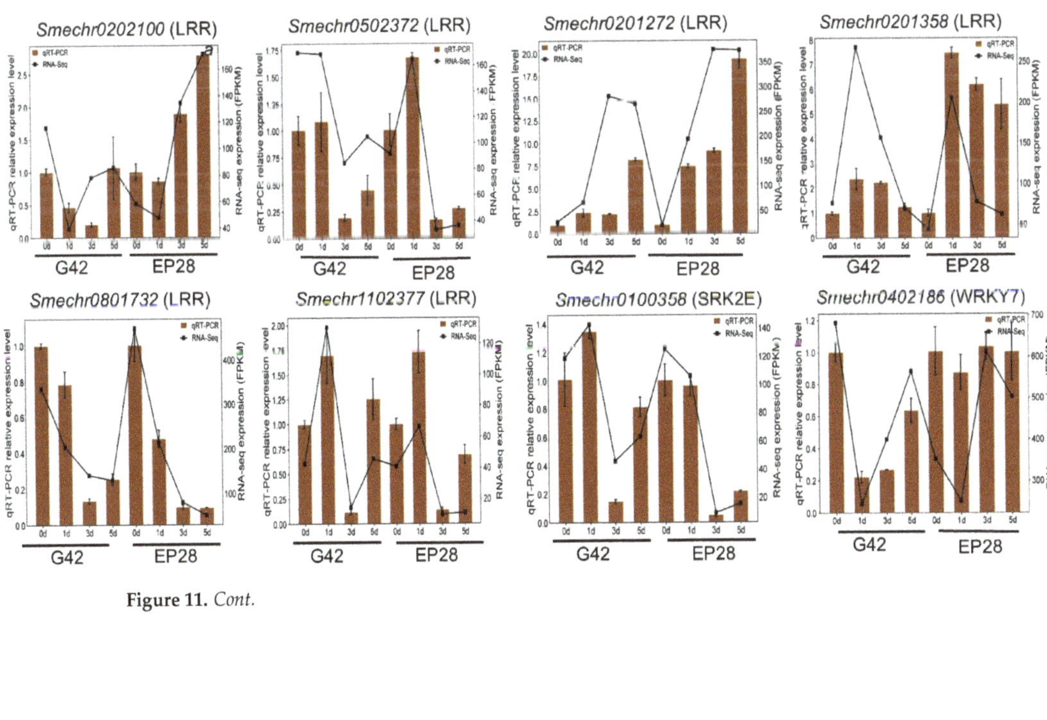

Figure 11. *Cont.*

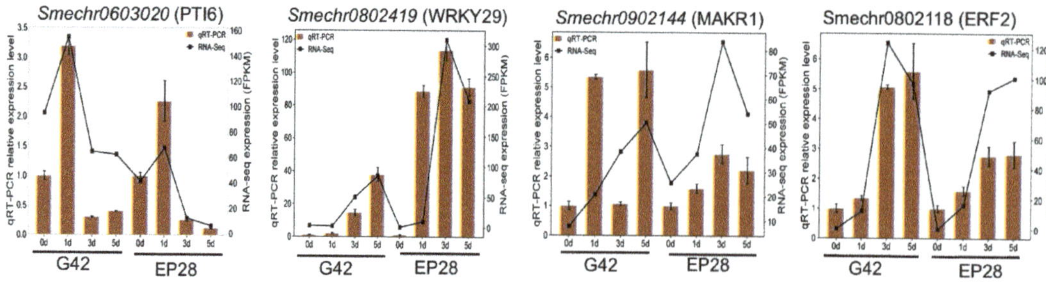

Figure 11. qRT-PCR validation of DEGs related to disease resistance. The twelve genes included members of LRR, WRKY, pathogenesis-related genes transcriptional activator (PTI6), membrane kinase regulator (*MAKR1*) and ethylene-responsive TF. The error bars represent ± SE of three replicates ($p \leq 0.05$).

3.9. VIGS-Mediated Silencing of SmPTI6 Enhances Eggplant Fruit Susceptibility to P. capsici Infestation

Several upregulated genes following *P. capsici* inoculation are promising candidates for studying the resistance mechanism against this oomycete in eggplant fruit. Consequently, we selected pathogenesis-related gene transcriptional activator (*SmPTI6*) for functional analysis using VIGS following its upregulation in the resistant eggplant genotype and downregulation in the susceptible genotype. Further, to justify the reason for choosing *SmPTI6* as our gene for VIGS analysis is the basis that Pti6 from tomato, when expressed in *Arabidopsis thaliana*, induced the expression of a diverse range of pathogenesis-related (PR) genes and served significant and unique functions in the defense mechanisms against *Pseudomonas syringae* pv. *tomato* [35]. However, no reports of the role of PTI6 regarding pathogen resistance have been reported in eggplant. Therefore, we conducted VIGs to determine the feasibility of *SmPTI6* involvement in *P. capsici* resistance. Seven days post-silencing, the fruits were inoculated with a mycelium agar plug of actively growing *P. capsici* colonies, and lesions were evaluated at 0, 3 and 5 dpi. The findings indicated that the *SmPTI6*-silenced fruits of G42 eggplant genotype exhibited disease symptoms following inoculation with *P. capsici* in contrast to the TRV2 empty vector plants (Figure 12A). In the fruits of the *SmPTI6*-silenced plants, the lesions became more pronounced and severe by 5 dpi. Conversely, the TRV2 empty vector plants showed no signs of disease. These observations suggest that the silencing of the *SmPTI6* gene in resistant eggplant fruit may compromise its resistance to *P. capsici*. Quantitative reverse transcription polymerase chain reaction (qRT-PCR) analysis demonstrated a significant downregulation of *SmPTI6* in the silenced fruits when compared to the TRV2:00 empty vector, with notable reductions in *SmPTI6* expression at both 3 and 5 dpi in the TRV:*SmPTI6* samples (Figure 12B).

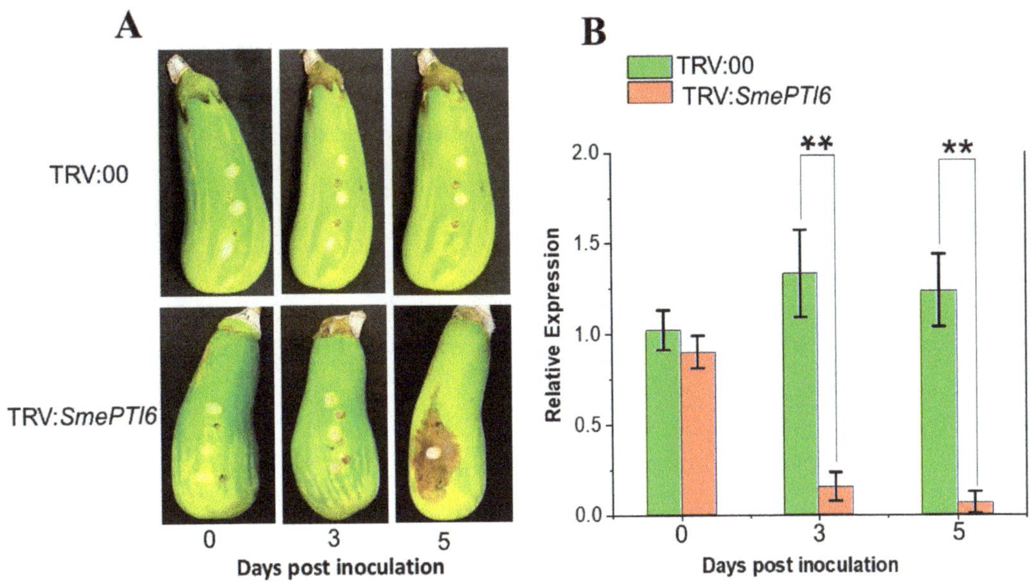

Figure 12. Effect of virus-induced gene silencing of *SmPTI6* on the resistance of G42 eggplant fruits to *Phytophthora capsici*. (**A**) Eggplant fruits were inoculated with *P. capsici* 7 days after agrobacterial infection, and infection progression was observed from 0 to 5 days dpi. TRV:00 is the control, injected with a mixture of agrobacteria containing empty vectors. TRV:*SmPTI6* are the VIGS fruits injected with a mix of agrobacteria harboring pTRV1 or pTRV2-*SmPTI6*. (**B**) VIGS reduced *SmPTI6* expression. The data are presented as the mean ± SD (n = 3). Statistical significance was determined by one-way ANOVA followed by Tukey's test. Asterisks denote significant differences (** $p < 0.01$).

4. Discussion

Phytophthora fruit rot (*Phytophthora capsici* L.) is a disease that poses a significant threat to various members of the solanaceous group, specifically crops like peppers, tomatoes and eggplants [6]. A comprehensive transcriptomic analysis of eggplant fruits inoculated with *P. capsici* generated a considerable dataset, which allowed for a detailed analysis of the differences between the resistant variety (G42) and the susceptible variety (EP28). This investigation focused on pinpointing the genes that contribute to resistance against *P. capsici* infections. In both genotypes, a high number of DEGs were exhibited at 3 dpi and then slightly reduced at 5 dpi, with EP28 showing the highest number of DEGs across all time points (Figure 2A). The KEGG analysis revealed the expression of several pathways, including the signal transduction pathways for plant hormones, the phenylpropanoid synthesis and phenylalanine metabolic pathways, the pathways governing plant–pathogen interactions and the mitogen-activated protein kinase (MAPK) pathway. Furthermore, we discovered that disease-resistance genes (R genes) such as *Smechr0500287*, *Smechr0500297* and *Smechr0402283*, among others, are linked to proteins related to specific courses such as pathogen recognition, immune response activation transcription factors and protein kinases. This suggests that these pathways contain genes contributing to disease resistance during pathogen attacks, enhancing the plant's ability to resist diseases. Plant disease resistance is a multifaceted systemic response, characterized by varying reactions from different resistance genes [36].

In response to pathogen stress, plants engage a sophisticated immune system characterized by two primary layers: pattern-triggered immunity (PTI) and effector-triggered immunity (ETI). These immune mechanisms can trigger the expression of genes involved in defensive responses, the hypersensitive response (HR) and programmed cell death (PCD) [37,38]. The combined action of PTI and ETI fosters a robust resistance against a variety of pathogens, mediated by calcium ion signaling, the production of reactive oxygen species (ROS), the stimulation of mitogen-activated protein kinases (MAPKs) and the reactivation of specific genes that respond to pathogen attacks. The infection induced by *P. capsici* initiates the propagation of calcium ion signals downstream via the cyclic nucleotide-gated channels known as CNGCs. Notably, there is a continuous upregulation of certain CNGC members, including *Smechr0700086* and *Smechr03024478*, particularly in the G42 context, which enhances the transmission of calcium signals. In response to this signaling, Rboh-related genes such as *Smechr0101010*, *Smechr0303236* and *Smechr0500887* were upregulated. This cascade of events results in alterations in ROS levels and ultimately triggers the hypersensitive response (HR). These findings suggest that the genes exhibiting marked upregulation could participate in mediating responses induced by *P. capsici* infections in plants (Figure 7).

Calcium-binding proteins are integral to the plant's response to pathogenic invasions. In stressful environments, numerous Ca^{2+} sensors or binding proteins can perceive changes in the cytoplasmic Ca^{2+} concentration, which regulates downstream gene expression to enhance plant resistance [39]. This study revealed that certain calcium-binding proteins, such as *Smechr1001819, CML15* (*Smechr0103759*), *CML49* (*Smechr1002778*), *CML19* (*Smechr1201626*) and *CML21* (*Smechr0400107*), were significantly upregulated in EP28 at 3 and 5 days post-inoculation (dpi) compared to G42. In G42, the calcium-binding proteins *CML16* (*Smechr1102012*), *CML48* (*Smechr1201717*), *CML50* (*Smechr1102453*) and *CML41* (*Smechr0300575*) were upregulated, while *CML18* (*Smechr0200838*) and *CML19* (*Smechr1201626*) were downregulated 3 and 5 dpi. The difference in expression pattern of calcium-binding protein between susceptible and resistant genotypes is likely from the fact that, in the susceptible genotype, some genes may be overexpressed to suppress or modulate immune responses, possibly to prevent excessive cell death or stress, while the overexpressed in resistant genotype is to enhance pathogen recognition, defense gene activation and cell wall reinforcement [40]. Previous studies demonstrated that in Irish potatoes, the calcium-independent protein kinase *StCDPK5* directly influences the formation of reactive oxygen species (ROS) by promoting the phosphorylation of respiratory burst oxidase homologs (Rboh) [41]. In tea plants, however, it was observed that the upstream cyclic nucleotide-gated channel (CNGC) experiences a significant downregulation during calcium transfer. At the same time, the majority of downstream Crassulacean acid metabolism/Calmodulin-like proteins (CaM/CMLs) genes exhibit a pronounced upregulation in expression during both the initial and later stages of the response to blister blight disease caused by *Exobasidium vexans* Massee [42]. In our study, CDPK genes showed significant differential expression between G42 and EP28 at different times, signifying their important role in response to *P. capsici* infections. For instance, six CDPK genes (*Smechr0900732, Smechr1002959, Smechr0800117, Smechr0100010, Smchr10002342* and *Smechr1101287*) were downregulated in EP28 compared to G42, signifying their role in defense against *P. capsici* in the resistant genotype.

The synthesis and transmission of ROS signals are vital components of the plant's defensive response to pathogenic threats. The homologous protein Rboh, responsible for modulating ROS levels, is regulated by calcium ions and the action of fungal effectors. The expression of Rboh-related genes, *Smechr0101010, Smechr0303236* and *Smechr0500887*, were markedly increased in the ROS pathway during the infection caused by *P. capsici*, particularly at the 3 and 5 dpi stages in EP28 compared to G42. The elevated synthesis and accumulation of ROS through the action of Rboh enable plant tissues to trigger their defense responses when faced with pathogenic threats [43,44]. Alternatively, genes associated with increased ROS are overexpressed in the susceptible genotype, which could suggest that an accumulation of ROS favors the development of the pathogen. This is likely because ROS can act as a signing molecule to pathogen growth by causing oxidative damage to host cells, suppressing immune cell function, or ROS-induced host cell lysis can release nutrients that pathogens scavenge for growth [45]. RPM1-induced protein kinase RIPK is instrumental in regulating RbohD-mediated reactive oxygen species ROS signaling across a range of immune responses, such as pathogen-triggered immunity (PTI), effector-triggered immunity (ETI), disease resistance (DTI, and systemic acquired resistance (SAR) [46]. Classic genetic analyses have frequently indicated that resistance (R) genes are predominantly found at single loci that confer resistance to various pathogens. The predominant category of identified R proteins comprises those featuring a nucleotide binding site (NBS) alongside leucine-rich repeat domains (LRR proteins) [47,48]. Proteins encoded by NBS-LRR play a crucial role in plant resistance by identifying receptors present in the pathogens. Certain R genes associated with specific diseases are capable of detecting the effectors linked to pathogen-associated molecular patterns (PTI) and effector-triggered immunity (ETI) [37,49]. The functional role of NBS-LRR-related genes has been previously elucidated in some vegetable species in response to fungal pathogenic attacks. The expression of *RsTNL03* and *RsTNL09* enhances the resistance of radish to *Fusarium oxysporum*, whereas *RsTNL06* plays a negative regulatory role [50]. Additionally, the transcriptional silencing of *MaNBS89* resulted in more severe leaf damage in banana plants (*Musa acuminata*) compared to the control group [51]. In our study, the observed upregulated NBS-LRR genes in the resistant genotype signify their potential roles in response to *P. capsici* infections. Their overexpression suggests an attempt to recognize *P. capsici* effectors and trigger resistance mechanisms in the resistant genotype compared to the susceptible type. The expression of plant-resistance genes serves as a negative regulator of effector-triggered immunity (ETI) while simultaneously having the capacity to genetically activate pattern-triggered immunity (PTI) through mediator proteins [52]. In this study, PTI-related genes, *Smechr0603020* (PTI6), were triggered in reaction to *P. capsici* infection showing its significance in response to pathogenic attack (Figure 7).

The role of mitogen-activated protein kinases (MAPKs) is significant in numerous (R)-mediated defense responses to pathogens affecting plants [53]. MAPK cascades are integral to various signaling pathways that operate downstream of receptor kinases, particularly in the biosynthesis of phytoalexins within plant systems [54]. The *StMKK5-StSIPK* module in potato (*Solanum tuberosum*) significantly contributes to the enhancement of resistance against phytophthora pathogens by activating salicylic acid and ethylene signaling pathways [55]. In rice (Oryza sativa), the regulatory mechanisms involving *OsMAPKKK16*, *OsMAPKKK18* and *OsMAPKKK19* and the *OsBSK1-2-OsMAPKKK16-18-19-OsMKK4-5* module have been identified as critical for the plant's response to rice blast [56]. Furthermore, research indicates that HvMPK4 phosphorylates HvWRKY1, amplifying the suppression of barley's immune response to the powdery mildew pathogen, *Blumeria graminis* f. sp. hordei. Additionally, the activation of the MAPK signaling pathway in maize in response to *Fusarium verticillioides* has been documented [57]. It has been observed that *OsMPK12* regulates plant defense mechanisms against rice blast [58]. OsMPK6 regulates bacterial blight resistance in rice [59]. Additionally, the MAPK signaling cascades (*MsDef1* and *MtDef4*) have been identified as regulators of host plant resistance to *Fusarium graminearum* [60]. In soybeans, the *GmMKK4–GmMPK6* and *GmERF113* cascade has been shown to promote the expression of genes associated with defense, thereby enhancing resistance to *Phytophthora sojae* [61]. In our study, most of the MAPK DEGs identified were upregulated in G42. Moreover, this study revealed that the MAPK cascade reaction led to the expression of downstream genes, resulting in the activation of PR proteins such as PRM1 (plant-resistant modulators) (*Smechr0502610*, *Smechr0500762* and *Smechr0603080*), PTO (*Smechr0502664*), PT1 (*Smechr0502649* and *Smechr0902695*) and PTI6 (*Smechr0603020*) that could ultimately contribute to the enhancement of resistance to pathogens.

Hormonal interactions are fundamental to the ability of plants to defend themselves against various pathogens [62]. Several hormones signaling pathways, including ethylene, abscisic, auxins and gibberellins expressed differently in both G42 and EP28 eggplant genotypes after *P. capsici* infections. The role of auxin in the stress response of plants has been recognized and displays complex plant–pathogen interaction patterns [63,64]. In this study, three auxin-related genes (*Smechr0602219*, *Smechr1201844* and *Smechr0100147*) were significantly upregulated in G42 compared to EP28. The results imply that signaling pathways linked to auxin mediate significant responses to *P. capsici* infections. The role of abscisic acid (ABA) in plant pathology is characterized as a negative regulator of disease resistance, where elevated levels of its expression are linked to an increased sensitivity to various diseases [65]. In this study, the expression of ABA-related genes changed after inoculation, showing upregulation in EP28, suggesting the relevance of ABA signaling pathway genes involvement in eggplant susceptibility response to *P. capsici* pathogen. Maize J1259 was found to activate two ABA signal transduction pathways in response to *Fusarium verticillioides* infection [57]. ERF, an ethylene response factor, contributes positively to the regulation of resistance gene expression through binding to the promoters of genes related to ethylene resistance [66], thus regulating plant immune responses. Ethylene-responsive transcription factors such as *Smechr092335*, *Smechr0300032* and *Smechr0500152* were upregulated in G42, while *Smechr010171* and *Smechr0300030* were upregulated in EP28, hence implying that ethylene-mediated defense response was involved in the response of eggplant to *P. capsici*. Ethylene is also associated with plant–pathogen interactions and transcription factors, thus displaying its diverse role in response to pathogen attack.

The regulation of the plant immune defense system in response to biotic and abiotic stresses is significantly influenced by transcription factors, which can either activate or inhibit these protective mechanisms [67]. In total, 703 transcription factors were recognized, categorized into 37 unique families (Figure 10A, Table S6). The top 15 abundant transcription factor families included ERF/ETH (Ethylene;121), MYB (myeloblastosis-related proteins; 93), NAM/NAC 64, bHLH (Basic helix–loop–helix; 58), HSF (Heat shock TFs; 49, WRKY; 48), NTFY (nuclear transcription factor Y subunit; 34), HD-ZIP (homeobox-leucine zipper protein; 32), GATA 30, bZIP (basic leucine zipper; 29), Mad-box 16, ASIL (Arabidopsis 6b-interacting protein 1-like; 15), Dof (Dof domain, zinc finger family protein; 13) and TCP (Teosinte branched1/Cincinnata/proliferating cell factor; 9) (Figure 10A). Additionally, transcription factors were found to be enriched in plant hormone signal transduction, MAPK signaling pathway, and plant–pathogen interaction. The results indicated that transcription factors from different families continue to be significantly expressed in response to *P. capsici* infections, suggesting a response to the pathogen and a crucial regulatory role in the process of disease resistance. The MYB gene family is characterized as the largest group of transcription factors found in plants. MYB-associated genes are significantly involved in regulating the flavonoid metabolism pathway, hormone signaling mechanisms, and stress response processes [68]. Furthermore, MYB genes play pivotal functions in plant disease resistance [69]. Importantly, one of the most represented TFs in our study was observed in the MYB family with 93 DEGs. MYB TFs recorded differential expressions between the two eggplant genotypes and different stages, with the majority showing upregulation in G42 at 3 and 5 dpi (Figure 10B). The findings suggest that MYB transcription factors mediate plant defense mechanisms in response to *P. capsici* infection. This aligns with previous research demonstrating their involvement in apple resistance to *Alternaria alternata* [70]. Additionally, among the various families of transcription factors in plants, bHLH transcription factors are distinguished as the second largest group [70]. In this study, 58 differentially expressed genes associated with bHLH transcription were obtained. It has been established that bHLH genes play a significant role in the defense response of potato plants to common scab, which is induced by *Streptomyces scabies* [71]. Therefore, the considerable accumulation of bHLH observed in this study may be indicative of the plant's resistance mechanisms against *P. capsici*.

WRKY proteins interact with the conserved W-box motif, which is present in the promoters of numerous genes associated with plant defense mechanisms [72]. In addition, the pronounced expression of WRKY-dependent components is of considerable interest in facilitating the generation of proteins that enhance plant disease resistance [67]. The enhancement of resistance to bacterial and fungal infections in wild tobacco (*Nicotiana benthamiana*) has been attributed to the overexpression of the cotton gene *GhWRKY39* [73]. In rice (*Oryza sativa*), the *WRKY45* gene functions as a positive regulator of resistance against the hemibiotrophic fungus *Magnaporthe grisea*, which causes rice blast [74]. In addition, in pepper (*Capsicum annum*), the upregulation of *CaWRKY01-10* and *CaWRKY08-4* genes is instrumental in providing resistance to *Phytophthora capsici* by directly activating a cluster of genes involved in defense responses [75]. In this study, WRKY TFs were significantly upregulated in G42 and downregulated in EP28. Additionally, two WRKY TFs (*WRKY7* and *WRKY24*) identified in the "plant-pathogen interaction pathway" (Figure 10E) showed contrasting expressions, suggesting that WRKY TFs have an influence on the expression of disease-resistance-related genes in eggplant to *P. capsici* infections. The difference in the expression pattern of WRKY TFs suggests that each WRKY gene plays a significant role in either suppressing or enhancing disease resistance, depending on the genotype.

We further conducted VIGs to determine the function of *SmPTI6* in eggplant response to *P. capsici* infection. Our data showed that silencing of *SmPTI6* significantly enhanced infection of eggplant plants to *P. capsici* (Figure 12A). The decreased expression of *SmPTI6* in silenced resistant fruits compared to the control empty vector samples signifies a successful VIGS process (Figure 12B). These data demonstrate that *SmPTI6* positively regulates eggplant response to *P. capsici*. A previous report showed that overexpression of Pti6 in tomato promotes plant defense [35]. PTI-related proteins (Pti4, Pti5 and Pti6) from tomato were characterized by their interaction with the PTO disease-resistance gene product, a serine-threonine protein kinase. These proteins are part of the ethylene-response factor (ERF) family, which consists of transcription factors exclusive to plants, and they specifically bind to the GCC-box cis-element located in the promoters of various pathogenesis-related (PR) genes [35], hence enhancing disease resistance.

Finally, we have provided a comprehensive summary of the molecular defense responses in G42 and EP28 eggplant genotypes against *P. capsici* infection, integrating gene expression dynamics, pathway interactions and regulatory networks (Figure 13). The data is structured to highlight not only the differences between the two varieties but also the functional relationships between key components, as indicated by directional arrows. The figure illustrates that the resistant eggplant mounts a robust, multi-layered defense against *P. capsici*, involving rapid gene expression changes, immune signaling (MAPK), hormone modulation and TF-mediated reprogramming, all dynamically interacting to mount an effective defense. In contrast, the susceptible genotype may lack this coordinated response, or disrupted interactions may lead to disease progression. Further, the upregulation of disease-resistance genes in G42 justifies their resistant capability to *P. capsici*, while the opposite is manifested in EP28. These insights could guide breeding strategies to enhance *P. capsici* in eggplants.

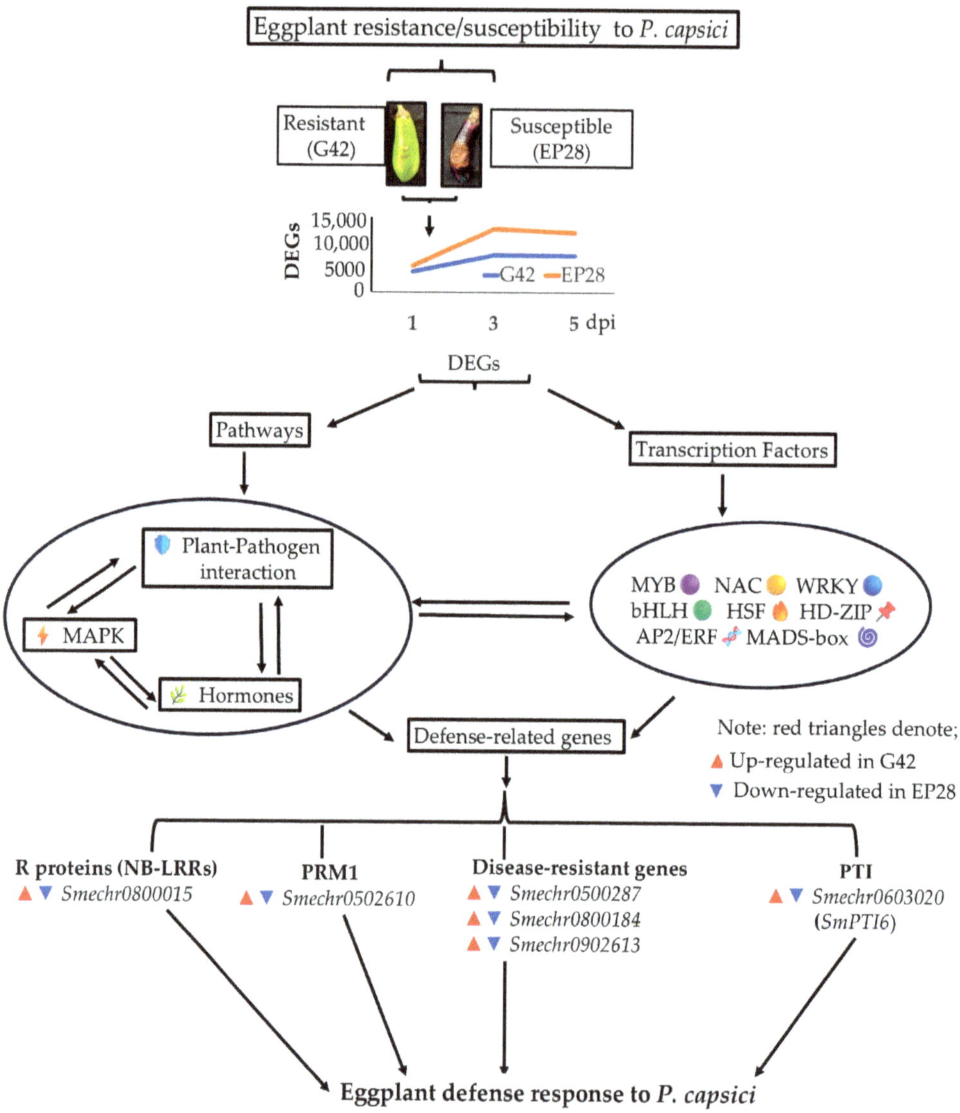

Figure 13. Summary of molecular defense responses in resistant (G42) and susceptible (EP28) eggplant varieties against *P. capsici* infection. The directional arrows show functional relationships or interactions between key components.

5. Conclusions

The comparative analysis of transcriptomes and gene expression patterns between G42 and EP28 genotypes infected with *P. capsici* has provided a thorough theoretical background for elucidating the defense mechanisms engaged in the response of eggplant to

P. capsici infection. The infection by *P. capsici* triggered the activation of various resistance pathways, with distinct genes contributing to the defense response and the associated signal transduction processes. VIGS silencing of *SmPTI6* significantly increased susceptibility to *P. capsici* infections, thus providing a candidate gene for studying the resistance mechanism to *P. capsici* in eggplant fruits. Analyzing the transcriptome contributes to a more profound understanding of the molecular mechanisms that are fundamental to biological functions regarding the resistance of eggplant to Phytophthora fruit rot. The findings from this research may also facilitate the identification of potential resistance genes that could be utilized for the genetic enhancement of germplasm resources.

Supplementary Materials: The following supporting information can be downloaded at: https://www.mdpi.com/article/10.3390/horticulturae11091026/s1, Figure S1: Fruit treatments in the humid chamber; Figure S2: Pearson correlation coefficients between samples. A0–A5 are different time points for G42, while C0–C5 are different time points for EP28; Table S1: Primers used in this study for qRT-PCR analysis; Table S2: Summary of mapped reads; Table S3: Mapping region statistics; Table S4: Top 25 DEGs upregulated in the resistant genotype G42 by *Phytophthora capsici* inoculation and relative up- or downregulation in EP28; Table S5: Differentially expressed genes involved in different pathways; Table S6: Transcription factors related genes involved in *P. capsici* infections.

Author Contributions: Conceptualization, Y.Z., X.Z. and H.O.O.; data curation, H.O.O. and X.Z.; formal analysis, H.O.O., X.Z., Y.Y. and J.L.; funding acquisition, Y.Z.; investigation, H.O.O., X.Z., S.L. and J.L.; methodology, H.O.O., X.Z. and S.L.; project administration, J.L. and Y.Z.; resources, X.Z., Y.Y., J.L. and Y.Z.; software, H.O.O., X.Z., S.L. and Y.Y.; supervision, Y.Z.; validation, X.Z. and Y.Y.; visualization, H.O.O. and X.Z.; writing—original draft, H.O.O.; writing—review and editing, X.Z., S.L., Y.Y., J.L. and Y.Z. All authors have read and agreed to the published version of the manuscript.

Funding: We acknowledge the funding support provided by the Key Research and Development Program of Jiangsu Province of China (funding number: BE2023349) and the Independent Innovation Foundation of Agricultural Sciences of Jiangsu Province, China (funding number: CX (24)3015).

Data Availability Statement: Data are contained within the article and Supplementary Materials.

Acknowledgments: We appreciate the technical support, access to and operation of laboratory equipment by the technical staff at the Vegetable Institute of the Jiangsu Academy of Agricultural Sciences.

Conflicts of Interest: The authors declare no conflicts of interest. The funders had no role in the design of the study; in the collection, analyses, or interpretation of data; in the writing of the manuscript; or in the decision to publish the results.

References

1. Fallahi, F.; Abdossi, V.; Bagheri, M.; Ghanbari Jahromi, M.; Mozafari, H. Genetic diversity analysis of Eggplant Germplasm from Iran: Assessments by morphological and SSR markers. *Mol. Biol. Rep.* **2022**, *49*, 11705–11714. [CrossRef]
2. FAOSTAT. *Statistical Database of the Food and Agriculture Organization of the United Nations*; Food and Agriculture Organization of the United Nations, Ed.; Online database; FAO: Rome, Italy, 2022. Available online: http://www.fao.org/faostat/en/#data (accessed on 10 September 2024).
3. Sharma, M.; Kaushik, P. Biochemical composition of eggplant fruits: A review. *Appl. Sci.* **2021**, *11*, 7078. [CrossRef]
4. Garcia-Estrada, R.S.; Cruz-Lachica, I.; Osuna-García, L.A.; Márquez-Zequera, I. First report of eggplant fruit rot caused by *Phytophthora nicotianae* in Mexico. *Plant Dis.* **2021**, *105*, 513. [CrossRef]
5. Kumar, S.; Singh, A.; Gupta, M.; Bist, C.M.S.; Gupta, B.; Sharma, S. Isolation, identification and multi-locus sequence typing of *Phytophthora capsici* from capsicum fields and its cross-infectivity in different crop species. *Physiol. Mol. Plant Pathol.* **2024**, *134*, 102413. [CrossRef]

6. Quesada-Ocampo, L.M.; Parada-Rojas, C.H.; Hansen, Z.; Vogel, G.; Smart, C.; Hausbeck, M.K.; Lamour, K. *Phytophthora capsici*: Recent progress on fundamental biology and disease management 100 years after its description. *Annu. Rev. Phytopathol.* **2023**, *61*, 185–208. [CrossRef] [PubMed]
7. Arkhipov, A.; Carvalhais, L.C.; Schenk, P.M. PGPR control *Phytophthora capsici* in tomato through induced systemic resistance, early hypersensitive response and direct antagonism in a cultivar-specific manner. *Eur. J. Plant Pathol.* **2023**, *167*, 811–832. [CrossRef]
8. Ghaderi, F.; Askari, S.; Abdollahi, M. Response of eggplant genotypes to *Phytophthora capsici* the causal agent of black stem disease in Kohgiluyeh va Boyer Ahmad, Iran. *Seed Plant J.* **2012**, *28*, 215–226. [CrossRef]
9. Kaniyassery, A.; Thorat, S.A.; Kiran, K.R.; Murali, T.S.; Muthusamy, A. Fungal diseases of eggplant (*Solanum melongena* L.) and components of the disease triangle: A review. *J. Crop Improvement.* **2023**, *37*, 543–594. [CrossRef]
10. Quesada-Ocampo, L.M.; Hausbeck, M.K. Resistance in tomato and wild relatives to crown and root rot caused by *Phytophthora capsici*. *Phytopathology* **2010**, *100*, 619–627. [CrossRef]
11. Mohammadbagheri, L.; Nasr-Esfahani, M.; Al-Sadi, A.M.; Khankahdani, H.H.; Ghadirzadeh, E. Screening for resistance and genetic population structure associated with *Phytophthora capsici*-pepper root and crown rot. *Physiol. Mol. Plant Pathol.* **2022**, *122*, 101924. [CrossRef]
12. Yu, X.Q.; Niu, H.Q.; Liu, C.; Wang, H.L.; Yin, W.; Xia, X. PTI-ETI synergistic signal mechanisms in plant immunity. *Plant Biotechnol. J.* **2024**, *22*, 2113–2128. [CrossRef] [PubMed]
13. Tripathi, A.; Pandey, V.K.; Jha, A.K.; Srivastava, S.; Jakhar, S.; Singh, G.; Choudhary, P. Intricacies of plants' innate immune responses and their dynamic relationship with Fungi: A Review. *Microbiol. Res.* **2024**, *285*, 127758. [CrossRef]
14. Han, Z.; Schneiter, R. Dual functionality of pathogenesis-related proteins: Defensive role in plants versus immunosuppressive role in pathogens. *Front. Plant Sci.* **2024**, *15*, 1368467. [CrossRef]
15. Qingshuo, G.; Shasha, L.; Zuhua, H.; Xiangzong, M.; Yiwen, D. NLRs in plant immunity: Structural insights and molecular mechanisms. *Crop Design* **2025**, *4*, 100103. [CrossRef]
16. Jiang, Y.; Du, J.; Latif, M.Z.; Yue, Y.; Li, Y.; Lu, C.; Li, Y.; Yin, Z.; Ding, X. Antimicrobial peptides: An important link in the game theory between plants and pathogens. *J. Adv. Res.* **2025**, *S2090–S1232*, 00492–00498. [CrossRef]
17. Saltos, L.A.; Monteros-Altamirano, Á.; Reis, A.; Garcés-Fiallos, F.R. *Phytophthora capsici*: The diseases it causes and management strategies to produce healthier vegetable crops. *Hortic. Bras.* **2022**, *40*, 5–17. [CrossRef]
18. Gaikwad, P.N.; Sharma, V.; Singh, J.; Sidhu, G.S.; Singh, H.; Omar, A.A. Biotechnological advancements in Phytophthora disease diagnosis, interaction and management in citrus. *Sci. Hortic.* **2023**, *310*, 111739. [CrossRef]
19. Shands, A.C.; Xu, G.; Belisle, R.J.; Seifbarghi, S.; Jackson, N.; Bombarely, A.; Manosalva, P.M. Genomic and transcriptomic analyses of *Phytophthora cinnamomi* reveal complex genome architecture, expansion of pathogenicity factors, and host-dependent gene expression profiles. *Front. Microbiol.* **2024**, *15*, 1341803. [CrossRef]
20. Naveed, Z.A.; Ali, G.S. Comparative transcriptome analysis between a resistant and a susceptible wild tomato accession in response to *Phytophthora parasitica*. *Int. J. Mol. Sci.* **2018**, *19*, 3735. [CrossRef]
21. Mansfeld, B.N.; Colle, M.; Zhang, C.; Lin, Y.C.; Grumet, R. Developmentally regulated activation of defense allows for rapid inhibition of infection in age-related resistance to *Phytophthora capsici* in cucumber fruit. *BMC Genom.* **2020**, *21*, 628. [CrossRef]
22. Shen, D.; Chai, C.; Ma, L.; Zhang, M.; Dou, D. Comparative RNA-Seq analysis of *Nicotiana benthamiana* in response to Phytophthora parasitica infection. *Plant Growth Regul.* **2016**, *80*, 59–67. [CrossRef]
23. Hao, C.; Xia, Z.; Fan, R.; Tan, L.; Hu, L.; Wu, B.; Wu, H. De novo transcriptome sequencing of black pepper (*Piper nigrum* L.) and an analysis of genes involved in phenylpropanoid metabolism in response to *Phytophthora capsici*. *BMC Genom.* **2016**, *17*, 822. [CrossRef] [PubMed]
24. Fan, R.; Tao, X.Y.; Xia, Z.Q.; Sim, S.; Hu, L.S.; Wu, B.D.; Hao, C.Y. Comparative transcriptome and metabolome analysis of resistant and susceptible piper species upon infection by the oomycete *Phytophthora capsici*. *Front. Plant Sci.* **2022**, *13*, 864927. [CrossRef] [PubMed]
25. Naegele, R.; Hill, T.A.; Ashrafi, H.; Reyes Chin-Wo, S.; Van Deynze, A.; Hausbeck, M.K. QTL mapping of fruit rot resistance to the plant pathogen *Phytophthora capsici* Leonian in a recombinant inbred line *Capsicum annuum* L. population. *Phytopathology* **2014**, *104*, 479–483. [CrossRef] [PubMed]
26. Naegele, R.P.; Boyle, S.; Quesada-Ocampo, L.M.; Hausbeck, M.K. Genetic diversity, population structure, and resistance to *Phytophthora capsici* of a worldwide collection of eggplant germplasm. *PLoS ONE* **2014**, *9*, e95930. [CrossRef]
27. Naegele, R.P.; Hausbeck, M.K. Phytophthora root rot resistance and its correlation with fruit rot resistance in *Capsicum annuum*. *HortScience* **2020**, *55*, 1931–1937. [CrossRef]
28. Bolger, A.M.; Lohse, M.; Usadel, B. Trimmomatic: A flexible trimmer for Illumina sequence data. *Bioinformatics* **2014**, *30*, 2114–2120. [CrossRef]

29. Florea, L.; Salzberg, S.L. Thousands of exons skipping events differentiate among splicing patterns in sixteen human tissues. *F1000Research* **2013**, *2*, 188. [CrossRef]
30. Love, M.I.; Huber, W.; Anders, S. Moderated estimation of fold change and dispersion for RNA-Seq data with DESeq2. *Genome Biol.* **2014**, *15*, 550. [CrossRef]
31. Livak, K.J.; Schmittgen, T.D. Analysis of relative gene expression data using real-time quantitative PCR and the $2^{-\Delta\Delta CT}$ method. *Methods* **2001**, *25*, 402–408. [CrossRef]
32. Gao, X.Q.; Shan, L.B. Functional genomic analysis of cotton genes with *Agrobacterium*-mediated virus-induced gene silencing. *Methods Mol. Biol.* **2013**, *975*, 157–165. [CrossRef]
33. Zhou, H.; Bai, S.; Wang, N.; Sun, X.; Zhang, Y.; Zhu, J.; Dong, C. CRISPR/Cas9-Mediated Mutagenesis of *MdCNGC2* in Apple Callus and VIGS-Mediated silencing of *MdCNGC2* in fruits improve resistance to *Botryosphaeria dothidea*. *Front. Plant Sci.* **2020**, *11*, 575477. [CrossRef]
34. Chen, C.J.; Chen, H.; Zhang, Y.; Thomas, H.R.; Frank, M.H.; He, Y.H.; Xia, R. TBtools: An integrative toolkit developed for interactive analyses of big biological data. *Mol. Plant.* **2020**, *13*, 1194–1202. [CrossRef] [PubMed]
35. Wang, Y.; Feng, G.; Zhang, Z.; Liu, Y.; Ma, Y.; Wang, Y.; Niu, X. Overexpression of Pti4, Pti5, and Pti6 in tomato promote plant defense and fruit ripening. *Plant Sci.* **2021**, *302*, 110702. [CrossRef] [PubMed]
36. Gou, M.; Balint-Kurti, P.; Xu, M.; Yang, Q. Quantitative disease resistance: Multifaceted players in plant defense. *J. Integr. Plant Biol.* **2023**, *65*, 594–610. [CrossRef] [PubMed]
37. Jones, J.D.; Dangl, J.L. The plant immune system. *Nature* **2006**, *444*, 323–329. [CrossRef]
38. Noman, A.; Aqeel, M.; Qari, S.H.; Al Surhanee, A.A.; Yasin, G.; Alamri, S.; Al-Saadi, A.M. Plant hypersensitive response vs pathogen ingression: Death of few gives' life to others. *Microb. Pathog.* **2020**, *145*, 104224. [CrossRef] [PubMed]
39. Hamel, L.P.; Sheen, J.; Séguin, A. Ancient signals: Comparative genomics of green plant CDPKs. *Trends Plant Sci.* **2014**, *19*, 79–89. [CrossRef]
40. Sampaio, J.R.; Oliveira, W.D.d.S.; Nascimento, F.d.S.; Junior, L.C.d.S.; Rebouças, T.A.; Moreira, R.F.C.; Ramos, A.P.D.S.; Santos-Serejo, J.A.d.; Amorim, E.P.; Ferreira, C.F. Calcium-Binding Protein and Polymorphism in *Musa* spp. Somaclones Resistant to *Fusarium oxysporum*. *Curr. Issues Mol. Biol.* **2024**, *46*, 12119–12132. [CrossRef]
41. Kobayashi, M.; Ohura, I.; Kawakita, K.; Yokota, N.; Fujiwara, M.; Shimamoto, K.; Yoshioka, H. Calcium-dependent protein kinases regulate the production of reactive oxygen species by potato NADPH oxidase. *Plant Cell* **2007**, *19*, 1065–1080. [CrossRef]
42. Zhou, X.L.; Hoang, N.H.; Tao, F.; Fu, T.T.; Guo, S.J.; Guo, C.M.; Buensanteai, K. Transcriptomics and phytohormone metabolomics provide comprehensive insights into the response mechanism of tea against blister blight disease. *Sci. Hortic.* **2024**, *324*, 112611. [CrossRef]
43. Ahammed, G.J.; Li, Z.; Chen, J.; Dong, Y.; Qu, K.; Guo, T.; Li, X. Reactive oxygen species signaling in melatonin-mediated plant stress response. *Plant Physiol. Biochem.* **2024**, *207*, 108398. [CrossRef]
44. Wang, R.; Li, J.; Liang, Y. Role of ROS signaling in the plant defense against vascular pathogens. *Curr. Opin. Plant Biol.* **2024**, *81*, 102617. [CrossRef]
45. Camejo, D.; Guzmán-Cedeño, Á.; Moreno, A. Reactive oxygen species, essential molecules, during plant-pathogen interactions. *Plant Physiol. Biochem.* **2016**, *103*, 10–23. [CrossRef] [PubMed]
46. Li, P.; Zhao, L.; Qi, F.; Htwe, N.M.P.S.; Li, Q.; Zhang, D.; Liang, Y. The receptor-like cytoplasmic kinase RIPK regulates broad-spectrum ROS signaling in multiple layers of plant immune system. *Mol. Plant* **2021**, *14*, 1652–1667. [CrossRef]
47. Lin, L.; Chen, Q.; Yuan, K.; Xing, C.; Qiao, Q.; Huang, X.; Zhang, S. E3 ubiquitin ligase PbrATL18 is a positive factor in pear resistance to drought and *Colletotrichum fructicola* infection. *Hortic. Plant J.* **2024**, *10*, 698–712. [CrossRef]
48. Wei, W.; Wu, X.; Garcia, A.; McCoppin, N.; Viana, J.P.G.; Murad, P.S.; Clough, S.J. An NBS-LRR protein in the Rpp1 locus negates the dominance of Rpp1-mediated resistance against *Phakopsora pachyrhizi* in soybean. *Plant J.* **2023**, *113*, 915–933. [CrossRef]
49. Liu, R.; Tan, X.; Wang, Y.; Lin, F.; Li, P.; Rahman, F.U.; Zhang, Y. The cysteine-rich receptor-like kinase CRK10 targeted by *Coniella diplodiella* effector CdE1 contributes to white rot resistance in grapevine. *J. Exp. Bot.* **2024**, *75*, 3026–3039. [CrossRef] [PubMed]
50. Ma, Y.; Chhapekar, S.S.; Lu, L.; Oh, S.; Singh, S.; Kim, C.S.; Choi, S.R. Genome-wide identification and characterization of NBS-encoding genes in *Raphanus sativus* L. and their roles related to *Fusarium oxysporum* resistance. *BMC Plant Biol.* **2021**, *21*, 47. [CrossRef]
51. Liu, S.W.; Huang, H.Q.; Li, Z.Y.; Ahmad, M.; Zhuo, M.X.; Li, C.Y.; Li, Y.D. NBS-LRR genes of *Musa acuminata* is involved in disease resistance to Fusarium wilt. *Sci. Hortic.* **2024**, *336*, 113361. [CrossRef]
52. Ma, M.; Li, M.; Zhou, R.; Yu, J.B.; Wu, Y.; Zhang, X.; Liang, X. CPR5 positively regulates pattern-triggered immunity via a mediator protein. *J. Integr. Plant Biol.* **2023**, *65*, 1613–1619. [CrossRef] [PubMed]
53. Wang, W.; Chen, S.; Zhong, G.; Gao, C.; Zhang, Q.; Tang, D. Mitogen-activated protein kinase3 enhances disease resistance of edr1 mutants by phosphorylating MAPKKK5. *Plant Physiol.* **2024**, *194*, 578–591. [CrossRef] [PubMed]

54. Zhang, M.; Su, J.; Zhang, Y.; Xu, J.; Zhang, S. Conveying endogenous and exogenous signals: MAPK cascades in plant growth and defense. *Curr. Opin. Plant Biol.* **2018**, *45 Pt A*, 1–10. [CrossRef] [PubMed]
55. Yang, H.; Chen, X.; Yang, R.; Cheng, J.; Chen, Y.; Joosten, M.H.A.J.; Du, Y. The potato StMKK5-StSIPK module enhances resistance to *Phytophthora* pathogens through activating the salicylic acid and ethylene signalling pathways. *Mol. Plant Pathol.* **2023**, *24*, 399–412. [CrossRef]
56. Li, S.; Xiang, X.; Diao, Z.; Xia, N.; Lu, L.; Zhang, J.; Tang, D. The *OsBSK1-2*-MAPK module regulates blast resistance in rice. *Crop J.* **2024**, *12*, 110–120. [CrossRef]
57. Zeng, H.; Cai, H.; Bai, T.; Ren, X.; Ci, J.; Yang, W. Transcriptome analysis of maize resistance to *Fusarium verticillioides*. *J. Plant Interact.* **2024**, *19*, 2393803. [CrossRef]
58. Lu, L.; Zhang, J.; Zheng, X.; Xia, N.; Diao, Z.; Wang, X.; Li, S. *OsMPK12* positively regulates rice blast resistance via *OsEDC4*-mediated transcriptional regulation of immune-related genes. *Plant Cell Environ.* **2024**, *47*, 3712–3731. [CrossRef] [PubMed]
59. Wang, D.; Wang, H.; Liu, Q.; Liu, Y.; Wang, Y.; Qin, R.; Zhou, H.; Xu, X. Reduction of *OsMPK6* Activity by a R89K Mutation Induces Cell Death and Bacterial Blight Resistance in Rice. *Plant Cell Rep.* **2021**, *40*, 835–850. [CrossRef]
60. Ramamoorthy, V.; Zhao, X.; Snyder, A.K.; Xu, J.R.; Shah, D.M. Two mitogen-activated protein kinase signaling cascades mediate basal resistance to antifungal plant defensins in *Fusarium graminearum*. *Cell. Microbiol.* **2007**, *9*, 1491–1506. [CrossRef]
61. Gao, H.; Jiang, L.; Du, B.; Ning, B.; Ding, X.; Zhang, C.; Zhang, S. GmMKK4-activated GmMPK6 stimulates GmERF113 to trigger resistance to *Phytophthora sojae* in soybean. *Plant J.* **2022**, *111*, 473–495. [CrossRef]
62. Aerts, N.; Pereira Mendes, M.; Van Wees, S.C. Multiple levels of crosstalk in hormone networks regulating plant defense. *Plant J.* **2021**, *105*, 489–504. [CrossRef]
63. Radouane, N.; Goura, K.; Lahmamsi, H.; Kenfaoui, J.; Farhaoui, A.; Belabess, Z.; Lahlali, R. Phytohormone signaling and plant–Pathogen interaction. In *Plant Pathogen Interaction*; Springer Nature: Singapore, 2024; pp. 185–220. [CrossRef]
64. Roychowdhury, R.; Hada, A.; Biswas, S.; Mishra, S.; Prusty, M.R.; Das, S.P.; Sarker, U. Jasmonic acid (JA) in plant immune response: Unravelling complex molecular mechanisms and networking of Defense signaling against pathogens. *J. Plant Growth Regul.* **2024**, *44*, 89–114. [CrossRef]
65. Jiang, Y.; Yue, Y.; Wang, Z.; Lu, C.; Wang, Z.; Yin, Z.; Ding, X. A novel ABA structural analogue enhanced plant resistance by inducing plant immunity and inactivating ABA signaling pathway. *Adv. Agrochem* **2024**, *3*, 64–73. [CrossRef]
66. Tang, Q.; Wei, S.; Zheng, X.; Tu, P.; Tao, F. APETALA2/ethylene-responsive factors in higher plant and their roles in the regulation of plant stress response. *Crit. Rev. Biotechnol.* **2024**, *44*, 1533–1551. [CrossRef]
67. Liu, F.; Xi, M.; Liu, T.; Wu, X.; Ju, L.; Wang, D. The central role of transcription factors in bridging biotic and abiotic stress responses for plants' resilience. *New Crops* **2024**, *1*, 100005. [CrossRef]
68. Pratyusha, D.S.; Sarada, D.V.L. MYB transcription factors—Master regulators of phenylpropanoid biosynthesis and diverse developmental and stress responses. *Plant Cell Rep.* **2022**, *41*, 2245–2260. [CrossRef] [PubMed]
69. Yu, Y.; Zhang, S.; Yu, Y.; Cui, N.; Yu, G.; Zhao, H.; Fan, H. The pivotal role of MYB transcription factors in plant disease resistance. *Planta* **2023**, *258*, 16. [CrossRef]
70. Zeng, X.; Wu, C.; Zhang, L.; Lan, L.; Fu, W.; Wang, S. Molecular mechanism of resistance to *Alternaria alternata* apple pathotype in apple by alternative splicing of transcription factor MdMYB6-like. *Int. J. Mol. Sci.* **2024**, *25*, 4353. [CrossRef]
71. Tao, H.; Wang, S.; Li, X.; Li, X.; Cai, J.; Zhao, L.; Cai, Y. Biological control of potato common scab and growth promotion of potato by *Bacillus velezensis* Y6. *Front. Microbiol.* **2023**, *14*, 1295107. [CrossRef] [PubMed]
72. Saha, B.; Nayak, J.; Srivastava, R.; Samal, S.; Kumar, D.; Chanwala, J.; Giri, M.K. Unraveling the involvement of WRKY TFs in regulating plant disease defense signaling. *Planta* **2024**, *259*, 7. [CrossRef]
73. Shi, W.; Liu, D.; Hao, L.; Wu, C.A.; Guo, X.; Li, H. GhWRKY39, a member of the WRKY transcription factor family in cotton, has a positive role in disease resistance and salt stress tolerance. *Plant Cell Tissue Organ Cult.* **2014**, *118*, 17–32. [CrossRef]
74. Nakayama, A.; Fukushima, S.; Goto, S.; Matsushita, A.; Shimono, M.; Sugano, S.; Jiang, C.J.; Akagi, A.; Yamazaki, M.; Inoue, H.; et al. Genome-Wide Identification of WRKY45-Regulated Genes That Mediate Benzothiadiazole-Induced Defense Responses in Rice. *BMC Plant Biol.* **2013**, *13*, 150. [CrossRef] [PubMed]
75. Cheng, W.; Wang, N.; Li, Y.; Zhou, X.; Bai, X.; Liu, L.; Chu, M. *CaWRKY01-10* and *CaWRKY08-4* confer pepper's resistance to *Phytophthora capsici* Infection by Directly Activating a Cluster of Defense-Related Genes. *J. Agric. Food Chem.* **2024**, *72*, 11682–11693. [CrossRef] [PubMed]

horticulturae

MDPI

Article

Exploring Polyploidization in *Nigella sativa* L.: An Applicable Strategy Towards Crop Improvement

Mohammed E. El-Mahrouk [1], Mossad K. Maamoun [2], Sobhia Saifan [3], Yousry A. Bayoumi [1], Hassan El-Ramady [4] and Neama Abdalla [5,*]

1 Department of Horticulture, Faculty of Agriculture, Kafrelsheikh University, Kafr El-Sheikh 33516, Egypt; threemelmahrouk@yahoo.com (M.E.E.-M.); yousry.bayoumi@agr.kfs.edu.eg (Y.A.B.)
2 Department of Breeding and Genetics of Vegetables, Aromatic & Medicinal Plants, Horticultural Research Institute, Agriculture Research Center, Giza 12619, Egypt; mossad_maamoun@yahoo.com (M.K.M.)
3 Department of Agricultural Biotechnology and Genetic Engineering, Faculty of Agricultural Technology, Al Ahliyya Amman University, Amman 19328, Jordan; s.saifan@ammanu.edu.jo (S.S.)
4 Soil and Water Department, Faculty of Agriculture, Kafrelsheikh University, Kafr El-Sheikh 33516, Egypt; hassan.elramady@agr.kfs.edu.eg (H.E.-R.)
5 Plant Biotechnology Department, Biotechnology Research Institute, National Research Centre, 33 El Buhouth St., Dokki, Giza 12622, Egypt
* Correspondence: neama_ncr@yahoo.com (N.A.)

Abstract: A plant breeding program needs helpful markers, especially morphological ones, which can allow breeders to dispense with other markers, including cytological traits and flow cytometry. These markers can assist plant breeders in distinguishing diploid and tetraploid plants during the seedling stage. Therefore, the present study aimed to investigate and validate effective methodologies for the early identification of artificially induced polyploids in black cumin. Thus, we established an extensional program for black cumin breeding including producing seeds, active compounds, and flowers as ornamental plants. Field experiments on tetraploids and diploids were carried out to evaluate the morphological and yield traits of both plants. Also, some cytological studies and Gas Chromatography (GC) analysis were conducted to achieve these goals. The results showed the possibility of realizing diploid and tetraploid plants in early growing black seeds in the field (mainly after the first cotyledon leaves). This crucial outcome can support plant breeders in identifying polyploidy during the seedling stage without referring to biochemical markers, flow cytometry, and cytological traits. All morphological and yield-related traits were superior in diploid plants compared to tetraploids. The results showed that diploid and tetraploid plants exhibited plant heights of 116 cm and 95 cm, numbers of secondary branches of 112 and 22, numbers of flowers of 111.7 and 24.75, and shoot fresh weights of 610 g and 147.5 g, respectively. Furthermore, the number of seeds per capsule, seed yield per plant, and oil percentage in diploids were 97.5 seeds, 24 g, and 22.94%, compared with 35.25 seeds, 4.62 g, and 17.76% in tetraploids, respectively. This work used the cotyledon leaf shape as a morphological marker to distinguish the tetraploid and diploid plants, as diploids are typically taller with pointed cotyledons, whereas tetraploids are shorter with rounded cotyledon tips. This study will create great opportunities for plant breeders to save time and costs during their programs. Further studies on such suggested black cumin breeding programs are needed on diploids, triploids, and tetraploids.

Keywords: diploids; morphological markers; *Nigella sativa* L.; polyploidy; tetraploids; thymoquinone

1. Introduction

Nigella sativa L., commonly known as black cumin or black seed, is an annual herbaceous plant belonging to the family Ranunculaceae. It has a diploid chromosome number of 2n = 2x = 12. This plant species has several culinary uses and has been recognized in traditional medicine for treating many ailments such as eczema, headache, inflammation, anorexia, rheumatism, and paralysis [1–4]. It has been widely used in different types of product such as essential oil, extract, powder, and paste [5,6]. It is native to Southern Europe, Southwest Asia, and North Africa. It is cultured in the Middle East, the Mediterranean, and Southern European regions [7]. Black cumin seeds are often used as food preservatives, spices, and flavoring additives for adding a distinctive aroma and taste to bread, savory dishes, and pickles due to their low level of toxicity [8,9]. Moreover, the seeds are known for their nutritional value due to their high content of iron, copper, phosphorus, calcium, zinc, folic acid, thiamin, niacin, and pyridoxine [10,11]. The pharmaceutical and commercial value of black cumin is determined by its seed yield, essential oil content, and concentration of bioactive compounds [12]. Due to the health benefits of black cumin, it is necessary to develop plants with desirable traits through induced genetic variations and breeding programs, as the existing germplasm may not satisfy future demands [12].

Polyploidy or whole genome duplication is the heritable state in which the cells carry more than two complete chromosomal sets [13]. The somatic cells of polyploid organs have more than two complete/paired sets of homologous chromosomes [14]. Polyploidy, or the doubling of the chromosome number, is one of the plant breeding methods for genetic improvement of plants. Autotetraploid induction for plants has been studied for over 85 years [15]. Tetraploidy of black cumin was chemically induced by colchicine treatment of different plant organs, i.e., seed and seedling [16], seedling [17], and apical meristematic tips of young seedlings bearing only two cotyledonary leaves [18], or by another method that involves using 2,4-dinitroaniline treatment on seeds [19].

Previous reports indicated that polyploidy extended flowering duration, increased organ size, and improved resistance to abiotic stresses and diseases [20]. For black cumin, induced tetraploid plants using colchicine had more branches and flowers, enhanced size and frequency of stomata, and an increased number of septa per fruit and seeds per septum [16]. Furthermore, leaf thickness, flower and anther size, and petals were increased in tetraploid black cumin plants [21]. Autotetraploid induction improved the fruit quality of kiwifruit (*Actinidia chinensis* Planch.) in weight, size, and diameter [22]. Tetraploids of the *Anemone sylvestris* L. flowering plant are characterized by strong, vigorous growth, and they flower early compared to diploid plants [23]. In addition, artificial polyploidy enhanced the biosynthesis of important secondary metabolites in many polyploidy plants compared to their diploid parents [24]. Tetraploid clones of *Artemisia annua* L. (synonym: *Artemisia chamomilla* C.Winkl.) increased artemisinin production six-fold compared to the diploid parent [25]. Tetraploid plants of radish (*Raphanas sativus* L.) have larger vegetative and reproductive organs, enhanced antioxidant enzyme activity, and higher soluble contents and quality [26]. One autotetraploid black cumin plant had some useful traits compared to diploids and other tetraploids [18]. Polyploidy induction enhances plant breeding potential by improving ornamental traits, stress tolerance, yield, and fertility restoration, and enabling the development of seedless cultivars.

On the other hand, induction of artificial polyploidy can negatively affect plant growth and fruit size for commercial production [27]. The growth of diploid plants of apple was more vigorous than tetraploid ones, and the stomatal conductance and photosynthetic rate of tetraploid plants were slightly lower [28]. Vegetative growth and yield of diploid black cumin plants were higher than tetraploid plants [21]. Seeds of tetraploid plants of black cumin were sterile due to widespread chromosome irregularities, producing varying numbers of quadrivalents (0–4), trivalents (0–2), and univalents (0–10) [17]. The most prominent morphological changes in black cumin tetraploid and its progenies were increased flower and capsule sterility, and reduced seed number per capsule and seed fertility [18].

Early identification of polyploids is a crucial issue in developing plant breeding programs. It was reported that approximately 70% of all angiosperms have undergone one or more polyploidy events [29]. One common method for identifying polyploids is morphological based-markers like size, shape, and color of plant organs (e.g., cotyledons, leaves, and flowers). It is a practical method for distinguishing polyploid plants, especially at the early stages of growth. Tetraploid radish seedlings were significantly taller (by 36%), had longer (61%) and wider (43%) leaves, and exhibited larger floral organs as compared to diploid plants [26]. Similarly, notable differences in plant height, leaf length, petiole characteristics, and leaflet width were observed between diploid and tetraploid plants [30]. Physiological markers are also considered as a reliable tool for polyploid identification. These include stomatal characteristics such as size, density, and number of chloroplasts per guard cell. It was demonstrated that stomatal size tends to increase in polyploid plants, while stomatal density generally decreases, providing a novel method for detecting polyploidy [29]. Later studies confirmed that polyploidy leads to increased stomatal size and chloroplast number, accompanied by reduced stomatal density [26,31,32]. Another precise method is chromosome counting during mitotic or meiotic division. Tetraploidy was confirmed in radish by counting chromosomes in pollen mother cells at metaphase I during meiosis [26]. Similarly, tetraploidy in black cumin was verified through mitotic chromosome analysis [18,19]. Flow cytometry is also widely used for determining ploidy levels by comparing nuclear DNA content between diploid and tetraploid plants. This technique has been successfully applied in various species, including radish [26], black cumin [33], purple coneflower [34], and Urochloa [35].

A limited number of studies have addressed the comparative evaluation of diploid and tetraploid black cumin, primarily focusing on mature plants. These studies have relied on various approaches, including morphological and physiological measurements, cytological assessments, and flow cytometry analysis [17–19,33]. However, these studies provide little information on early developmental stages, which are critical for accelerating selection processes in breeding programs. In the present study, we extend the scope of evaluation by comparing diploid and tetraploid black cumin at both the early cotyledon stage and the mature plant stage. This dual stage approach allows for the identification of early morphological and physiological indicators that can reliably distinguish polyploid plants. Such early detection methods are particularly valuable for plant breeders, as they reduce the time and resources required to confirm ploidy status at later growth stages. The main objective of this study was to explore and validate practical methodologies for the early identification of artificially induced polyploids in black cumin. In addition, this study sought to assess economic traits such as plant vigor, leaf and flower morphology, potential yield components, and thymoquinone (TQ) content of diploid versus tetraploid plants in order to evaluate their agronomic significance.

2. Materials and Methods

This research was performed in 2021–2022 at the Physiology and Breeding of Horticultural Crops Laboratory, Horticulture Department, Faculty of Agriculture, Kafrelsheikh University, Egypt.

2.1. Plant Materials

In the present study, seeds from a tetraploid breeder line (induced using 2,4-dinitroaniline at 10 mg L^{-1} for 24 h) and a selected diploid line obtained from the breeding program (El-Mahrouk et al. [19]) at the Faculty of Agriculture, Kafrelsheikh University, Egypt, were obtained, identified, and compared.

2.2. Cytological and Flow Cytometry Analysis

Flower buds of diploid and tetraploid plants were treated with freshly prepared fixation solution (three parts absolute alcohol/one part glacial acetic acid) for 24 h. Samples were preserved in 70% ethanol and placed in a cool place for scanning. Anthers were excised on the slide and stained with acetocarmin, then sealed and observed under a Leica Aristoplan light microscope (Neu-Isenburg, Germany) with Leica DC 300 F digital imaging using an oil immersion lens [36]. Ploidy level was detected by an Attune flow cytometer (Applied Biosystem, Foster City, CA, USA). Young leaves from seedlings (30 days old) of diploid and tetraploid plants were used as samples according to Galbraith et al. [37]. Leaf tissue (50 mg) was chopped and macerated in lysis buffer (1.0 mL) to release intact nuclei by a razor blade in Galbraith buffer [45 mM $MgCl_2$; 30 mM sodium citrate; 20 mM MOPS [3-(N-morpholino)-propane sulfonic acid]; 0.1% (*w*/*v*) Triton X-100; pH 7.0] for 1 h. The cell suspensions were filtered by a 0.45 μm nylon filter to eliminate cell debris for 5 min at room temperature. Then, the cell nuclei were stained with 10 μL 4′,6′-diamino-2phenylidole (DAPI) solution (solution A of high-resolution kit type P, Partec) for 30 min on ice in the dark. Nearly 10,000 nuclei were analyzed using a logarithmic scale. Histograms were analyzed using Attune cytometric software version 2.1.

2.3. Establishment of Seedlings

Seeds of tetraploid plants were cultivated separately in isolated field to avoid cross pollination with other black cumin plants to maintain pure tetraploid seeds. Under greenhouse conditions, the seeds were germinated in peat moss and vermiculite mixture (1:1; *v*/*v*). The medium was fertilized with 1.0 g L^{-1} of water-soluble compound fertilizer in a ratio of 19:19:19 (Rosasol; Rosier, Moustier, Belgium) and sterilized with a solution of fungicide (1.0 g L^{-1} Rizolex; Kafr El-Zayat Company, El-Gharbia, Egypt). The pH of the medium was adjusted to 6.0 ± 1 with calcium carbonate (Jenway 3510, Staffordshire, UK). Soaked seeds (for 24 h in distilled water) were sown in expanded polyurethane foam trays (3.0 × 3.0 cm; 209 cells/tray), with one seed in each cell and one for each diploid and tetraploid line. Trays were put in the greenhouse at 25 ± 2 °C and light intensity of 300 μm m^{-2} s^{-1} after sowing. The trays were also manually watered weekly using 10 L watering cans with 2.0 L for each tray. All trays were covered with plastic sheets (35 μm) until the first germinated seeds became visible, and then the plastic covers were removed. Germination indices including final germination percentage (FGP), germination rate index (GRI), corrected germination rate index (CGRI), and time to 50% germination (GT50) were recorded. At this stage, the morphological shape of cotyledons and chromosomal number in mitosis division were recorded. The seeds were considered germinated when their cotyledons became visible above the surface of the growth medium. Germination parameters were calculated as follows:

(a) Final germination percentage (FGP)

$$= [\text{Number of germinated seeds after 30 days from sowing} / \text{number of sown seeds}] \times 100$$

$$\text{(b) Germination rate index (GRI)} = \left[\left(\frac{G1}{1}\right) + \left(\frac{G2}{2}\right) + \left(\frac{Gx}{X}\right)\right]$$

where G = germination on each alternate day after placement; 1, 2, x = corresponding day of germination [38,39].

$$\text{(c) Corrected germination rate index (CGRI)} = \left(\frac{GRI}{FGP}\right) \times 100$$

$$\text{(d) GT50} = \text{number of days lapsed to reach 50\% of FGP}$$

2.4. *Soil Analysis and Cultivation of Black Cumin Seedlings*

The soil analysis of the experimental site is presented in Table 1. Soil pH was measured in a 1:2.5 ratio (soil/deionized water suspension) using a calibrated pH meter 3510 (Jenway, Stafford Shire, UK). Soil salinity [electrical conductivity (EC)] was measured in a 1:5 ratio (soil/deionized water) using an EC Meter (MI 170; Treviglio, Italy). Soluble ions in saturated extracts were measured according to the methods of Jackson [40]. Total carbonate was determined using a volumetric calcimeter [41]. Organic matter content was determined using the dichromate oxidation method [41]. Available nitrogen (NH_4^+) was determined using the micro Kjeldahl method [42], and available phosphorus (P_2O_5) was determined [43]. Calcium (Ca^{2+}) and magnesium (Mg^{2+}) were also measured [40]. Sodium (Na^+) and potassium (K^+) were extracted according to the methods described by Black [44], and concentrations were determined using a Flame photometer PFP7 (Jenway, Stafford Shire, UK). Chloride (Cl^-) was determined by titration with a standard solution of silver nitrate [40]. Both diploid and tetraploid seedlings of black cumin were transplanted in clay soil during two successive seasons, 2021 and 2022, in the experimental farm of the Horticulture Department, Faculty of Agriculture, Kafrelsheikh University. Each genotype was planted in five rows (6.0 × 0.8 m), each row considered as one replicate. The distance between plants was 30 cm. The experiment was maintained for 7 months. All cultural practices for best plant growth were followed according to Rajeswara et al. [45].

Table 1. Soil chemical analysis of experimental farm of the Horticulture Department, Faculty of Agriculture, Kafrelsheikh University.

EC ($dS\ m^{-1}$) (1:5)	pH (1:2.5)	Available Nutrients ($mg\ kg^{-1}$)			Soluble Cations ($mmol_c\ L^{-1}$)				Soluble Anions ($mmol_c\ L^{-1}$)				OM (%)
		N	P	K	Mg^{2+}	Ca^{2+}	K^+	Na^+	CO_3^{2-}	HCO_3^-	Cl^-	SO_4^{2-}	
3.96	8.13	32.65	10.9	380	7.1	12.47	0.24	19.77	0.0	3.33	18.4	17.9	1.45

2.5. *Assessment of Plant Morphology, Flowering, and Yield Traits Under Field Conditions*

The morphological traits of tetraploid plants and their corresponding diploid plants growing under the same field conditions were employed to verify the polyploidy variations. Ten seedlings from each genotype were used for determining the shape and area of cotyledon leaves. Twenty-five plants from every genotype were randomly selected from five replicates in order to assess the following traits. Morphological vegetative traits including plant height, number of main and secondary branches per plant, crown diameter (cm), shoot fresh and dry weights, root length (cm), and root fresh weight. Morphological flower traits in terms of the number of first days to flowering, number of flowers per plant, converted flowering parts, and stamen number and size were also investigated, as well as yield traits in terms of the number of seeds per capsule, seed yield per plant, and 100-seed weight. Oil percentage was determined according to Folch et al. [46].

2.6. Analysis of Thymoquinone Content

An analysis of thymoquinone content in the seed extracts was carried out in the High Institute for Public Health Central Lab, Alexandria University, using the Gas Chromatography GC-HP (Hewlett Packard) 6890, with FID detector (flame ionizing) and DB-23 Column (50% cyanopropyl–methylpolysiloxane), 30 m × 0.32 mm, ID = 0.25 μm film thickness, to determine TQ content. The carrier gas was nitrogen (1 mL min^{-1} gas flow). The seed oil of diploid and tetraploid plants (extracted from 9 seeds) was dissolved in HPLC methanol to prepare them for GC analysis. Samples of 1.0 μL were injected at a temperature of 220 °C. The oven temperature for the first 2 min was 100 °C and increased at 10 °C min^{-1} until 240 °C, then was held for 2 min. Injector and detector temperatures were set at 250 °C. Thymoquinone standard (2-isopropyl-5-methyl-l, 4 benzoquinone) 99% (Sigma Chemical Co. St Louis, MO, USA) was dissolved in HPLC methanol, whose dilution factor was 19.7 mg 0.2 mL^{-1} methanol. The concentration of thymoquinone of the seeds was calculated using the following equations:

$$TQ\ concentration\ (\text{mg/mL}) = ((area\ of\ sample \div area\ of\ standard)\ 98.5) \times dilution\ factor) / volume\ of\ used\ extract$$

$$TQ\ \text{consentration of 1 g} = \text{TQ concentration (mg/mL)} \times \text{n}$$

n number of miles extracted from 1 g plant material.

$$TQ\ concentration\ (\text{mg}/100\ \text{g}\ plant\ material) = TQ\ concentration\ of\ 1\ \text{g} \times 1000$$

2.7. Statistical Analyses

The experiments were set up in a completely randomized block design. Data were analyzed with one-way ANOVA and the unpaired *t*-test in SAS software (version 9.4; SAS Institute, Inc., Cary, NC, USA).

3. Results

3.1. Cytological and Flow Cytometry Analysis of Diploid vs. Tetraploid Plants

Concerning the cytological analysis, the normal meiosis for diploids (n = x = 6) and tetraploids (n = 2x = 12) of black cumin plants is presented in Figure 1. Due to its potential as an important proposed method for ploidy analysis of different polyploidy levels after applying 2,4-dinitroaniline, flow cytometry was used to confirm the production of tetraploid through treatment. The flow cytometry data are displayed as a histogram of relative fluorescence intensity, representing relative DNA content. The histogram peak of mean fluorescence intensity (MFI) was 4495 and 10,520 for diploid and tetraploid plants, respectively. By accounting for the total area under the curve or MFI, there seems to be a higher total area for the tetraploid plants compared with the diploids.

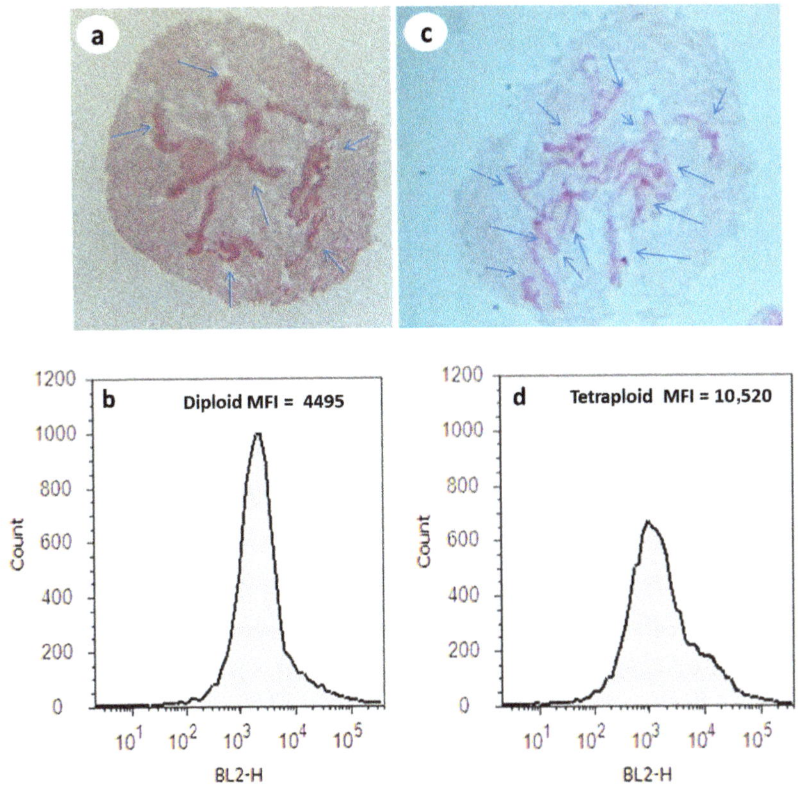

Figure 1. Cytological and flow cytometry analysis of meiosis of diploid and tetraploid black cumin plants: (**a**,**b**) meiosis and flow cytometry in diploid plant (n = x = 6); (**c**,**d**) meiosis and flow cytometry in tetraploid plant (n = 2x = 12). Magnification of 1a and 1c = 1000×. MFI = mean fluorescence intensity, where the *y*-axis is the counts (number of fluorescent nuclei at each intensity) and the *X*-axis is the MFI. (**a**,**c**) Each arrow refers to one chromosome.

3.2. Germination of Diploid and Tetraploid Black Cumin

The effect of ploidy level on germination parameters including final germination percentage (FGP), germination rate index (GRI), corrected germination rate index (CGRI), and median germination time (GT_{50}) in black cumin is presented in Figure 2. The data showed that germination started for diploid and tetraploid soaked seeds after 4 and 8 days, respectively. All these measured parameters, including FGP, GRI, and CGRI, showed a higher germination rate for the diploids of the studied plants compared to the tetraploids, except GT50 Days. The increasing rate of germination parameters was recorded at 125, 66.6, and 60% for FGP, GRI, and CGRI, respectively, in the diploid plant, whereas the GT50 was 8.9 and 14.2 days for the diploid and tetraploid plants, respectively. Also, the polyploidy level affected the number of days for germination of 50% of the seeds (GT50), and the positive effect was for the diploid plant rather than the tetraploid plant. Significant changes could also be observed in all the studied germination parameters between the diploid and tetraploid plants.

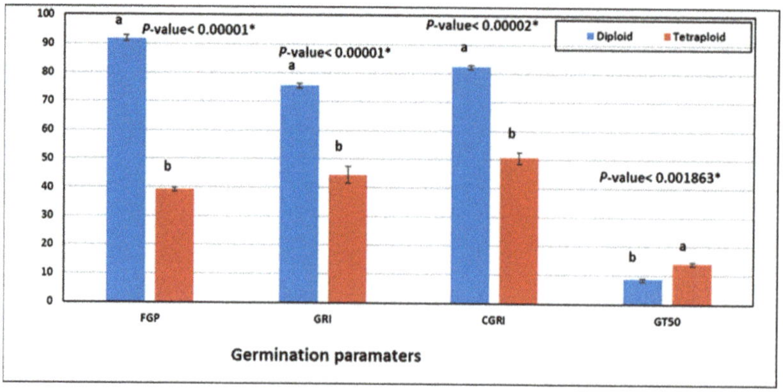

Figure 2. Seed germination of diploid and tetraploid black cumin: final germination percentage (FGP %), germination rate index (GRI %), corrected germination rate index (CGRI %), number of days lapsed to reach 50% of final germination percentage (GT50 Days) after 30 days in culture. The same letters mean non-significant impact, according to Duncan's multiple tests, at a 5% level. * Significant at $p < 0.05$ according to Student's unpaired t-test.

3.3. Morphology of Black Cumin Seedlings in Diploid Versus Tetraploid Plants

Various morphological traits of both diploid and tetraploid plants are presented in Table 2. All measured morphological characteristics were higher in diploid plants than in tetraploids, except for converted flowering parts and leaf areas, which were greater in tetraploids (188 cm and 22.5 cm, respectively) compared to diploids (20.34 cm and 17.5 cm, respectively). The analyzed traits included plant height, number of main and secondary branches per plant, days to first flowering, number of flowers per plant, stamen number, shoot fresh and dry weight, crown diameter, root length, and root fresh weight. The results indicated that most of these traits were several times higher in diploids than in tetraploids, ranging from three- to ninefold differences, particularly in stamen number, number of flowers per plant, and shoot fresh and dry weight.

Table 2. Key morphological traits of diploid and tetraploid black cumin plants.

Trait (Unit)	Diploid	Tetraploid	p-Value
Plant height (cm)	116.50 ± 0.866	95.00 ± 2.887	0.001021 *
The main number of branches	10.50 ± 0.289	8.75 ± 0.144	0.002804 *
Number of secondary branches per plant	112.00 ± 2.309	22.00 ± 1.155	0.00001 *
First days to flowering (day)	132.70 ± 1.443	97.09 ± 3.453	0.00001 *
Number of flowers per plant	111.70 ± 1.732	24.75 ± 0.289	0.00001 *
Flower diameter (cm)	3.70 ± 0.058	3.27 ± 0.010	0.000918 *
Converted flowering parts (petaloid stamen)	20.34 ± 0.510	188.00 ± 1.020	0.00001 *
Stamen number	43.34 ± 0.820	9.44 ± 0.290	0.00021 *
Capsule diameter (cm)	1.49 ± 0.078	0.83 ± 0.880	0.002601 *
Shoot fresh weight (g)	610.00 ± 5.774	147.50 ± 1.443	0.00001 *
Shoot dry weight (g)	110.50 ± 2.598	37.00 ± 1.732	0.00001 *
Crown diameter (cm)	1.91 ± 0.052	1.16 ± 0.118	0.002142 *
Leaf area (mm)	17.50 ± 0.058	22.50 ± 0.289	0.000035 *
Root length (cm)	21.65 ± 0.202	19.00 ± 0.577	0.006164 *
Root fresh weight (g)	8.00 ± 0.577	5.95 ± 0.029	0.011939 *

* = significant at $p < 0.05$ according to Student's unpaired t-test.

Clear differences in leaf morphological features were observed between diploid and tetraploid plants, including variations in cotyledon and mature leaf shape (Figure 3a–f). The cotyledon leaf of a diploid is taller and more pointed than the shorter tetraploid, which has rounded leaf tops, whereas the mature leaves are lower in thickness and more trimmed than tetraploid ones. On the level of field plants, the plant size of the diploid plants at the flowering stage of both diploid and tetraploid plants was observed and considered. It was observed that diploid plants exhibited greater growth and branch intensity compared to tetraploid plants. The flower shapes of diploid and tetraploid plants and stamen size were compared. This is a morphological analysis of diploid and tetraploid plants performed by examining the impact of applied 2,4-dinitroanaline on the plant polyploidy. It is apparent that the flower size of the tetraploid plants was bigger than that of the diploid plants (Figure 4a–d), and the same trend for stamen size can be noticed (Figure 4e). Also, it is observed in tetraploid plants that a number of stamens converted to petals, affecting the flower size and fertility.

Figure 3. Morphology of diploid and tetraploid black cumin plants: (**a**,**b**) cotyledon shape of diploid and tetraploid plants, respectively; (**c**,**d**) mature leaf shape of diploid and tetraploid plants, respectively (bar = 1 cm); (**e**,**f**) vegetative growth at the maturity stage of diploid and tetraploid plants, respectively (bar = 10 cm).

Figure 4. Flowering of diploid and tetraploid black cumin plants: (**a**,**b**) flowering and flower shape of diploid plants grown in the field; (**c**,**d**) flowering and flower shape of tetraploid plants grown in the field; (**e**) stamen size of diploid versus tetraploid plants (magnification = 15×).

3.4. Yield Traits and Thymoquinone Content in Diploid vs. Tetraploid Plants

Figure 5 shows the yield traits of the diploid and tetraploid black cumin plants (i.e., number of seeds per capsule, seed yield per plant, 100-seed weight, and oil percent). All previous yield parameters recorded higher values for the diploid plants compared with the tetraploid plants except the 100-seed weight. The number of seeds per capsule, seed yield per plant, 100-seed weight, and oil percent were 96.5, 24 g, 0.205 g, and 22.5% of the diploid plants, respectively, and 36.5, 4.5 g, 0.385 g, and 17.67% of the tetraploid plants, respectively. Thus, the 100-seed weight of the tetraploid plants was double that of the diploid plants. Data presented in Figure 6 show the thymoquinone percentage in the diploid and tetraploid plants. The obtained results indicated that tetraploid plants are superior to diploid plants in thymoquinone content, with 13.58 and 11.45 mg/100 g seeds, respectively (Figure 6D).

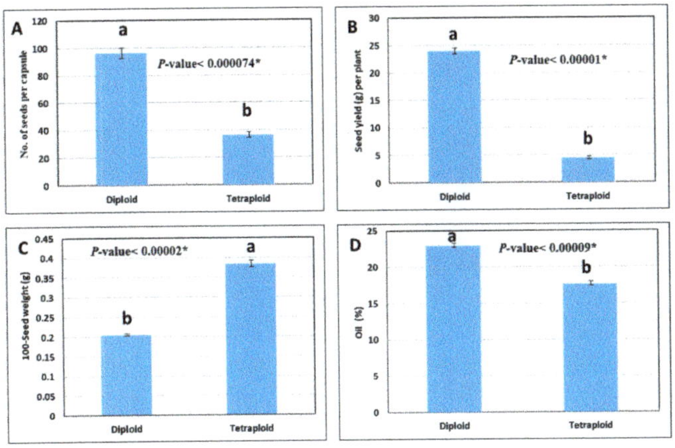

Figure 5. Yield traits of diploid and tetraploid black cumin plants: (**A**) number of seeds per capsule; (**B**) seed yield per plant; (**C**) 100-seed weight; (**D**) oil percentage per plant. The same letters mean non-significant impact, according to Duncan's multiple tests, at a 5% level. * Significant at $p < 0.05$ according to Student's unpaired t-test.

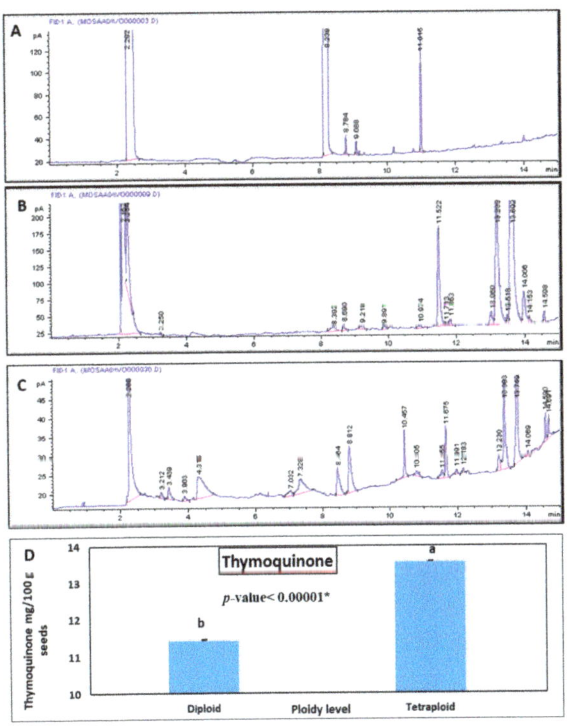

Figure 6. GC analysis of thymoquinone content in diploid versus tetraploid black cumin seeds: (**A**) thymoquinone standard; (**B**) diploid seeds; (**C**) tetraploid seeds; and (**D**) thymoquinone content in 100 g seeds of diploid versus tetraploid plants. Retention time (RT) for standard, diploid, and tetraploid seeds are 8.238, 8.392, and 8.464 min, respectively. The same letters mean non-significant impact, according to Duncan's multiple tests, at a 5% level. * Significant at $p < 0.05$ according to Student's unpaired t-test.

4. Discussion

Chromosome counting and flow cytometry could be used for genotyping-based ploidy determination in most accessions of Yams (*Dioscorea* spp.), allowing future rapid screening of ploidy levels [47]. Flow cytometry measures the relative fluorescence intensity that refers to the relative DNA content or genome size in the nuclei of the polyploidy samples compared to diploids. In our results the MFI of diploid plants was nearly half that of tetraploid. So this technique is widely used as the most convenient and rapid method for screening DNA ploidy level in plants [33,48,49]. Previous studies reported that flow cytometry has much higher accuracy than any other methodology for measuring ploidy, so it has become the best assay for ploidy level detection [50,51]. Therefore, diploid and tetraploid plants could be confirmed using flow cytometry assessment.

Ploidy level in black cumin was significantly associated with variations in seed germination. Diploid plants exhibited a higher germination percentage and required fewer days to reach 50% of final germination percentage (FGP) compared to tetraploid plants. Polyploidy may influence seed establishment, germination, and development [52]. Also, interactions between polyploidy level and the seed developmental environment may affect subsequent dormancy, and early growth traits, particularly under stress conditions [53]. A previous study indicated that tetraploid seeds had higher dormancy and a lower germination rate as compared with diploid seeds [53,54].

The present study highlights three morphological markers that can be effectively applied in black cumin breeding programs. Notably, cotyledon leaf shape proved to be a reliable indicator of ploidy level, as diploids consistently exhibited pointed tips, whereas tetraploids displayed rounded tips. These findings align with previous reports demonstrating that cotyledon morphology in black cumin is influenced by ploidy level, thereby supporting the use of this trait as an early, practical marker for distinguishing cytotypes [4]. This important feature can help black cumin breeders to detect the polyploidy level, and select tetraploid plants in the early growing stage, saving time and costs. The second marker was a little leaf-slitting and thicker leaves for the tetraploids, whereas the opposite was observed for the diploids. Previous studies demonstrated that triploids and tetraploids showed a wide range of variations in morphological traits, such as thicker and darker-green leaves [4,55]. The third marker was the flowers; there are obvious differences between diploids and tetraploids. These differences are due to the stamen number, which was more than fourfold for the diploids (43.34) what it was for the tetraploids (9.44). A possible explanation for using tetraploid black cumin plants in a breeding program as ornamental plants might be due to their converting stamens into petals, producing a big flower and increased longevity of the flowers. These findings for tetraploids are in agreement with the earlier flowering (after 97 days), lower number of flowers (24 per plant), and converted flowering parts (188), which increased the size of flowers. Another possible explanation for this result may be the increased longevity of these flowers, and the infertile flowers may be due to converting flowering parts into petaloid stamen.

Morphological traits of black cumin are crucial parameters, as they are closely linked to yield components and essential oil production. Tetraploid black cumin plants exhibit distinct morphological changes under field conditions. Several previous studies confirmed the potential value of black cumin traits (e.g., [56,57]). The differentiation in morphological characteristics between the diploid and tetraploid plants was investigated in both seedlings and mature plants. The start of flowering was earlier in January for tetraploids, whereas it was in February for diploids. The most interesting finding was that only the 100-seed weight of tetraploid plants was higher than that of diploid plants. The same finding was observed by Stevens et al. [53] on perennial grass. They found that tetraploid seeds were significantly heavier than diploid seeds by an average of 35%, whereas the yield of seeds associated with diploid plants was recorded as being fivefold that of tetraploids. A possible explanation for tetraploids' lower seed yield may be the conversion of flowering parts into petaloid stamen (infertile flowers). Another possible explanation for this result may be the big thickness of pollen tubes in tetraploids, which slow growth in the stigma tissues and then delay the fertilization of ovules, lowering the number of seeds in each ovary. Tetraploid plants exhibited stalk-lodging at maturity and low levels of segregation in the F2 and upcoming generations [55]. Tetraploids of *Jatropha curcas* L. possessed a decreased number of flowers per inflorescence, fruits per fluorescence, and seeds per fruit. This condition is due to lower fertility of pollen of the tetraploid plants and/or the slow rate of the photosynthesis process [58]. Tetraploids of *Echinacea purpurea* (L.) Moench exhibited larger flowers and seeds [34], and significantly larger pollen grains were noticed in tetraploids of *Ocimum basilicum* L. [59]. On the other hand, black cumin producers can benefit from the tetraploid genotype by intensifying planting to obtain a high yield near the diploid plant.

Although the polyploidy may improve some traits, sometimes it can affect others negatively. In this context, flowering and bolting genes of radish were reported to have lower expression in tetraploid compared to diploid plants; however, tetraploid plants were greater in size in terms of both vegetative and reproductive organs [26]. Tetraploidy induction (2n = 4x = 24) was reported on wallflower (*Erysimum cheiri* (L.) Crantz), which enhanced the morphological, physiological, and biochemical traits of this plant [60]. In our present study, it was found that the tetraploid plants of black cumin had a negative effect on the traits related to vegetative and yield components. A possible explanation for these results may be the tetraploids, which often can alter rates of vegetative growth relative to their diploid progenitors. The degradation of chloroplast in tetraploid leaves was accelerated, the rate of photosynthesis was decreased, and then the synthesis of carbohydrates was decreased [61]. Previous studies indicated that polyploid plants have slower growth and development [62], because the difficulties happen in the cell cycle in addition to slow cell division [63], which resulted in decreased cell numbers and smaller organs.

A program of plant breeding of a certain crop like medicinal plants of promising value has many strategies depending on its purpose. A program for cumin plants is preferable to using tetraploids for producing high essential oil compared to diploids [64]. The producing tetraploids of some plants have many desirable characteristics compared to diploid plants, such as potato tubers of big size [65,66]. Black cumin crops can be propagated for three purposes: producing seeds, a high content of active compounds in oil, and the desired flowers. Therefore, the breeders of black cumin production face many challenges depending on the purpose of such production, which may include (1) producing the seeds by propagation of diploid plants, (2) producing the active compounds through a callus of tetraploids, and (3) producing the flowers, as ornamental plants, via triploid and tetraploid plants.

Producing active compounds (mainly thymoquinone) in black cumin seeds is a crucial issue for the breeders of this crop. This production can be achieved monthly through the callus protocol of tetraploids [7]. In Egypt, the production of black seeds is mainly in May, and postharvest, the seeds should be under cooling and modified conditions to keep such compounds active and ready for cultivation after storing for one year. In the current study, the 100-seed weight was reported to be high in tetraploids compared to diploid plants of black cumin, but at the same time, these findings suggest the lowering seed yield of tetraploids, as mentioned before. The oil percent in black seeds was higher in diploids than in tetraploid plants, along with the yield of seeds, which has the same trend. The variability in active compounds content in different varieties of the plant grown in different regions has been studied [67,68]. The significant differences in the environmental conditions at various sites, and genetic variation of the plant material, are what affect the phytochemical content. Additionally, advanced analytical methods enabled us to explore black cumin's environmental, genotypic, and ontogenetic variability regarding its thymoquinone content [69]. Thymoquinone in black cumin depends on the plant's genetic and environmental conditions. Our results indicated that tetraploid plants had more concentration of thymoquinone than diploid plants. The same finding was reported for tetraploid lines of black cumin where they contained higher concentrations of thymoquinone in the seeds' oil by a ratio of approximately 46% [55]. Several studies assumed that synthetic polyploidization might enhance the biosynthesis of primary and secondary metabolites due to chromosome duplication that could affect the biological activities of the polyploid plants [70–72].

We compared in our published studies the production of diploid and tetraploid plants of black cumin and their comparison at the mature stage under field conditions (e.g., [19,21,33]). In contrast, this study is the first report on such a comparison at the early growth stage (seedling). The findings of the current research on tetraploid plants confirmed that these plants are a treasure for black cumin breeders because of their superiority in one or more of their genotypes. Therefore, it is necessary to identify them in their different growing stages and by several ways. The most obvious finding from the current investigation is that the differentiation between diploid and tetraploid plants could be distinguished during the seedling stage.

The production story began in 2015 with El-Mahrouk et al. [19], followed by cultivation of tetraploid plants for more than 5 years until 2021 to confirm this production of tetraploid plants. During the two successive seasons, the tetraploid and diploid seeds were sown while recording important guidelines for the studied breeding program. The first result is distinguishing between diploids and tetraploid seeds after germination under field conditions. The second one clarifies the suggested three strategies of black cumin production, including producing seeds, active compounds, and flowers of this plant. The third one is how to orient the plant breeding of black cumin by focusing on the morphological and cytological attributes and selecting the right time for cultivating practices from germination to harvesting. This study opens considerable opportunities for more research on black cumin plant breeding programs.

5. Conclusions

Black cumin is considered a highly potent medicinal plant with unique active phytochemicals for its therapeutic potential, and several benefits. Black cumin crops can be mainly propagated for three purposes (producing seeds, high content of active compounds in oil, and the desired flowers). The present study provides an important reference for black cumin breeders. Cotyledons can be an excellent marker for distinguishing between tetraploid and diploid plants due to the possibility of identifying polyploid plants at an early stage of growth (cotyledon stage). This approach can help breeders to avoid undesirable hybridizations that result in sterile plants without any crop and to determine the intended purpose of black cumin cultivation. Furthermore, vegetative trait markers could help in reducing the high cost of genetic markers or flow cytometry. Our study showed that diploid plants are superior in most morphological traits (i.e., plant height; number of branches, flowers, and stamens; and vegetative fresh weight) and yield traits (i.e., number of seeds per capsule, seed weight per plant, and oil percentage) when compared with tetraploid plants. Despite the negative effects of polyploidy in black cumin on both vegetative and yield traits, it had a positive and promotional effect on the ratio of thymoquinone content and 100-seed weight. This may indicate the possibility of producing tetraploid plants commercially through intensive planting. This study contributes valuable information to ongoing efforts in black cumin improvement. The findings support the integration of morphological markers as a cost-effective and accessible tool in polyploid breeding programs aimed at enhancing the medicinal and economic value of this important species.

Author Contributions: Conceptualization and visualization, M.E.E.-M. and M.K.M.; methodology, M.E.E.-M., M.K.M. and N.A.; software, Y.A.B. and S.S.; validation, M.K.M. and M.E.E.-M.; formal analysis, M.E.E.-M., N.A., S.S., M.K.M. and Y.A.B.; investigation, resources and data curation, M.E.E.-M., M.K.M. and N.A.; writing—original draft preparation, M.E.E.-M., H.E.-R. and N.A.; writing—review and editing, M.E.E.-M., H.E.-R. and N.A.; project administration, H.E.-R. and Y.A.B. All authors have read and agreed to the published version of the manuscript.

Funding: This research received no external funding.

Institutional Review Board Statement: Not applicable.

Data Availability Statement: The original contributions presented in this study are included in the article. Further inquiries can be directed to the corresponding author(s).

Acknowledgments: The authors thank the staff members of the Physiology and Breeding of Horticultural Crops Laboratory, Dept. of Horticulture, Fac. of Agric., Kafrelsheikh University, Kafr El-Sheikh, Egypt, for conducting the biochemical assays and other parameters.

Conflicts of Interest: The authors declare no conflicts of interest.

References

1. Ahmad, A.; Husain, A.; Mujeeb, M.; Khan, S.A.; Najmi, A.; Siddique, N.A.; Damanhouri, Z.A.; Anwa, F. A review on therapeutic potential of *Nigella sativa*: A miracle herb. *Asian Pac. J. Trop. Biomed.* **2013**, *3*, 337–352. [CrossRef]
2. Adam, S.H.; Abu, I.F.; Kamal, D.A.M.; Febriza, A.; Kashim, M.I.A.M.; Mokhtar, M.H. A Review of the Potential Health Benefits of *Nigella sativa* on Obesity and Its Associated Complications. *Plants* **2023**, *12*, 3210. [CrossRef]
3. Hussain, S.; Rukhsar, A.; Iqbal, M.; Ain, Q.; Fiaz, J.; Akhtar, N.; Afzal, M.; Ahmad, N.; Ahmad, I.; Mnif, W.; et al. Phytochemical profile, nutritional and medicinal value of *Nigella sativa*. *Biocatal. Agric. Biotechnol.* **2024**, *60*, 103324. [CrossRef]
4. Verma, S.; Hariwal, M.; Kumar, S. Exploratory analysis of agro-morphological characteristics in *Nigella sativa* L. plant genotypes to determine mutagen colchicine ameliorative/non-ameliorative impacts. *Sci. Rep.* **2024**, *14*, 24521. [CrossRef] [PubMed]
5. Hannan, M.A.; Rahman, M.A.; Sohag, A.A.M.; Uddin, M.J.; Dash, R.; Sikder, M.H.; Rahman, M.S.; Timalsina, B.; Munni, Y.A.; Sarker, P.P.; et al. Black Cumin (*Nigella sativa* L.): A Comprehensive Review on Phytochemistry, Health Benefits, Molecular Pharmacology, and Safety. *Nutrients* **2021**, *13*, 1784. [CrossRef]
6. Oubannin, S.; Jadouali, S.M.; Atifi, H.; Bijla, L.; Ibourki, M.; Gagour, J.; Bouzid, H.A.; Aabd, N.A.; Bouyahya, A.; Harhar, H.; et al. Antioxidant activity, physico-chemical properties, and bioactive compounds of *Nigella sativa* seeds and oil impacted by microwave processing technique. *Heliyon* **2024**, *10*, e37603. [CrossRef]
7. Higazy, A.E.; El-Mahrouk, M.E.; El-Banna, A.N.; Maamoun, M.K.; El-Ramady, H.; Abdalla, N.; Dobránszki, J. Production of Black Cumin via Somatic Embryogenesis, Chemical Profile of Active Compounds in Callus Cultures and Somatic Embryos at Different Auxin Supplementations. *Agronomy* **2023**, *13*, 2633. [CrossRef]
8. Dubey, P.N.; Singh, B.; Mishra, B.K.; Kant, K.; Solanki, R. Nigella (*Nigella sativa*): A high value seed spice with immense medicinal potential. *Indian J. Agric. Res.* **2016**, *86*, 967–979. [CrossRef]
9. Lal, G.; Meena, S.S.; Lal, S. Nigella (*Nigella sativa* L.), a novel herb can cure many diseases: A review. *Int. J. Seed Spices* **2020**, *10*, 1–10.
10. Yimer, E.M.; Tuem, K.B.; Karim, A.; Ur-Rehman, N.; Anwar, F. *Nigella sativa* L. (Black Cumin): A Promising Natural Remedy for Wide Range of Illnesses. *Evid. Based Complement. Altern. Med.* **2019**, *2019*, 1528635. [CrossRef]
11. Li, H.; Da Silva, N.A.; Liu, W.; Xu, J.; Dombi, G.W.; Dain, J.A.; Li, D.; Chamcheu, J.C.; Seeram, N.P.; Ma, H. Thymocid®, a Standardized Black Cumin (*Nigella sativa*) Seed Extract, Modulates Collagen Cross-Linking, Collagenase and Elastase Activities, and Melanogenesis in Murine B16F10 Melanoma Cells. *Nutrients* **2020**, *12*, 2146. [CrossRef]

12. Datta, A.K.; Saha, A.; Bhattacharya, A.; Mandal, A.; Paul, R.; Sengupta, S. Black Cumin (*Nigella sativa* L.)—A Review. *J. Plant Dev. Sci.* **2012**, *4*, 1–43.
13. Heslop-Harrison, J.S.; Schwarzacher, T.; Liu, Q. Polyploidy: Its consequences and enabling role in plant diversification and evolution. *Ann. Bot.* **2023**, *131*, 1–10. [CrossRef]
14. Morris, J.P.; Baslan, T.; Soltis, D.E.; Soltis, P.S.; Fox, D.T. Integrating the Study of Polyploidy Across Organisms, Tissues, and Disease. *Annu. Rev. Genet.* **2024**, *58*, 297–318. [CrossRef]
15. Blakeslee, A.F.; Avery, A.G. Methods of inducing doubling of chromosomes in plants: By treatment with colchicine. *J. Hered.* **1937**, *28*, 393–411. [CrossRef]
16. Biswas, A.K.; Chatterjee, A.K. Studies on the induction of ploidy in some species. *Bull. Bot. Soc. Bengal* **1971**, *25*, 19–21.
17. Biswas, A.K.; Datta, A.K. Studies on induced autotetraploids in *Nigella sativa* L. *Cell Chromosome Res.* **1982**, *5*, 81–83.
18. Saha, A.; Datta, A.K. Induced autotetraploidy in black cumin (*Nigella sativa* L). *Indian J. Genet. Plant Breed.* **2002**, *62*, 275–276.
19. El-Mahrouk, M.E.; Maamoun, M.K.; Dewir, Y.H.; Omran, S.A.; EL-Banna, A.N. Morphological and molecular characterization of induced mutants in *Nigella sativa* L. using irradiation and chemical mutagens. *Egypt. J. Plant Breed.* **2015**, *19*, 257–272.
20. Dhooghe, E.; Van Laere, K.; Eeckhaut, T.; Leus, L.; Van Huylenbroeck, J. Mitotic chromosome doubling of plant tissues in vitro. *Plant Cell Tissue Organ Cult.* **2011**, *104*, 359–373. [CrossRef]
21. El-Mahrouk, M.E.; Maamoun, M.K.; Abu El-Leel, O.F.; Dewir, Y.H.; El-Banna, A.N.; Naidoo, Y.; Datta, S.K. Morpho-agronomical and Biochemical Traits Screening and Genetic Variability in Selected Black Cumin (*Nigella sativa*) Mutant Lines. *Sains Malays.* **2020**, *49*, 503–515. [CrossRef]
22. Wu, J.; Ferguson, R.A.; Murray, B.G.; Jia, Y.; Datson, P.M.; Zhang, J. Induced polyploidy dramatically increases the size and alters the shape of fruit in *Actinidia chinensis*. *Ann. Bot.* **2012**, *109*, 169–179. [CrossRef] [PubMed]
23. Zahumenická, P.; Fernández, E.; Šedivá, J.; Žiarovská, J.; Ros-Santaella, J.L.; Martínez-Fernández, D.; Russo, D.; Milella, L. Morphological, physiological and genomic comparisons between diploids and induced tetraploids in *Anemone sylvestris* L. *Plant Cell Tissue Organ Cult.* **2018**, *132*, 317–327. [CrossRef]
24. Dhawan, O.P.; Lavania, U.C. Enhancing the productivity of secondary metabolites via induced polyploidy: A review. *Euphytica* **1996**, *87*, 81–89. [CrossRef]
25. De Jesus-Gonzalez, L.; Weathers, P.J. Tetraploid *Artemisia annua* hairy roots produce more artemisinin than diploids. *Plant Cell Rep.* **2003**, *21*, 809–813. [CrossRef]
26. Pei, Y.; Yao, N.; He, L.; Deng, D.; Li, W.; Zhang, W. Comparative study of the morphological, physiological and molecular characteristics between diploid and tetraploid radish (*Raphunas sativus* L.). *Sci. Hortic.* **2019**, *257*, 108739. [CrossRef]
27. Hancock, J.F. The colchicine story. *Hortic. Sci.* **1997**, *32*, 1011–1012. [CrossRef]
28. Hias, N.; De Dauw, K.; Davey, M.W.; Leus, L.; Van Labeke, M.C.; Van Huylenbroeck, J.; Keulemans, J. Influence of ploidy level on the physiological response of apple to water deficit. *Acta Hortic.* **2017**, *1177*, 333–338. [CrossRef]
29. Masterson, J. Stomatal size in fossil plants: Evidence for polyploidy in majority of angiosperms. *Science* **1994**, *264*, 421–423. [CrossRef]
30. Guerra, D.; Wittmann, M.T.S.; Schwarz, S.F.; de Souza, P.V.D.; Mateus Pereira Gonzatto, M.P.; Weiler, R.L. Comparison between diploid and tetraploid citrus rootstocks: Morphological characterization and growth evaluation. *Bragantia* **2014**, *73*, 1–7. [CrossRef]
31. Sattler, M.C.; Carvalho, C.R.; Clarindo, W.R. The polyploidy and it skeyrolein plant breeding. *Planta* **2016**, *243*, 281–296. [CrossRef]
32. Rao, S.; Kang, X.; Li, J.; Chen, J. Induction, identification and characterization of tetraploidy in *Lycium ruthenicum*. *Breed. Sci.* **2019**, *69*, 160–168. [CrossRef]
33. El-Mahrouk, M.E.; Maamoun, M.K.; EL-Banna, A.N.; Omran, S.A.; Dewir, Y.H.; El-Hendawy, S. In Vitro Gynogenesis and Flow Cytometry Analysis of the Regenerated Haploids of Black Cumin (*Nigella sativa*). *Hortic. Sci.* **2018**, *53*, 681–686. [CrossRef]
34. Abdoli, M.; Moieni, A.; Badi, H.N. Morphological, physiological, cytological and phytochemical studies in diploid and colchicine-induced tetraploid plants of *Echinacea purpurea* (L.). *Acta Physiol. Plant.* **2013**, *35*, 2075–2083. [CrossRef]
35. Tomaszewska, P.; Pellny, T.K.; Hernández, L.M.; Mitchell, R.A.C.; Castiblanco, V.; de Vega, J.J.; Schwarzacher, T.; Heslop-Harrison, P. Flow Cytometry-Based Determination of Ploidy from Dried Leaf Specimens in Genomically Complex Collections of the Tropical Forage Grass Urochloa s. l. *Genes* **2021**, *12*, 957. [CrossRef] [PubMed]
36. Anamthawat, J.K. Preparation of chromosomes from plant leaf meristems karyotype analysis and in situ hybridization. *Methods Cell Sci.* **2003**, *25*, 91–95. [CrossRef]
37. Galbraith, D.W.; Harkins, K.R.; Maddox, J.M.; Ayres, N.M.; Sharma, D.P.; Firoozabady, E. Rapid flow cytometric analysis of the cell cycle in intact plant tissues. *Science* **1983**, *220*, 1049–1051. [CrossRef] [PubMed]
38. Esechie, H. Interaction of salinity and temperature on the germination of sorghum. *J. Agron. Crop Sci.* **1994**, *172*, 194–199. [CrossRef]

39. Hsu, F.H.; Nelson, C.J.; Matches, A.G. Temperature effects on germination of perennial warm-season forage grasses. *Crop Sci.* **1985**, *25*, 212–220.
40. Jackson, M.L. *Soil Chemical Analysis*; Prentice Hall of India, Private Limited: New Delhi, India, 1973.
41. Nelson, D.W.; Sommers, L.E. Total carbon, organic carbon, and organic matter. In *Methodsofsoil Analysis. Part 3. Chemical Methods*; Black, C.A., Ed.; Soil Science of America and American Society of Agronomy: Madison, WI, USA, 1996; pp. 961–1010.
42. Bremner, J.M.; Mulvaney, C.S. Nitrogen total. In *Methods of Soil Analysis. Agron. No. 9 Part II; Chemical and Microbiological Properties*, 2nd ed.; Page, A.L., Ed.; American Society of Agronomy: Madison, WI, USA, 1982; pp. 595–624.
43. Olsen, S.R.; Sommers, L.E. Phosphorus. In *Methods of Soil Analysis. Agron. No. 9, Part 2; Chemical and Microbiological Properties*, 2nd ed.; Page, A.L., Ed.; American Society of Agronomy: Madison, WI, USA, 1982; pp. 403–430.
44. Black, C.A. *Methods of Soil Analysis Part I No. 9*; American Society of Agronomy: Madison, WI, USA, 1965.
45. Rajeswara, R.B.R.; Singh, K.; Kaul, P.N.; Bhattacharya, A.K. The effect of plant spacing and application of N and P fertilizers on the productivity and nutrient uptake of davana (*Artemisia pallens* Wall.). *Int. J. Trop. Agric.* **1989**, *7*, 229–236.
46. Folch, J.; Lees, M.; Sloane Stanley, G.H. A simple method for the isolation and purification of total lipides from animal tissues. *J. Biol. Chem.* **1957**, *226*, 497–509. [CrossRef]
47. Gatarira, C.; Sládeková, L.; Němečková, A.; Šimoníková, D.; Paliwal, R.; Asfaw, A.; Abberton, M.; Badara, G.; Asiedu, R.; Čížková, J.; et al. Cytological and Molecular Characterization for Ploidy Determination in Yams (*Dioscorea* spp.). *Agronomy* **2021**, *11*, 1897. [CrossRef]
48. Loureiro, J.; Rodriguez, E.; Doležel, J.; Santo, C. Comparison of four nuclear isolation buffers for plant DNA flow cytometry. *Ann. Bot.* **2006**, *98*, 679–689. [CrossRef]
49. Doležel, J.; Greilhuber, J.; Suda, J. Estimation of nuclear DNA content in plants using flow cytometry. *Nat. Protoc.* **2007**, *2*, 2233–2244. [CrossRef]
50. Couto, E.G.; Davide, L.M.; Bustamante, F.; VonPinho, R.G.; Silva, T.N. Identification of haploid maize by flow cytometry morphological and molecular markers. *Cienc. Agrotecnol.* **2013**, *37*, 25–31. [CrossRef]
51. Kielkowska, A.; Adamus, A. In vitro culture of unfertilized ovules in carrot (*Daucus carota* L.). *Plant Cell Tissue Organ Cult.* **2010**, *102*, 309–319. [CrossRef]
52. Chan, J.C.S.; Ooi, M.K.J.; Guja, L.K. Polyploidy but Not Range Size Is Associated with Seed and Seedling Traits That Affect Performance of Pomaderris Species. *Front. Plant Sci.* **2022**, *12*, 779651. [CrossRef] [PubMed]
53. Stevens, A.V.; Nicotra, A.B.; Godfree, R.C.; Guja, L.K. Polyploidy affects the seed, dormancy and seedling characteristics of a perennial grass, conferring an advantage in stressful climates. *Plant Biol.* **2020**, *22*, 500–513. [CrossRef]
54. Hacker, J.B. Polyploid distribution and seed dormancy in relation to provenance rainfall in the *Digitaria milanjiana* complex. *Aust. J. Bot.* **1988**, *36*, 693–700. [CrossRef]
55. Dixit, V.; Verma, S.; Chaudhary, B.R. Changes in ploidy and its effect on thymoquinone concentrations in *Nigella sativa* L. seeds. *J. Hortic. Sci. Biotechnol.* **2015**, *90*, 537–542. [CrossRef]
56. Hosseini, S.S.; Nadjafi, F.; Asareh, M.H.; Rezadoost, H. Morphological and yield related traits, essential oil and oil production of different landraces of black cumin (*Nigella sativa*) in Iran. *Sci. Hortic.* **2018**, *233*, 1–8. [CrossRef]
57. Telci, İ.; Özek, T.; Gül, F.; Yur, S.; Özek, G.; Demirtaş, İ.; Günay, E.; Aslancan, H.; Kacar, O. Diversity of black cumin genotypes and their classification based on functional properties. *Biochem. Syst. Ecol.* **2024**, *113*, 104802. [CrossRef]
58. Niu, L.; Tao, Y.; Chen, M.; Fu, Q.; Dong, Y.; He, H.; Xu, Z. Identification and characterization of tetraploid and octoploid *Jatropha curcas* induced by colchicine. *Caryologia* **2016**, *69*, 58–66. [CrossRef]
59. Omidbaigi, R.; Mirzaeea, M.; Hassani, M.E.; Sedghi-Moghadam, M. Induction and identification of polyploidy in basil (*Ocimum basilicum* L.) medicinal plant by colchicine treatment. *Int. J. Plant Prod.* **2010**, *4*, 87–98.
60. Fakhrzad, F.; Jowkar, A.; Shekafandeh, A.; Kermani, M.J.; Moghadam, A. Tetraploidy induction enhances morphological, physiological and biochemical characteristics of wallflower (*Erysimum cheiri* (L.) Crantz). *Sci. Hortic.* **2023**, *308*, 111596. [CrossRef]
61. Xu, C.; Zhang, Y.; Han, Q.; Kang, X. Molecular Mechanism of Slow Vegetative Growth in Populus Tetraploid. *Genes* **2020**, *11*, 1417. [CrossRef]
62. Ranney, T.G. Polyploidy: From Evolution to New Plant Development. *Proc. Int. Plant Propagators Soc.* **2006**, *56*, 604–607.
63. Comai, L. The advantages and disadvantages of being polyploid. *Nat. Genet.* **2005**, *6*, 836–846. [CrossRef] [PubMed]
64. Sanaei-Hoveida, Z.; Mortazavian, S.M.M.; Norouzi, M.; Sadat-Noori, S.A. Elevating morphology and essential oil in cumin genotypes through polyploidy induction. *Sci. Hortic.* **2024**, *329*, 113031. [CrossRef]
65. Muthoni, J.; Shimelis, H.; Melis, R. Production of hybrid potatoes: Are heterozygosity and ploidy levels important? *Aust. J. Crop Sci.* **2019**, *13*, 687–694. [CrossRef]
66. Pandey, J.; Scheuring, D.C.; Koym, J.W.; Vales, M.I. Genomic regions associated with tuber traits in tetraploid potatoes and identification of superior clones for breeding purposes. *Front. Plant Sci.* **2022**, *13*, 952263. [CrossRef]

67. Kumar, S.; Dobos, G.J.; Rampp, T. The significance of ayurvedic medicinal plants. *Evid.-Based Complement. Altern. Med.* **2017**, *22*, 494–501. [CrossRef] [PubMed]
68. Ravi, Y.; Vethamoni, P.I.; Saxena, S.N.; Raveendran, M.; Velmurugan, S.; Santhanakrishnan, P. Extraction and estimation of thymoquinone (a highly valued metabolite) from *Nigella sativa* L. *Med. Plants-Int. J. Phytomed. Relat. Ind.* **2022**, *14*, 492–498.
69. Ravi, Y.; Vethamoni, I.P.; Saxena, S.N.; Velmurugan, S.; Santanakrishnan, V.P.; Raveendran, M.; Bariya, H.; Harsh, M. Guesstimate of thymoquinone diversity in *Nigella sativa* L. genotypes and elite varieties collected from Indian states using HPTLC technique. *Open Life Sci.* **2023**, *18*, 20220536. [CrossRef]
70. Jadaun, J.S.; Yadav, R.; Yadav, N.; Bansal, S.; Sangwan, N.S. Influence of Genetics on the Secondary Metabolites of Plants. In *Natural Secondary Metabolites: From Nature, Through Science, to Industry*; Springer International Publishing AG: Cham, Switzerland, 2023; pp. 403–433.
71. Jmii, G.; Gharsallaoui, S.; Mars, M.; Haouala, R. Polyploidization of *Trigonella foenum-graecum* L. enhances its phytotoxic activity against *Cyperus rotundus* L. *S. Afr. J. Bot.* **2023**, *153*, 336–345. [CrossRef]
72. Gupta, N.; Bhattacharya, S.; Dutta, A.; Tauchen, J.; Landa, P.; Urbanová, K.; Houdková, M.; Fernández-Cusimamani, E.; Leuner, O. Synthetic polyploidization induces enhanced phytochemical profile and biological activities in *Thymus vulgaris* L. essential oil. *Sci. Rep.* **2024**, *14*, 5608. [CrossRef] [PubMed]

horticulturae

MDPI

Article

Identification of DMP Family Members in *Solanaceous* Vegetables Potentially Involved in Haploid Induction

Xuan Deng [1,†], Wenjian Zhong [2,†], Bo Liu [1], Xinyan Shen [1], Zhiyong Ren [2], Yongen Lu [1], Xin Wang [1,*] and Bo Ouyang [1,*]

1 National Key Laboratory for Germplasm Innovation and Utilization of Horticultural Crops, Huazhong Agricultural University, Wuhan 430070, China; dx2022hzau@163.com (X.D.); 18731968101@163.com (B.L.); xyseiichi@163.com (X.S.); luyongen@mail.hzau.edu.cn (Y.L.)

2 Institute of Economic Crops, Hubei Academy of Agricultural Sciences, Wuhan 430064, China; rzy@hbaas.com (Z.R.)

* Correspondence: xinwang@mail.hzau.edu.cn (X.W.); bouy@mail.hzau.edu.cn (B.O.)

† These authors contributed equally to this work.

Abstract: Haploid breeding technology offers a promising means of significantly shortening the breeding cycle by rapidly generating homozygous inbred lines. Previous studies have shown that *DMP8* is involved in haploid induction across various plant species. In this study, we performed whole-genome identification and bioinformatics analyses to investigate the evolutionary relationships, gene structures, conserved domains, and expression patterns of DMP gene family members in tomato (*Solanum lycopersicum*), pepper (*Capsicum annuum*) and eggplant (*S. melongena*). A total of seven, six, and eight DMP genes were identified in the genomes of tomato, pepper, and eggplant, respectively. All encoded proteins contained the DUF679 domain, and the DMP family members were clustered into three distinct groups. Collinearity analysis revealed species-specific expansions of DMP genes in the *Solanaceae* family. Phylogenetic analysis indicated that *CaDMP8* and *SmDMP8* are homologous to *SlDMP8*, with conserved gene and protein structures, suggesting that *CaDMP8* and *SmDMP8* are potential targets for developing haploid induction lines. Expression pattern analysis demonstrated that *SlDMP4* and *SlDMP8* are highly expressed in tomato flower tissues, suggesting their potential functional synergy. This study provides the first comprehensive insight into the evolutionary characteristics and functional diversification of the DMP gene family in *Solanaceous* vegetables. The findings offer a theoretical foundation for the targeted editing of *DMP8* homologs to create haploid induction lines, which is critical for accelerating the genetic improvement of *Solanaceous* crops.

Keywords: DUF679 domain; bioinformatics analysis; evolutionary relationship; expression pattern; DMP8

1. Introduction

Double haploid breeding is an essential tool for accelerating plant breeding and has been widely applied in maize. The maize line Stock6 can generate 2–3% haploids when crossed with different female parents, and was therefore designated as a haploid inducer line [1]. Pure lines (2n), referred to as double-haploid lines (DHs) are obtained through chromosome doubling. Elite DHs exhibit no genetic and phenotypic segregation in their progeny. Compared with traditional breeding, which typically requires 6–8 generations of self-pollination to achieve pure lines, haploid breeding significantly shortens the breeding cycle, making it highly valuable.

Haploids can be generated using either *in vitro* or *in vivo* approaches. The *in vitro* method typically involves tissue or plant culture [2], but its success is influenced by factors such as genotype, developmental stage, culture conditions, physiological state, and stress pretreatment of plant material. Attempts to establish anther and microspore culture in tomato have largely been unsuccessful, as this species exhibits strong recalcitrance and requires substantial optimization [3]. By contrast, successful anther culture has been reported in pepper and eggplant [4]. Nevertheless, the efficiency of this approach is highly genotype dependent, and the technical complexity of the procedure renders it more akin to a specialized craft than a routine protocol [5]. The *in vivo* approach, on the other hand, utilizes interspecific or intraspecific hybridization to induce haploid formation. In this case, haploid induction typically results from the elimination of the parental genome during the first division of the fertilized egg or through parthenogenesis of the egg cell, although the induction frequency remains low [6]. To elucidate the mechanism of haploid induction in maize, several research groups have cloned a key QTL gene, *ZmPLA1/MTL/NLD*, which controls haploid induction [7–9]. The *MTL* gene is conserved in monocots, and haploid induction can be successfully achieved through knockout approaches [10–13]. However, no homologous genes for this gene have been identified in dicots. Subsequently, another haploid induction-related QTL was cloned [14], and its corresponding gene, *ZmDMP8*, encodes a DUF679 domain protein (Domain of unknown function 679 membrane proteins). Interestingly, homologs of *DMP8* have been detected across a wide range of flowering plants. Haploid inducer lines have been obtained by knocking out the homologous genes of *ZmDMP8* in various dicot species, such as *Arabidopsis*, *Medicago truncatula*, watermelon, and tobacco [15–18]. In tomato, *SlDMP8*-knockout plants produce aborted seeds and a small proportion of maternal haploids in cross- or self-pollinated progenies, with an average induction rate of 1.9% [19].

The DMP family has been documented in at least 24 species, including maize, tomato, pepper, cotton, and others [20]. Studies have shown that DMP homologous proteins are present exclusively in higher plants, mosses and Chlamydomonas, with no related amino acid motifs detected in mammals, fungi, and prokaryotes. All DMP proteins contain a conserved DUF679 domain, and the domain distribution within the same group is highly conserved, suggesting that their functions are likely consistent across species. The DMP8 protein contains the typical DUF679 conserved domain, and the transmembrane motifs (2–5) are all located within the DMP domain, indicating that its core function may be conserved throughout evolution. *AtDMP8* is expressed in both reproductive organs (such as flowers and seeds) and vegetative tissues (such as roots), with particularly high expression during seed development stage. This is consistent with the presence of endosperm development-related elements (*GCN4*) and seed-specific elements (*RY*) in its promoter, suggesting a role in seed development or germination. Cotton's *GhDMP8A* is expressed in both reproductive tissues (petals, anthers) and vegetative tissues, with RT-qPCR confirming its high expression in petals, implying a potential role in floral organ development or reproductive regulation. Beyond haploid induction, the DMP gene family has also been implicated in various physiological processes. Ten DMP family members have been identified in *Arabidopsis*, where they are involved in diverse physiological functions [20]. Among them, *AtDMP1* and *AtDMP2* participate in plant senescence regulation [21], *AtDMP4* is involved in programmed cell death [22], and *AtDMP8* and *AtDMP9* are implicated in haploid induction [16]. The functions of other members remain to be elucidated. In addition to *Arabidopsis*, *AsDMP1*, *AsDMP19* and *AsDMP22* have been identified in oats as potential regulators of seed maturation [23]. Some DMP members in cotton are induced by stress treatments [24], and similar findings have been reported in sugar beet, where certain DMP genes regulate stress responses [25].

In this study, we comprehensively analyzed and compared the DMP gene family in three *Solanaceous* species: tomato, eggplant and pepper. We examined their gene structures, conserved domains, evolutionary relationships, and expression patterns, and analyzed the three-dimensional structure of DMP family proteins. Our primary focus was on DMP members potentially involved in haploid induction. The results not only provide a foundation for the editing and pyramiding of haploid-related genes but also have significant implications for identifying family members involved in haploid induction and establishing a streamlined haploid breeding approach for *Solanaceous* vegetables.

2. Materials and Methods

2.1. *Identification of DMP Gene Family Members*

To identify DMP family genes in tomato, pepper, and eggplant, we used the amino acid sequences of *Arabidopsis* DMP proteins (AtDMP8 and AtDMP9) as query sequences to perform BLASTP (v2.17.0) searches in the protein databases of tomato (ITAG2.4), pepper (L_Zunla-1_v2.0), and eggplant (Eggplant_V4.1). A preliminary screening threshold was set to retain candidate sequences with an E-value < 1×10^{-10}, using additional default parameters. For more precise identification of DMP family members, the Hidden Markov Model (HMM) file for the DUF679 domain (PF05078.15) was obtained from the Pfam database. HMMER 3.0 software [26] was then used to conduct HMM searches on the whole-genome protein sequences of the three species. Only those sequences containing the complete DUF679 domain were retained, while sequences with missing or incomplete domains were discarded.

The candidate genes identified through this combined screening strategy were named based on their sequence homology with *Arabidopsis* DMP proteins. Detailed annotation information for each gene, including gene ID, chromosome location, gene length, and number of exons, was extracted from the corresponding genome annotation files (ITAG2.4 for tomato, L_Zunla-1_v2.0 for pepper, and Eggplant_V4.1 for eggplant).

2.2. *Construction of Phylogenetic Tree*

A phylogenetic tree was constructed for the DMP family members from the three Solanaceous species and *Arabidopsis*. Multiple sequence alignment was conducted using MUSCLE v3.8.425 software [27] with default parameters, and the alignment results were saved in FASTA format. The aligned sequences were then imported into MEGAX 11 software to construct a phylogenetic tree using the Neighbor-Joining method. A bootstrap analysis with 1000 replicates was performed to assess branch reliability [28]. After construction, the phylogenetic tree was exported in Newick format and uploaded to the iTOL online platform (https://itol.embl.de (accessed on 28 September 2025)) for visualization and enhancement, including branch color adjustments, labeling, and optimization of branch thickness [29]. A high-resolution PNG image was then downloaded for further analysis.

To explore the evolutionary relationships of DMP genes, a genome collinearity analysis was further performed using the One Step MCScanX function in TBtools [30]. The genome FASTA files and GFF3 annotation files were input and analyzed, including automatic gene alignment, block collinearity detection, and gene duplication types (tandem, dispersed, or whole-genome duplication). The generated result files were imported into TBtools for visualization.

2.3. Chromosome Localization, Gene Structure and Motif Analysis

The Gene Location Visualize (Advanced) module of TBtools software (v2.0) was used for chromosome location analysis. Chromosome length files (ChrLen.tab) and gene structure annotation files (GFF format) were imported into the software to generate the gene distribution map on chromosomes. For motif analysis, the Simple MEME Wrapper tool in TBtools was used to conduct motif discovery for all DMP family protein sequences. The parameters were set to ZOOPS mode, with a maximum of 10 motifs and a motif width ranging from 6 to 50 amino acids. After the analysis, the Visualize MEME Motif Pattern function in TBtools was used to visualize the identified motifs. The final figure was arranged, labeled, and composed using Adobe Illustrator 2025.

2.4. Prediction of Cis-Elements in Promoters

The 2000 bp upstream sequence of each DMP coding region was retrieved from the FASTA file of the reference genome and the corresponding GFF3 annotation file using TBtools software. The extracted sequences were then quality-checked using the "Fasta Extract or Filter (Quick)" tool, and subsequently converted to uppercase with the "Sequence Manipulate (RevComp)" tool to satisfy the input requirements of downstream analysis. The processed promoter sequence files were uploaded to the PlantCARE online database (http://bioinformatics.psb.ugent.be/webtools/plantcare/html/ (accessed on 5 October 2025)) for cis-element prediction. After obtaining the prediction results, the cis-element information was organized, and elements related to the biological processes of interest were selected, including hormone response elements (e.g., ABRE, G-box), stress response elements (e.g., DRE, LTRE), and light-response elements (e.g., GATA, GT1). Finally, the distribution of the selected cis-elements was visualized using the "Simple BioSequence Viewer" or "Gene Structure View (Advanced)" module in TBtools. The organized cis-element information table was imported into TBtools, and the "Promoter cis-element visualization" function was used to generate distribution maps of each cis-element.

2.5. Analysis of Protein Physicochemical Properties and Subcellular Localization Prediction

The physicochemical properties of the DMP proteins, including protein molecular weight (kDa), isoelectric point (pI), and instability index, were analyzed using the Expasy-ProtParam (https://web.expasy.org/protparam/ (accessed on 5 October 2025)). Hydrophobicity was evaluated using ProtScale (https://web.expasy.org/protscale/ (accessed on 5 October 2025)), and transmembrane domains were predicted with TMHMM-2.0 (https://services.healthtech.dtu.dk/services/TMHMM-2.0/ (accessed on 12 October 2025)). Subcellular localization predictions were made using WoLF PSORT (https://wolfpsort.hgc.jp/ (accessed on 5 October 2025)).

2.6. Protein Tertiary Structure Prediction

The three-dimensional structures of the DMP proteins were predicted using SWISS-MODEL (https://swissmodel.expasy.org/ (accessed on 12 October 2025)). The amino acid sequences were input, and the optimal template was selected with a confidence level greater than 30%. The model quality was evaluated using the GMQE (Global Model Quality Estimate) value (0–1), where higher values indicate better quality. Template matching was assessed by the QMEAN value (−4 to 0), with values closer to 0 indicating better alignment between the predicted protein and the template protein.

After modeling, the three-dimensional structure model of the selected DMP protein was exported as a PNG image. Finally, the three-dimensional structure images of all DMP proteins were arranged and labeled uniformly using Adobe Illustrator 2025, and a final composite image was generated.

2.7. Tissue Expression Analysis of DMP Family Genes Based on Transcriptome Data

The gene expression data of DMP genes from different tissues of tomato were obtained from the Solanaceae website (https://solgenomics.net/ (accessed on 28 September 2025)). Dynamic developmental expression data of the pepper DMP family genes were retrieved from the PepperHub database [31], and expression level data for eggplant DMP family genes were extracted from publicly available transcriptome data for different tissues [32].

The raw FPKM (or TPM) values were converted to $\log_2$(FPKM + 1) to reduce variance and avoid the effect of zero values on logarithmic calculations. Hierarchical clustering analysis of the gene expression matrix was then performed based on tissue type and developmental stage, using Euclidean distance as the similarity measure and the Ward.D2 method for cluster merging [33]. The clustering results were visualized in a heatmap, with a blue-white-red gradient color scheme to clearly present the upregulation and downregulation patterns of gene expression [34]. The final heatmap was imported into Adobe Illustrator 2025 for layout optimization and labeling.

2.8. Verification of the Tissue Expression Profile of Tomato DMP Genes

RT-qPCR was performed to validate the tissue expression of the seven DMP genes in tomato. RNA was extracted from roots, stems, leaves, fruits, and flowers of Micro-Tom plants 90 days after sowing using TRIzol reagent (Invitrogen, Carlsbad, CA, USA), following the manufacturer's instructions. Total RNA (1 μg) was reverse-transcribed using the HiScript II 1st Strand cDNA Synthesis Kit with genomic DNA wiper (Vazyme, Nanjing, China). Primers were designed using Primer5 and the NCBI Primer-BLAST tool (https://www.ncbi.nlm.nih.gov/tools/primer-blast (accessed on 30 July 2024)) (Table S1). qPCR reactions (10 μL each) contained 4 μL diluted cDNA, 5 μL SYBR Premix Ex Taq (Takara, Dalian, China), and 0.5 μL of each primer (10 mM). Amplification was performed on a QuantStudio 7 system (Thermo Fisher Scientific, Waltham, MA, USA) under the following program: 95 °C for 1 min, followed by 40 cycles of 95 °C for 10 s, 55 °C for 15 s, and 72 °C for 20 s. Three biological and three technical replicates were analyzed for each sample. Relative expression levels were calculated using the $2^{-\Delta\Delta Ct}$ method with *Actin* as the internal reference gene [35].

3. Results

3.1. DMP Gene Family Members in the Three Solanaceous Vegetable Genomes

A total of seven, six, and eight DMP genes were identified in the genomes of tomato, pepper, and eggplant, respectively. Phylogenetic analysis including *Arabidopsis* DMP family genes clustered these Solanaceous genes into distinct clades, and gene names were assigned accordingly. The basic characteristics of these DMP genes, including gene ID, chromosome location, gene length, and exon number, were obtained from the chromosome length and gene annotation files (Table 1). Gene lengths varied from 537 bp (*CaDMP1* and *SmDMP1*) to 1641 bp (*SmDMP8*). Chromosome localization showed that DMP family genes were scattered across various chromosomes, without a clear clustering pattern. Regarding gene structure, 17 of the 21 genes contained a single exon, while three homologous genes (*SlDMP8*, *CaDMP8*, and *SmDMP8*) and *CaDMP5* contained two exons. The genes with introns, such as the DMP8 homologs, were longer (average 1486 bp) than those without introns, which averaged 832 ± 215 bp.

Table 1. Characteristics of DMP family genes in tomato, pepper, and eggplant.

Gene Name	Gene ID	Chromosome Location	Length (bp)	Exon
SlDMP1	Solyc01g080490.2.1	Chr01: 79,721,811–79,722,647	837	1
SlDMP3	Solyc01g103580.2.1	Chr01: 92,152,884–92,153,909	1026	1
SlDMP4	Solyc10g008970.1.1	Chr10: 2,997,582–2,998,241	660	1
SlDMP6	Solyc02g077640.1.1	Chr02: 42,537,780–42,538,409	630	1
SlDMP7	Solyc12g005490.1.1	Chr12: 292,388–293,065	678	1
SlDMP8	Solyc05g007920.2.1	Chr05: 2,350,007–2,351,451	1445	2
SlDMP10	Solyc10g050200.1.1	Chr10: 48,118,856–48,119,491	636	1
CaDMP1	Capana01g003066	Chr01: 203,232,367–203,232,903	537	1
CaDMP3	Capana08g001834	Chr08: 137,105,053–137,105,745	693	1
CaDMP5	Capana08g001833	Chr08: 137,013,757–137,014,443	686	2
CaDMP7	Capana09g000598	Chr09: 25,314,397–25,315,185	789	1
CaDMP8	Capana04g002148	Chr04: 176,729,920–176,731,434	1515	2
CaDMP10	Capana08g001404	Chr08: 130,303,193–130,303,879	687	1
SmDMP1	SMEL4.1_01g024990.1.01	Chr01: 31,942,637–31,943,173	537	1
SmDMP3	SMEL4.1_01g000730.1.01	Chr01: 651,021–651,725	705	1
SmDMP4	SMEL4.1_04g004560.1.01	Chr04: 5,949,557–5,950,189	633	1
SmDMP6	SMEL4.1_02g015810.1.01	Chr02: 63,371,892–63,372,521	630	1
SmDMP7A	SMEL4.1_10g003550.1.01	Chr10: 4,252,481–4,253,140	660	1
SmDMP7B	SMEL4.1_10g003560.1.01	Chr10: 4,255,085–4,255,744	660	1
SmDMP8	SMEL4.1_10g015200.1.01	Chr10: 70,725,962–70,727,602	1641	2
SmDMP10	SMEL4.1_12g004810.1.01	Chr12: 8,986,251–8,986,886	636	1

3.2. Phylogenetic Relationships of DMP Family Genes

To investigate the evolutionary relationships among DMP family members in tomato, pepper, and eggplant, a phylogenetic tree was constructed. The DMP genes were classified into three groups based on their branching and clustering patterns (Figure 1). Group I contains only *SlDMP10* and *SmDMP10*. Group II consists of *DMP8* homologs from all species, along with *AtDMP9* from *Arabidopsis*. Group III is the largest group and can be further divided into two subgroups. Subgroup 1 includes *AtDMP10* and *CaDMP10*. Subgroup 2 can be further divided into two clades: the first clade includes DMP1 homologs from all Solanaceae species, as well as *AtDMP1* and *AtDMP2* genes from *Arabidopsis*, while the second clade splits into two sub-branches. The first sub-branch contains *AtDMP3* and *AtDMP5*, along with DMP3 homologs from Solanaceae species and *CaDMP5* from pepper. The second sub-branch contains DMP7 homologs from all species, and DMP6 and DMP4 homologs from all species except for pepper.

Gene duplication events were also assessed through collinearity analysis (Figure 2). In tomato, a pair of DMP genes exhibited collinearity: *SlDMP6* on chromosome 2 and *SlDMP4* on chromosome 10. No collinearity was observed among the six DMP genes in pepper. In eggplant, *SmDMP6* on chromosome 2 and *SmDMP4* on chromosome 4 were collinear.

The collinearity of DMP genes across the three species was further analyzed (Figure 3). The results showed that DMP1, DMP3, DMP7, and DMP8 homologous genes exhibit collinearity in all three species. In addition, DMP4 and DMP6 homologs showed collinearity between tomato and eggplant.

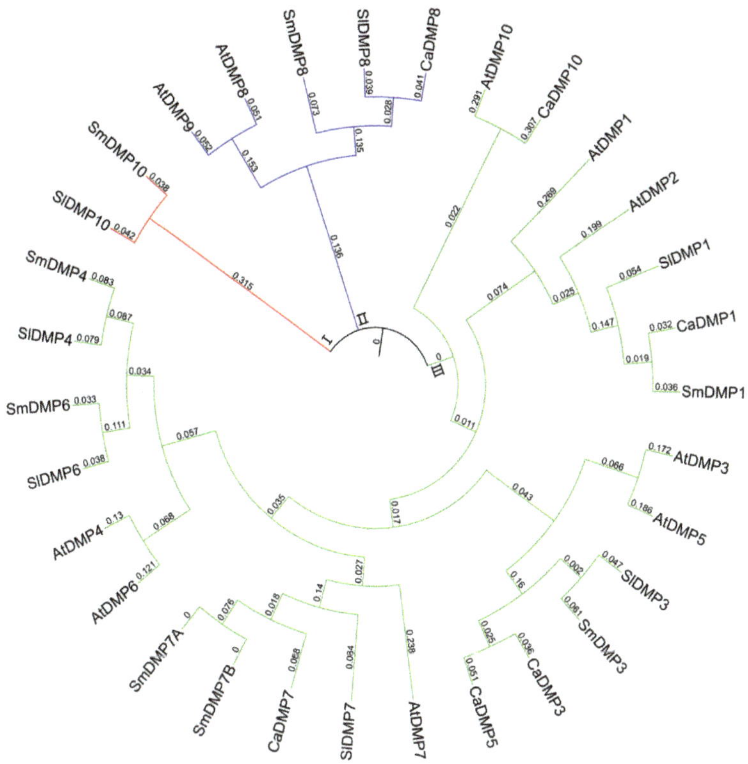

Figure 1. Phylogenetic tree of DMP family members in *Arabidopsis*, tomato, pepper, and eggplant. The tree was constructed using MEGA 11 with the Neighbor-Joining (NJ) method (bootstrap values of 1000). The three groups are labeled with different colors, and the support for each branch is represented by bootstrap values.

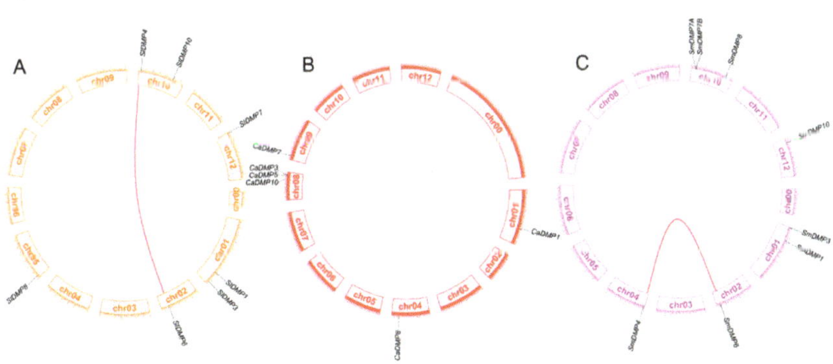

Figure 2. Collinearity of DMP genes in tomato (**A**), pepper (**B**), and eggplant (**C**). Connecting lines represent collinear gene pairs.

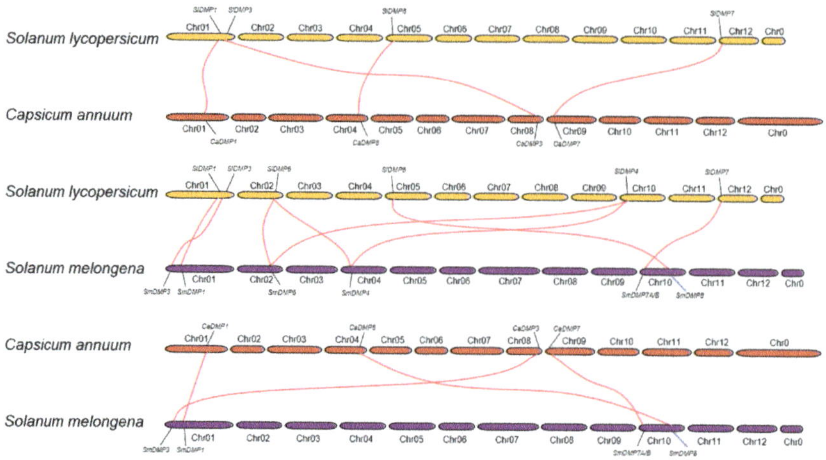

Figure 3. Collinearity of DMP genes among tomato, pepper, and eggplant. Lines represent collinear gene pairs.

3.3. Chromosomal Distribution and Gene Structure of DMP Genes

The chromosomal distribution of DMP genes was analyzed based on genome annotation files (Figure 4). In tomato, the seven DMP genes were distributed across chromosomes 1, 2, 5, 10, and 12. Chromosomes 1 and 10 each contained two DMP genes (*SlDMP1* and *SlDMP3*; *SlDMP4* and *SlDMP10*), while *SlDMP8* was located on chromosome 5. In pepper, the six DMP genes were located on chromosomes 1, 4, 8, and 9, with chromosome 8 containing three DMP genes (*CaDMP3*, *CaDMP5* and *CaDMP10*), and *CaDMP8* located on chromosome 4. In eggplant, the eight DMP genes were distributed across chromosomes 1, 2, 4, 10, and 12, with chromosome 10 carrying three genes (*SmDMP7A*, *SmDMP7B* and *SmDMP8*), and chromosome 1 carrying two genes (*SmDMP1* and *SmDMP3*).

Gene structure analysis provided further insights into the functional characteristics of DMP genes (Figure 5). All 21 DMP genes contained Motif1 to Motif4, with the structure and distribution of these motifs being highly conserved across the three species. The motifs in the DMP1, DMP7, and DMP8 genes were particularly well-conserved, while those in the DMP4 and DMP6 genes were also highly consistent in tomato and eggplant. Regarding the exon-intron structure, most DMP genes contained only one exon, with the exception of DMP8 homologs and *CaDMP5*, which contained two exons (Figure 5B).

3.4. Cis-Elements in DMP Gene Promoters

Cis-elements in promoters play a key role in gene regulation and expression. Therefore, the 2000 bp upstream regions of each DMP gene were analyzed for cis-elements (Figure 6). Among the detected cis-elements, drought-response elements were the most abundant. Specifically, of the 162 cis-elements identified in tomato, 81 were related to drought response; in pepper, 69 out of 147 cis-elements were related to drought response; and in eggplant, 71 out of 172 cis-elements were related to drought response. All DMP gene promoter regions contained light-response elements.

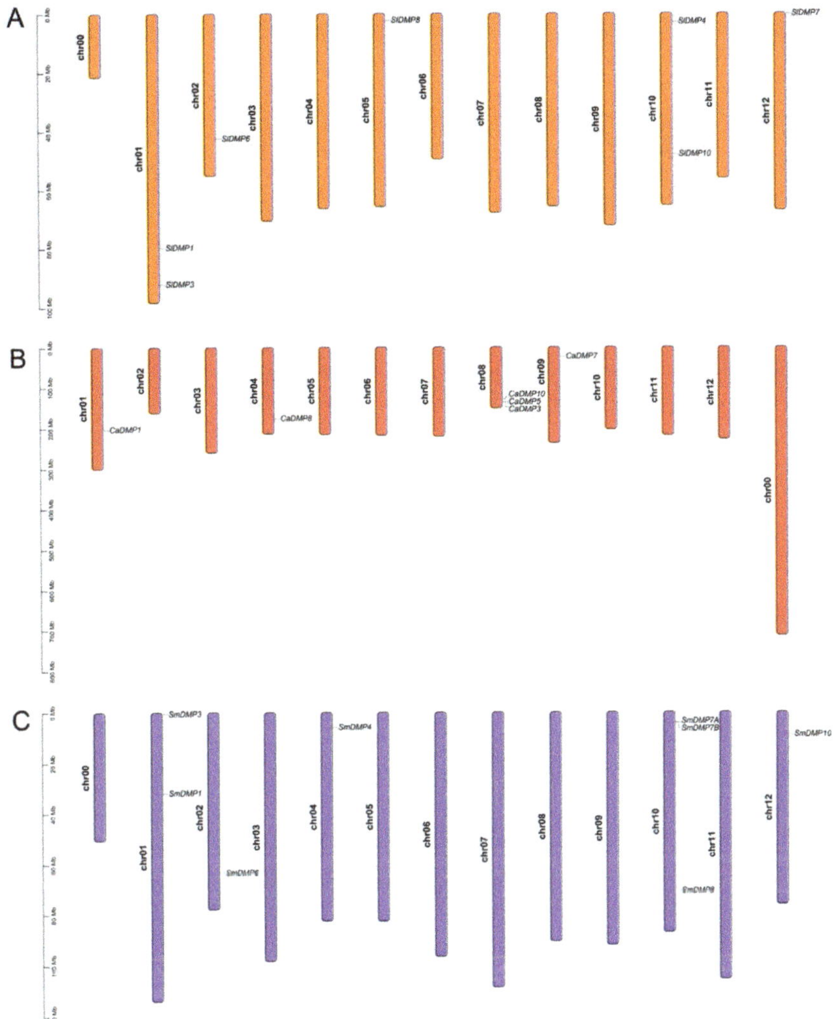

Figure 4. Chromosomal distribution of DMP genes in tomato (**A**), pepper (**B**), and eggplant (**C**). The scale bar indicates chromosome length (Mb).

Among the 21 DMP genes, 18, except for *SlDMP4*, *SmDMP5*, and *SmDMP6*, carried abscisic acid response elements in their promoters. Only the promoters of *SlDMP4*, *SlDMP6*, *CaDMP3*, and *SmDMP8* did not contain anaerobic induction elements, and the promoters of *SlDMP1*, *SlDMP4*, *SlDMP6*, *SmDMP1*, *SmDMP4*, *SmDMP7A*, and *CaDMP8* lacked biotic stress-response elements.

Numerous hormone-responsive cis-elements were also identified in the promoters of DMP genes. Fourteen promoters contained methyl jasmonate (MeJA) response elements, except for the promoters of *SlDMP10*, *CaDMP8*, *CaDMP10*, *SmDMP3*, *SmDMP7A*, *SmDMP7B*, and *SmDMP8*. Moreover, the promoters of *SlDMP4*, *SlDMP7*, *CaDMP8*, and *SmDMP10* contained gibberellin response elements. Thirteen promoters contained salicylic acid response elements (*SlDMP1*, *SlDMP4*, *SlDMP6*, *SlDMP10*, *CaDMP3*, *CaDMP7*, *SmDMP1*, *SmDMP3*, *SmDMP4*, *SmDMP6*, *SmDMP7A*, *SmDMP7B*, and *SmDMP8*).

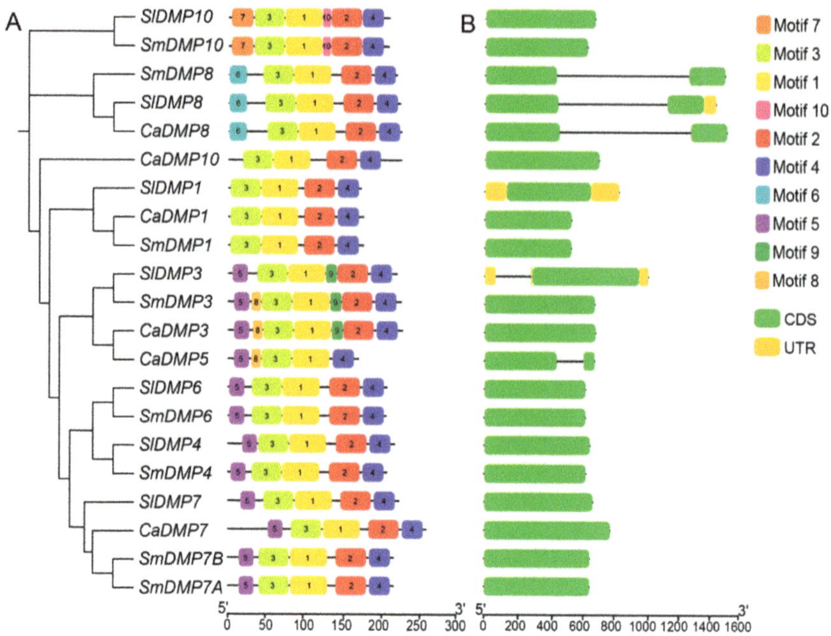

Figure 5. Gene structure of DMP genes from tomato, pepper, and eggplant. (**A**), Conserved motifs of *DMP* genes, organized based on the evolutionary tree. (**B**), Gene structures of DMP genes. The green regions represent the untranslated region (UTR), the yellow regions represent the coding sequence (CDS), and the black lines represent the non-coding regions.

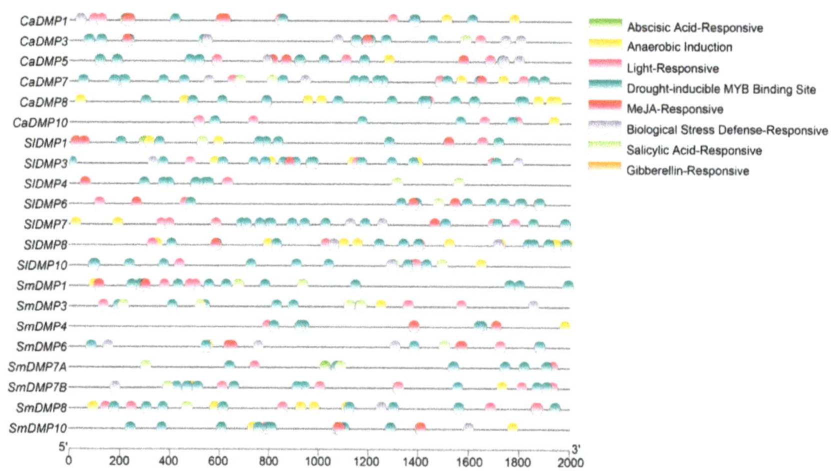

Figure 6. Cis-elements distribution in DMP promoters from tomato, pepper, and eggplant.

Notably, the promoters of *SlDMP8*, *CaDMP8*, and *SmDMP8* contained the highest number of drought-inducible cis-elements, followed by anaerobic induction elements and light-response elements.

3.5. Physicochemical Properties of DMP Family Members

The physicochemical properties of proteins were analyzed to provide insights into their functions (Table 2). Protein lengths ranged from 172 to 262 aa, with CaDMP5 being the shortest (172 aa) and CaDMP7 the longest (262 aa). Correspondingly, molecular weight ranged from 18.88 (CaDMP5) to 28.70 kDa (CaDMP7). Among the DMP homologs, SlDMP8, CaDMP8, and SmDMP8 contained 225, 228, and 222 amino acids, with molecular weights of 24.69, 24.85, and 24.41 kDa, respectively. Isoelectric point (pI) values ranged from 4.91 (SmDMP6) to 9.17 (CaDMP10). The DMP8 homologs exhibited relatively high pIs (8.73, 8.19, and 8.51 for SlDMP8, CaDMP8, and SmDMP8, respectively). The instability index ranged from 30.20 to 52.17. Based on the instability index threshold of 40 [36], ten DMP proteins were classified as stable, whereas the rest were unstable. Notably, all three DMP8 homologs were stable (36.04, 36.96, and 39.36, respectively). In addition, all proteins were predicted to be hydrophilic.

Table 2. Physicochemical properties of DMP family members in tomato, pepper, and eggplant.

Protein Name	Number of Amino Acids	Molecular Weight (kDa)	Isoelectric Point	Instability Index	Subcellular Localization *
SlDMP1	174	19.01	7.64	30.20	CM
SlDMP3	222	24.24	6.95	40.27	CM, ER
SlDMP4	219	24.64	8.87	38.17	CM
SlDMP6	209	22.91	5.28	34.5	CM
SlDMP7	225	24.94	7.55	40.88	CM
SlDMP8	225	24.69	8.73	36.04	CM, ER, Cyto
SlDMP10	221	24.07	5.49	44.43	CM
CaDMP1	178	19.51	8.2	32.62	CM
CaDMP3	230	25.31	8.3	39.17	CM
CaDMP5	172	18.88	7.76	45.77	ER
CaDMP7	262	28.70	7.48	45.80	CM
CaDMP8	228	24.85	8.19	36.96	Cyto, CM
CaDMP10	228	25.07	9.17	52.17	CP
SmDMP1	178	19.44	6.87	36.60	CM
SmDMP3	228	24.66	6.88	42.67	CM, ER
SmDMP4	210	23.60	8.09	36.89	CM
SmDMP6	209	22.98	4.91	40.25	CM
SmDMP7A	219	24.08	8.69	40.25	ER, CM
SmDMP7B	219	24.08	8.69	40.25	ER, CM
SmDMP8	222	24.41	8.51	39.36	Cyto, CM
SmDMP10	211	24.05	5.24	50.11	CM

* CM, Cell membrane; ER: Endoplasmic reticulum; Cyto, Cytoplasm; CP, Chloroplast.

The results of subcellular localization analysis showed that, among all DMP proteins, 19 were localized to the cell membrane, 6 to the endoplasmic reticulum, 3 to the cytoplasm, and 1 to the chloroplast. Among them, SlDMP3, SmDMP3, SmDMP7A, and SmDMP7B might be simultaneously localized to the cell membrane and endoplasmic reticulum; SlDMP8 might be simultaneously localized to the cell membrane, endoplasmic reticulum, and cytoplasm; and CaDMP8 and SmDMP8 might be simultaneously localized to the cytoplasm and cell membrane.

3.6. Three-Dimensional (3D) Structure of DMP Proteins

To explore higher-order structural characteristics, the 3D structures of the 21 DMP proteins were predicted (Figure 7). Overall, homologous proteins exhibited conserved structures across the three Solanaceous species. Specifically, DMP1, DMP3, and DMP8 homologs in tomato, pepper, and eggplant displayed highly similar 3D structures. The DMP4, DMP6, DMP7, and DMP10 homologs in tomato and eggplant also showed highly similar structures.

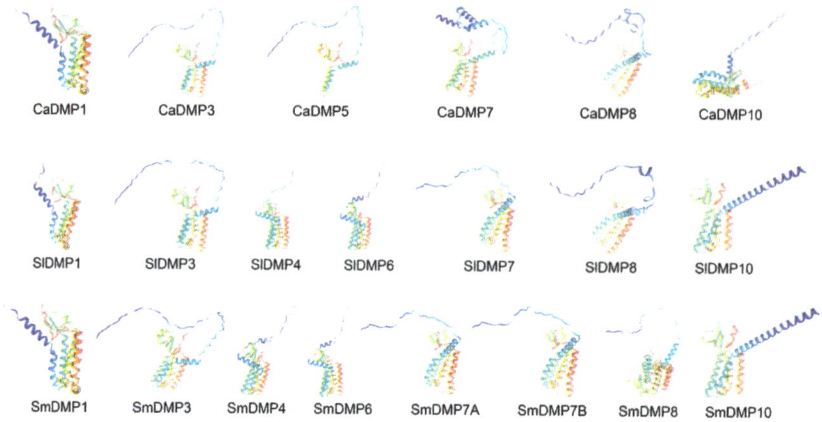

Figure 7. Three-dimensional structure of DMP family member proteins in tomato, pepper, and eggplant.

To further elucidate structural characteristics, transmembrane helices were predicted using TMHMM-2.0 (Figure S1). Most DMP proteins were predicted to contain four transmembrane helices. Exceptions included SlDMP6 (three), CaDMP3 (five), and CaDMP5 (two). In most proteins, the N-terminus was predicted to localize to the cytoplasmic side with probability > 0.8. In contrast, the N-terminal cytoplasmic probability of DMP1 homologs was <0.6. Collectively, all DMP proteins, except SlDMP1, CaDMP1, and SmDMP1, were classified as hydrophilic transmembrane proteins.

3.7. Tissue Expression Profiles of DMP Genes

Based on public data, tissue-specific expression heatmaps were generated for tomato, pepper, and eggplant (Figure 8). In tomato, *SlDMP4* and *SlDMP8* were strongly expressed in flower buds. *SlDMP1, SlDMP3,* and *SlDMP6* also showed relatively high expression in flower buds. The expression levels of *SlDMP10, SlDMP3,* and *SlDMP6* were higher in roots compared to other tissues, with SlDMP10 showing the most prominent expression. In pepper, *CaDMP7* and *CaDMP8* were highly expressed in flower buds and anthers, but were barely detectable in other tissues. *CaDMP1* was highly expressed in developing seeds and flower buds. *CaDMP3* showed relatively high expression in the stamen, placenta and young seeds of fruits, and in certain stages of flowers and seeds. *CaDMP10* had the highest expression in stems, followed by petals and ovaries in open flowers. In eggplant, *SmDMP3* was highly expressed in roots, while *SmDMP8* was predominantly expressed in leaves. In flower buds, *SmDMP1* and *SmDMP7A/7B* were strongly expressed, whereas *SmDMP4, SmDMP6* and *SmDMP10* were highly expressed in stems.

3.8. Verification of Tissue Expression of Tomato DMP Genes

RT-qPCR was performed to validate the tissue-specific expression patterns of the seven DMP genes in tomato (Figure 9).

The results showed that the qPCR data for *SlDMP1, SlDMP3,* and *SlDMP10* were highly consistent with the transcriptome data. Specifically, *SlDMP10* was significantly higher in root tissue; *SlDMP1* exhibited higher expression levels in flower buds, roots, and mature green fruits; and *SlDMP3* showed higher expression in roots, flower buds, and stems. The qPCR results for *SlDMP6* and *SlDMP8* were mostly consistent with the transcriptome data, but slight differences were observed. For *SlDMP6*, high expression was found not only in roots but also in fruits after the breaker stage, a feature not reflected in the transcriptome data. Similarly, for *SlDMP8*, high expression was detected in flower buds, roots, and fruits after the breaker stage, but this pattern was not observed in the transcriptome data.

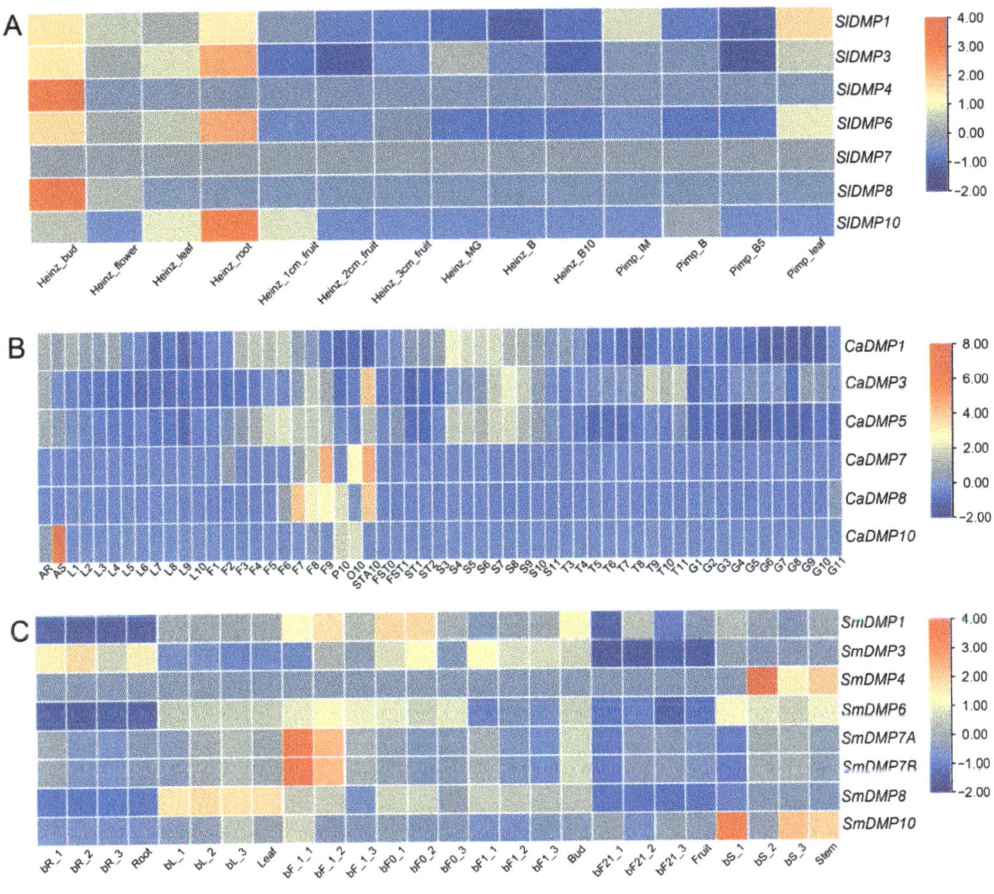

Figure 8. Tissue expression profiles of DMP genes in tomato (**A**), pepper (**B**), and eggplant (**C**). Heinz represents cultivated tomato (*Solanum lycopersicum* cv. Heinz 1706). B, breaker stage; B5, breaker plus 5 days; IM, immature green; MG, mature green. For details of pepper and eggplant tissues, please refer to published papers [31,32].

The qPCR results for *SlDMP4* and *SlDMP7* showed significant differences from the transcriptome data. *SlDMP4* was highly expressed in flower buds, stems, roots, and fruits after the breaker stage, while the transcriptome data suggested it was highly expressed only in flower buds. *SlDMP7* showed high expression in roots and was also detectable in flower buds, stems, and fruits after the breaker stage, a feature not reflected in the transcriptome data. Overall, the tissue expression profiles of tomato DMP genes showed both similarities and differences when compared to the data from the public database.

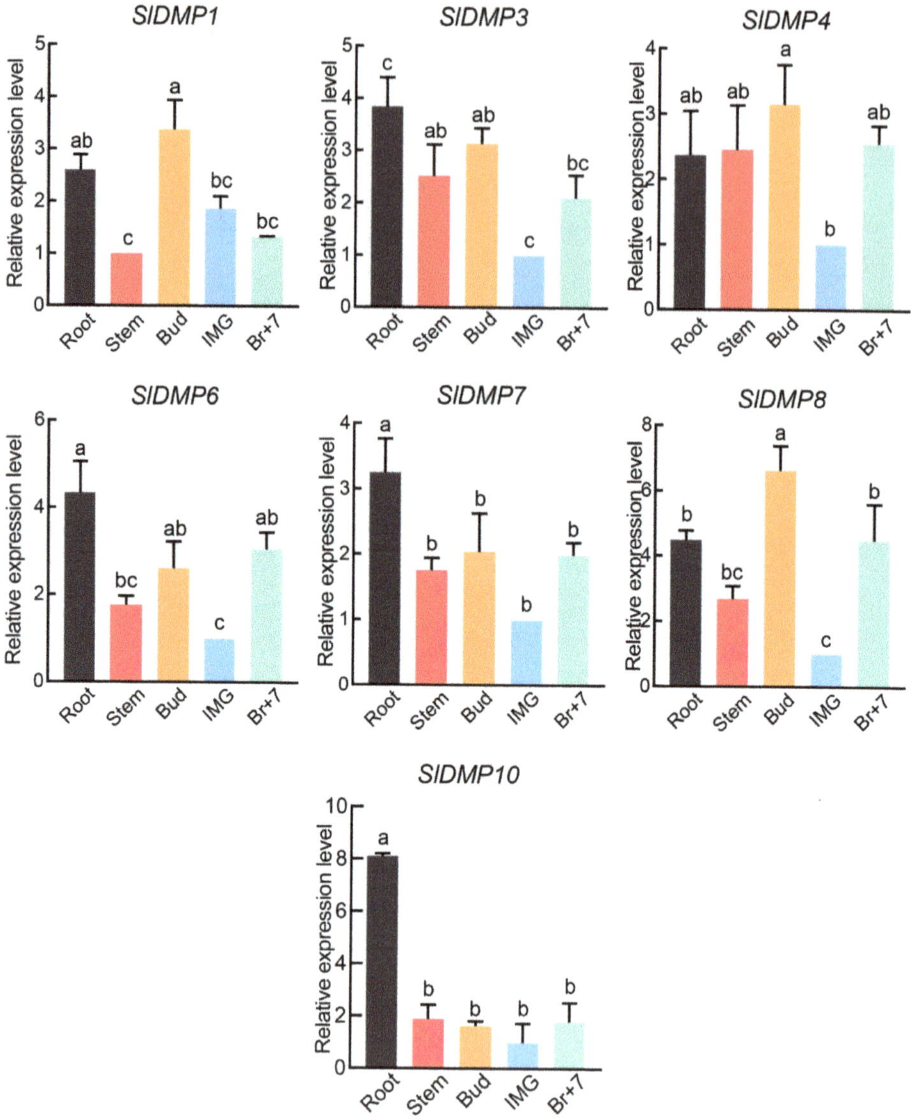

Figure 9. Tissue expression profiles of DMP genes in tomato plants based on RT-qPCR. The gene IDs for the DMP genes are Solyc01g080490 (*SlDMP1*), Solyc01g103580 (*SlDMP3*), Solyc10g008970 (*SlDMP4*), Solyc02g077640 (*SlDMP6*), Solyc12g005490 (*SlDMP7*), Solyc05g007920 (*SlDMP8*), and Solyc10g050200 (*SlDMP10*). Expression profiles were derived from RT-qPCR analyses of roots, stems, flowers, and fruits of MicroTom plants at 90 days of age. Relative expression levels of each DMP gene were normalized to the reference gene *Actin*. Data were $\log_2$-transformed and are presented as the mean ± standard deviation of three biological replicates. Statistical significance was determined by one-way ANOVA followed by post hoc multiple comparisons. Different letters above the bars indicate significant differences between groups ($p < 0.05$).

4. Discussion

We identified seven DMP family genes in tomato, six in pepper, and eight in eggplant by searching for genes containing the conserved DUF679 domain (PF05078.15). Consistent with a previous study, six DMP genes were identified in the pepper variety Zunla [37]. Although a prior study identified six DMP genes in tomato [20], we identified seven, with the additional gene being *SlDMP8* (Solyc05g007920). This discrepancy may be due to the use of different protein sequence databases (ITAG4.0 in their work versus ITAG2.4 in ours). We also noted that the number of DMP genes is similar across these three Solanaceous species. These numbers are much smaller than those identified in polyploid plants, such as peanut and cotton [24,38], suggesting that the number of DMP genes is largely determined by the species' ploidy level.

The phylogenetic tree showed that DMP8 members from the three *Solanaceous* vegetables cluster with *AtDMP8* and *AtDMP9*, which are known to be homologous and functionally redundant in *Arabidopsis* [16]. Haploid induction lines can be generated by knocking out *DMP8* in many species, including tomato [14,19,39], or both *DMP8* and *DMP9* in *Arabidopsis* and *Medicago truncatula* [16,17]. Only one highly homologous DMP8 gene was found in each of the tomato, pepper, and eggplant genomes. Theoretically, haploid inducers can also be generated in pepper and eggplant by knocking out the DMP8 homolog. Our recent breakthrough in pepper transformation makes it possible to generate a DMP8-based haploid inducer in this typically recalcitrant species [40].

We further investigated the expression patterns of DMPs in these *Solanaceous* species. Based on expression data from public databases, *SlDMP8* and *SlDMP4* were strongly expressed in buds, followed by *SlDMP1*, *SlDMP3*, and *SlDMP6* (Figure 8). This was supported by our RT-qPCR validation, especially for *SlDMP8*, *SlDMP4*, and *SlDMP1* (Figure 9). Previous studies reported that haploid induction lines obtained by knocking out *SlDMP8* had an induction rate ranging from 0.5% to 3.7% (average 1.9%) [19]. Given the expression patterns and functional connections in this gene family, we speculate that other members, such as *DMP4*, might also be involved in haploid induction. Previous studies have shown that the homologous gene *CaDMP5* (referred to as *CaDMP8* in this study) is highly expressed in anthers, which also supports its role in haploid induction [37]. In line with these findings, our expression data indicated that *CaDMP8* and *CaDMP7* were highly expressed in the stamens of pepper flowers, and *CaDMP8* showed a collinear relationship with *SlDMP8* in between the two species (Figures 3 and 8B). These results further support *CaDMP8* as a target for developing haploid induction lines in pepper and suggest that knocking out *CaDMP7* may further enhance the induction efficiency. For *SmDMP8*, its expression level was also relatively high in flower tissue (Figure 8C), which, together with its phylogenetic relationship with other DMP8 homologs, suggests it can also be targeted for the development of a haploid inducer line. In addition to intrinsic genetic influences, exogenous chemicals can also affect the creation of haploid lines. In *Brassica napus*, certain antimitotic substances have been shown to promote doubled-haploid regeneration [41], suggesting that applying antimitotic substances exogenously, along with knocking out target genes, may improve the efficiency of producing double-haploid lines.

To further analyze the regulatory mechanisms of DMP genes, we examined their gene structures and cis-elements in their promoter regions. We found that the gene structures of DMP8 in the three *Solanaceous* species were highly similar, with conserved motif arrangements and consistent exon-intron distributions, suggesting that the function of DMP8 is conserved across these species (Figure 5). The promoter regions of DMP8 contained cis-elements responsive to light and anaerobic conditions (Figure 6). Light-responsive cis-elements (LREs) are often found in the promoters of genes involved in photosynthesis, photomorphogenesis, and other light-regulated pathways [42–46], while anaerobic-responsive elements (AREs) are involved in responses to low oxygen and dehydration [47]. The presence of these promoter elements suggests that DMP8 may be involved in photosynthesis, anaerobic metabolism, and stress responses, and that environmental factors could affect its role in haploid induction

Protein structure plays a crucial role in determining protein function. As the structural characteristics of DMP proteins have rarely been discussed, we predicted the physicochemical properties and 3D structures of the DMP proteins in these three *Solanaceous* species (Figure 7 and Table 2). Overall, homologs of a particular DMP member, such as DMP1, DMP3 and DMP10, showed highly similar 3D structures (Figure 7). In tomato, the 3D structures of DMP4 and DMP6 proteins were highly similar (Figure 7), indicating their functions may also be conserved. In contrast, the protein structures of DMP3 and DMP8 showed some differences (Figure 7). Despite these structural differences, we found that the DMP3 and DMP8 proteins in the three species had similar molecular weights and amino acid counts (Table 2), which may suggest a degree of functional conservation. Analysis of transmembrane domains indicated that DMP family proteins in these species each contain 3–5 transmembrane helices, suggesting potential roles in channel formation and small-molecule transmembrane transport across membranes, though their specific functions remain unknown. Except for SlDMP1, CaDMP1, and SmDMP1, the N-terminal regions of all other DMP proteins are predicted to be on the cytoplasmic side of the membrane. This suggests that DMP1 family proteins may perform other, unidentified extracellular functions that are conserved across species. In summary, based on the identification of the DMP gene family in tomato, pepper, and eggplant, this study found that DMP8 genes are highly expressed in floral organs and exhibit conserved gene structures and protein characteristics across these species, suggesting their conserved role in haploid induction. Other DMP members may also be involve in haploid induction and may thus enhance the efficiency of DMP8. Future research should employ gene-editing technologies to directly validate the functions of DMP8 and other DMP members expressed in flower tissue within Solanaceous crops and further elucidate the specific molecular mechanisms through which they participate in reproductive development.

5. Conclusions

This study provides a comprehensive genome-wide identification of DMP genes in tomato, pepper, and eggplant, characterizing seven, six, and eight DMP genes, respectively. All these genes encode proteins containing the conserved DUF679 domain. Phylogenetic analysis clustered these genes into three distinct groups, revealing that Group II members, comprising evolutionarily conserved DMP8 homologs, share conserved and protein structures. Moreover, DMP8 homologs are predicted to localize to the cell membrane. Notably, tissue-specific expression profiling revealed that *SlDMP8* and *SlDMP4* are predominantly expressed in reproductive tissues, suggesting their functional involvement in haploid induction pathways. These findings identify *CaDMP8* and *SmDMP8* as key targets for developing haploid induction lines in pepper and eggplant. In addition, co-editing DMP4 may enhance induction efficiency in tomato. Together, these results provide a solid theoretical foundation for advancing doubled-haploid breeding in Solanaceous crops through targeted genome editing.

Supplementary Materials: The following supporting information can be downloaded at: https://www.mdpi.com/article/10.3390/horticulturae11111329/s1, Figure S1: Transmembrane structure prediction of DMP family members in tomato, pepper and eggplant; Table S1: RT-qPCR primer list.

Author Contributions: X.D., B.L. and X.S., performed wet lab; X.W., X.D. and W.Z. performed dry lab; X.D., W.Z., X.W. and B.O. wrote the manuscript; Z.R., Y.L. and X.W. revised the manuscript. All authors have read and agreed to the published version of the manuscript.

Funding: This research funded by the National Natural Science Foundation of China (32472743, U21A20230), and the High-Quality Development Project for the Seed Industry in Hubei Province (HBZY2023B004), China.

Data Availability Statement: All data supporting the findings of this study are available. The sequence data were sourced from public genome resources.

Acknowledgments: We thank our laboratory members for their support and the reviewers for their valuable comments on this manuscript.

Conflicts of Interest: The authors have not disclosed any competing interests.

References

1. Coe, E.H. A line of maize with high haploid frequency. *Am. Nat.* **1959**, *93*, 381–382. [CrossRef]
2. Gilles, L.M.; Martinant, J.P.; Rogowsky, P.M.; Widiez, T. Haploid induction in plants. *Curr. Biol.* **2017**, *27*, R1095–R1097. [CrossRef]
3. Seguí-Simarro, J.M.; Nuez, F. Embryogenesis induction, callogenesis, and plant regeneration by culture of tomato isolated microspores and whole anthers. *J. Exp. Bot.* **2007**, *58*, 1119–1132. [CrossRef]
4. Seguí-Simarro, J.M.; Corral-Martínez, P.; Parra-Vega, V.; González-García, B. Androgenesis in recalcitrant solanaceous crops. *Plant Cell Rep.* **2011**, *30*, 765–778. [CrossRef]
5. Islam, M.T.; Arif, M.R.; Hasan, M.T.; Robin, A.H.K. Anther culture in crop plants: Progress and perspectives. *Plant Breed. Biotechnol.* **2023**, *11*, 69–96. [CrossRef]
6. Khammona, K.; Dermail, A.; Suriharn, K.; Lübberstedt, T.; Wanchana, S.; Thunnom, B.; Poncheewin, W.; Toojinda, T.; Ruanjaichon, V.; Arikit, S. Accelerating haploid induction rate and haploid validation through marker-assisted selection for *qhir1* and *qhir8* in maize. *Front. Plant Sci.* **2024**, *15*, 1337463. [CrossRef] [PubMed]
7. Liu, C.X.; Li, X.; Meng, D.X.; Zhong, Y.; Chen, C.; Dong, X.; Xu, X.W.; Chen, B.J.; Li, W.; Li, L.; et al. A 4-bp insertion at *ZmPLA1* encoding a putative phospholipase a generates haploid induction in maize. *Mol. Plant* **2017**, *10*, 520–522. [CrossRef] [PubMed]
8. Gilles, L.M.; Khaled, A.; Laffaire, J.B.; Chaignon, S.; Gendrot, G.; Laplaige, J.; Bergès, H.; Beydon, G.; Bayle, V.; Barret, P.; et al. Loss of pollen-specific phospholipase NOT LIKE DAD triggers gynogenesis in maize. *EMBO J.* **2017**, *36*, 707–717. [CrossRef]
9. Kelliher, T.; Starr, D.; Richbourg, L.; Chintamanani, S.; Delzer, B.; Nuccio, M.L.; Green, J.; Chen, Z.Y.; McCuiston, J.; Wang, W.L.; et al. MATRILINEAL, a sperm-specific phospholipase, triggers maize haploid induction. *Nature* **2017**, *542*, 105–109. [CrossRef]
10. Liu, C.; Zhong, Y.; Qi, X.; Chen, M.; Liu, Z.; Chen, C.; Tian, X.; Li, J.; Jiao, Y.; Wang, D.; et al. Extension of the in vivo haploid induction system from diploid maize to hexaploid wheat. *Plant Biotechnol. J.* **2020**, *18*, 316–318. [CrossRef]
11. Yao, L.; Zhang, Y.; Liu, C.; Liu, Y.; Wang, Y.; Liang, D.; Liu, J.; Sahoo, G.; Kelliher, T. OsMATL mutation induces haploid seed formation in indica rice. *Nat. Plants* **2018**, *4*, 530–533. [CrossRef]
12. Wang, C.; Liu, Q.; Shen, Y.; Hua, Y.; Wang, J.; Lin, J.; Wu, M.; Sun, T.; Cheng, Z.; Mercier, R.; et al. Clonal seeds from hybrid rice by simultaneous genome engineering of meiosis and fertilization genes. *Nat. Biotechnol.* **2019**, *37*, 283–286. [CrossRef]
13. Liu, H.; Wang, K.; Jia, Z.; Gong, Q.; Lin, Z.; Du, L.; Pei, X.; Ye, X. Efficient induction of haploid plants in wheat by editing of TaMTL using an optimized Agrobacterium-mediated CRISPR system. *J. Exp. Bot.* **2020**, *71*, 1337–1349. [CrossRef]
14. Zhong, Y.; Liu, C.; Qi, X.; Jiao, Y.; Wang, D.; Wang, Y.; Liu, Z.; Chen, C.; Chen, B.; Tian, X.; et al. Mutation of *ZmDMP* enhances haploid induction in maize. *Nat. Plants* **2019**, *5*, 575–580. [CrossRef]
15. Chen, X.E.; Li, Y.X.; Ai, G.L.; Chen, J.F.; Guo, D.L.; Zhu, Z.H.; Zhu, X.J.; Tian, S.J.; Wang, J.F.; Liu, M.; et al. Creation of a watermelon haploid inducer line via *ClDMP3*-mediated single fertilization of the central cell. *Hortic. Res. Engl.* **2023**, *10*, uhad081. [CrossRef] [PubMed]
16. Zhong, Y.; Chen, B.; Li, M.; Wang, D.; Jiao, Y.; Qi, X.; Wang, M.; Liu, Z.; Chen, C.; Wang, Y.; et al. A *DMP*-triggered in vivo maternal haploid induction system in the dicotyledonous *Arabidopsis*. *Nat. Plants* **2020**, *6*, 466–472. [CrossRef]
17. Wang, N.; Xia, X.Z.; Jiang, T.; Li, L.L.; Zhang, P.C.; Niu, L.F.; Cheng, H.M.; Wang, K.J.; Lin, H. In planta haploid induction by genome editing of *DMP* in the model legume *Medicago truncatula*. *Plant Biotechnol. J.* **2022**, *20*, 22–24. [CrossRef]
18. Zhang, X.L.; Zhang, L.L.; Zhang, J.S.; Jia, M.A.; Cao, L.G.; Yu, J.; Zhao, D.G. Haploid induction in allotetraploid tobacco using *DMPs* mutation. *Planta* **2022**, *255*, 98. [CrossRef] [PubMed]

19. Zhong, Y.; Chen, B.; Wang, D.; Zhu, X.; Li, M.; Zhang, J.; Chen, M.; Wang, M.; Riksen, T.; Liu, J.; et al. In vivo maternal haploid induction in tomato. *Plant Biotechnol. J.* **2022**, *20*, 250–252. [CrossRef]
20. Ahmad, Z.; Tian, D.; Li, Y.; Aminu, I.M.; Tabusam, J.; Zhang, Y.; Zhu, S. Characterization, evolution, expression and functional divergence of the *DMP* gene family in plants. *Int. J. Mol. Sci.* **2024**, *25*, 10435. [CrossRef] [PubMed]
21. Kasaras, A.; Melzer, M.; Kunze, R. Arabidopsis senescence-associated protein *DMP1* is involved in membrane remodeling of the ER and tonoplast. *BMC Plant Biol.* **2012**, *12*, 54. [CrossRef] [PubMed]
22. Gao, Z.; Daneva, A.; Salanenka, Y.; Van Durme, M.; Huysmans, M.; Lin, Z.; De Winter, F.; Vanneste, S.; Karimi, M.; Van de Velde, J.; et al. KIRA1 and ORESARA1 terminate flower receptivity by promoting cell death in the stigma of Arabidopsis. *Nat. Plants* **2018**, *4*, 365–375. [CrossRef]
23. Ma, Y.; Liu, H.; Wang, J.; Zhao, G.; Niu, K.; Zhou, X.; Zhang, R.; Yao, R. Genomic identification and expression profiling of *DMP* genes in oat (*Avena sativa*) elucidate their responsiveness to seed aging. *BMC Genom.* **2024**, *25*, 863. [CrossRef]
24. Zhu, S.; Wang, X.; Chen, W.; Yao, J.; Li, Y.; Fang, S.; Lv, Y.; Li, X.; Pan, J.; Liu, C.; et al. Cotton DMP gene family: Characterization, evolution, and expression profiles during development and stress. *Int. J. Biol. Macromol.* **2021**, *183*, 1257–1269. [CrossRef]
25. Han, P.G.; Chang, Y.; Tang, K.G.; Wang, L.; Xiu, Z.J.; Yang, J.; Shi, H.B.; Liang, Y.H.; Sun, R.F.; Su, S.F.; et al. Genome-wide identification of the sugar beet (*Beta vulgaris* L.) *DMP* gene family and its potential role in abiotic stress. *Genet. Resour. Crop Evol.* **2025**, *72*, 3561–3579. [CrossRef]
26. Finn, R.D.; Clements, J.; Eddy, S.R. HMMER web server: Interactive sequence similarity searching. *Nucleic Acids Res.* **2011**, *39*, W29–W37. [CrossRef]
27. Edgar, R.C. MUSCLE: Multiple sequence alignment with high accuracy and high throughput. *Nucleic Acids Res.* **2004**, *32*, 1792–1797. [CrossRef]
28. Tamura, K.; Stecher, G.; Kumar, S. MEGA11 molecular evolutionary genetics analysis version 11. *Mol. Biol. Evol.* **2021**, *38*, 3022–3027. [CrossRef]
29. Letunic, I.; Bork, P. Interactive Tree Of Life (iTOL) v5: An online tool for phylogenetic tree display and annotation. *Nucleic Acids Res.* **2021**, *49*, W293–W296. [CrossRef] [PubMed]
30. Chen, C.J.; Wu, Y.; Li, J.W.; Wang, X.; Zeng, Z.H.; Xu, J.; Liu, Y.L.; Feng, J.T.; Chen, H.; He, Y.H.; et al. TBtools-II: A "one for all, all for one" bioinformatics platform for biological big-data mining. *Mol. Plant* **2023**, *16*, 1733–1742. [CrossRef] [PubMed]
31. Liu, F.; Yu, H.; Deng, Y.; Zheng, J.; Liu, M.; Ou, L.; Yang, B.; Dai, X.; Ma, Y.; Feng, S.; et al. PepperHub, an informatics hub for the chili pepper research community. *Mol. Plant* **2017**, *10*, 1129–1132. [CrossRef]
32. Chen, X.; Zhang, M.; Tan, J.; Huang, S.P.; Wang, C.L.; Zhang, H.Y.; Tan, T.M. Comparative transcriptome analysis provides insights into molecular mechanisms for parthenocarpic fruit development in eggplant (*Solanum melongena* L.). *PLoS ONE* **2017**, *12*, e0179491. [CrossRef]
33. Fionn, M.; Pierre, L. Ward's hierarchical agglomerative clustering method: Which algorithms implement ward's criterion? *J. Classif.* **2014**, *31*, 274–295. [CrossRef]
34. Gu, Z.; Eils, R.; Schlesner, M. Complex heatmaps reveal patterns and correlations in multidimensional genomic data. *Bioinformatics* **2016**, *32*, 2847–2849. [CrossRef]
35. Liu, G.Z.; Li, C.X.; Yu, H.Y.; Tao, P.W.; Yuan, L.; Ye, J.; Chen, W.F.; Wang, Y.; Ge, P.F.; Zhang, J.H.; et al. *GREEN STRIPE*, encoding methylated TOMATO AGAMOUS-LIKE 1, regulates chloroplast development and Chl synthesis in fruit. *New Phytol.* **2020**, *228*, 302–317. [CrossRef]
36. Martinez, G.S.; Dutt, M.; Kumar, A.; Kelvin, D.J. Multiple protein profiler 1.0 (MPP): A webserver for predicting and visualizing physiochemical properties of proteins at the proteome level. *Protein J.* **2024**, *43*, 711–717. [CrossRef]
37. Zhang, Y.M.; Zhang, D.D.; Li, X.R.; He, J.; Chen, Z.N.; Xu, N.; Zhong, Y.K.; Yao, S.Q.; Qu, L.B.; Li, B.; et al. Genome-wide identification, characterization, and expression analysis of the DMP gene family in pepper (*Capsicum annuum* L.). *Horticulturae* **2024**, *10*, 1073. [CrossRef]
38. Qu, P.; He, L.; Xue, L.; Liu, H.; Li, X.; Zhao, H.; Fu, L.; Han, S.; Dai, X.; Dong, W.; et al. Genome-wide identification, evolution, and expression analysis of the DMP gene family in peanut (*Arachis hypogaea* L.). *Int. J. Mol. Sci.* **2025**, *26*, 7243. [CrossRef] [PubMed]
39. Li, T.; Wang, C.; Pan, J.; Tabusam, J.; Li, Y.; Yao, J.; Chen, W.; Wang, Y.; Gao, W.; Rong, J.; et al. Exploring potential strategies for haploid induction based on double fertilization in plants. *Plant Biotechnol. J.* **2025**, *23*, 4000–4016. [CrossRef] [PubMed]
40. Tang, Y.; Shen, X.; Deng, X.; Song, Y.; Zhou, Y.; Lu, Y.; Li, F.; Ouyang, B. Establishment of an efficient Agrobacterium-mediated transformation system for chilli pepper and its application in genome editing. *Plant Biotechnol. J.* **2025**, *23*, 4752–4754. [CrossRef]
41. Klíma, M.; Shmeit, Y.H.; Kopecký, P.; Vítámvás, P.; Kosová, K.; Prášil, I.T.; Fernández-Cusimamani, E. Impact of selected antimitotic substances on doubled haploid and polyploid regeneration in microspore cultures of swede (*Brassica napus* ssp. napobrassica (L.) Hanelt). *Czech J. Genet. Plant Breed.* **2024**, *60*, 79–85. [CrossRef]

42. Ain-Ali, Q.U.; Mushtaq, N.; Amir, R.; Gul, A.; Tahir, M.; Munir, F. Genome-wide promoter analysis, homology modeling and protein interaction network of Dehydration Responsive Element Binding (DREB) gene family in *Solanum tuberosum*. *PLoS ONE* **2021**, *16*, e0261215. [CrossRef]
43. Basu, S.; Roychoudhury, A.; Sengupta, D.N. Deciphering the role of various cis-acting regulatory elements in controlling SamDC gene expression in rice. *Plant Signal. Behav.* **2014**, *9*, e28391. [CrossRef] [PubMed]
44. Cota-Ruiz, K.; Oh, S.; Montgomery, B.L. Phytochrome-dependent regulation of ZFP6 and ZFPH impacts photomorphogenesis in *Arabidopsis thaliana*. *Front. Plant Sci.* **2022**, *13*, 846262. [CrossRef] [PubMed]
45. Lopez-Ochoa, L.; Acevedo-Hernandez, G.; Martinez-Hernandez, A.; Arguello-Astorga, G.; Herrera-Estrella, L. Structural relationships between diverse cis-acting elements are critical for the functional properties of a rbcS minimal light regulatory unit. *J. Exp. Bot.* **2007**, *58*, 4397–4406. [CrossRef] [PubMed]
46. Sheshadri, S.A.; Nishanth, M.J.; Simon, B. Stress-mediated cis-element transcription factor interactions interconnecting primary and specialized metabolism in planta. *Front. Plant Sci.* **2016**, *7*, 1725. [CrossRef]
47. Dolferus, R.; Klok, E.J.; Ismond, K.; Delessert, C.; Wilson, S.; Good, A.; Peacock, J.; Dennis, L. Molecular basis of the anaerobic response in plants. *IUBMB Life* **2001**, *51*, 79–82. [CrossRef]

MDPI AG
Grosspeteranlage 5
4052 Basel
Switzerland
Tel.: +41 61 683 77 34

Horticulturae Editorial Office
E-mail: horticulturae@mdpi.com
www.mdpi.com/journal/horticulturae

www.ingramcontent.com/pod-product-compliance
Lightning Source LLC
Chambersburg PA
CBHW061559210926
51850CB00009B/2159

9783725871360